FAMILY VIOLENCE
IN THE UNITED STATES

FAMILY VIOLENCE
IN THE UNITED STATES
Defining, Understanding, and Combating Abuse

DENISE A. HINES
University of New Hampshire

KATHLEEN MALLEY-MORRISON
Boston University

SAGE Publications
Thousand Oaks ▪ London ▪ New Delhi

For information:

Sage Publications, Inc.
2455 Teller Road
Thousand Oaks, California 91320
E-mail: order@sagepub.com

Sage Publications Ltd.
1 Oliver's Yard
55 City Road
London EC1Y 1SP
United Kingdom

Sage Publications India Pvt. Ltd.
B-42, Panchsheel Enclave
Post Box 4109
New Delhi 110 017 India

Printed in the United States of America

Library of Congress Cataloging-in-Publication Data

Hines, Denise A.
Family violence in the United States: Defining, understanding,
and combating abuse / Denise A. Hines and Kathleen Malley-Morrison.
    p. cm.
Includes bibliographical references and index.
ISBN 0-7619-3085-X (cloth)—ISBN 0-7619-3086-8 (pbk.)
    1. Family violence—United States. I. Malley-Morrison, Kathleen. II. Title.
HV6626.2.H56 2005
362.82′92′0973—dc22

                        2004016590

04 05 06 07 10 9 8 7 6 5 4 3 2 1

Acquisitions Editor:	Jim Brace-Thompson
Editorial Assistant:	Karen Ehrmann
Production Editor:	Melanie Birdsall
Copy Editor:	Bill Bowers, Interactive Composition Corporation
Typesetter:	C&M Digitals (P) Ltd.
Proofreader:	Mary Meagher
Indexer:	Kathleen Paparchontis
Cover Designer:	Edgar Abarca

Contents

Preface and Acknowledgments

My friend went out with a boy who tried to control what she did. He would say he hated her.

I had a former female friend who tended to be very domineering. She would never resort to physical violence, but she used guilt-manipulation and yelling to control her boyfriends.

My best friend was in an abusive relationship. Her boyfriend used to push her around and try to control her appearance, activities, and friendships.

My boyfriend and I were fighting and he tried to go to sleep, so I punched him in the chest to wake him up.

These descriptions of aggressive relationships are quotes from college students, just like most of you who are reading this book right now. They describe the typical type of aggression occurring between partners in relationships—what some researchers call "minor violence" (Straus, 1990a) and others call "common couple violence" (Johnson, 1995). Indeed, most aggression in relationships, both intimate relationships and other family relationships, is not the type that we see on the news or in most of the case studies presented in this book. Such cases are extreme examples; they are meant to capture our attention and spur research into, and resources for, people involved in highly abusive situations.

People involved in extremely abusive family relationships are the ones most in need of intervention services such as battered women's shelters, parenting classes, and counseling. They are also the ones who tend to receive the most research attention. However, most cases of family aggression involve psychological aggression and occasional minor violence. In fact, according to dozens of studies on dating aggression in college students and young adults (e.g., Hines & Saudino, 2003; Makepeace, 1986; Morse, 1995), about one third of the students reading this book right now are involved in dating relationships in which at least some violence (e.g., slapping, pushing, shoving) has occurred. Some researchers argue that prevention services need to focus on these types of relationships because preventing minor violence in intimate partner and family relationships will result in a large improvement in social and psychological health (M. Straus, personal communication, March 4, 2004). Compared to people not involved in aggressive relationships, men and women in relationships in which minor violence has occurred suffer from more depression, psychological distress, and psychosomatic symptoms (e.g., Stets & Straus, 1990a). Because at least one third of the population have been involved in these types of relationships, eliminating even minor violence from close relationships, Straus argues, would result in vast improvements in mental health. We focus in this Preface on dating aggression because many of you are involved in such relationships; however, consider how much social and psychological health would improve if violence and aggression are eliminated from all forms of family and intimate relationships.

Family and intimate partner aggression, in the form of minor physical violence or verbal aggression, has touched most of us. Most of us have witnessed it, experienced it, and/or used it in our lifetimes. In order for us to eliminate aggressive and abusive behaviors from relationships, we must be willing to confront our own experiences with these behaviors. All aggression matters, whether it is extreme or minor, verbal or physical, committed by men or women, by ourselves or someone else. How often have you been victimized by someone else's cruel words or actions? More important, how often have you done or said something that could harm another person (e.g., boyfriend or girlfriend, brother or sister, mother or father) either emotionally or physically? Because most of us have been socialized to view aggression as something outside of ourselves, confronting our own abilities to behave aggressively is a crucial first step in understanding and eliminating aggression from our lives and the lives of others in this country (Mills, 2003).

Because the bulk of the research on family violence has focused on those most in need of intervention services, much of the research and many of the case examples presented in this book involve individuals subjected to severe forms of family violence. We also present as much research as possible on population-based studies, which include cases of minor violence and psychological aggression as well. As you read this book, consider the quotes at the beginning of this Preface and this discussion of the impact of even minor aggression on people's psychological and social health. Consider also your own life and the lives of your loved ones—how many of them have been involved in relationships in which aggression occurs? How much better do you think your life and/or their lives would be if aggression were not present in your or their closest relationships?

This book addresses all types of family aggression. For the most part, we limit our discussion to relationships in which some type of close, long-term commitment is involved (e.g., parent-child relationship, sibling relationship, husband-wife relationship). Thus, we generally do not include dating relationships, although there are exceptions to this pattern. For example, there is much more research on abuse against men in dating relationships than on abuse against husbands; therefore, in Chapter 8, "Husband Abuse," we include dating violence research in our discussion. Similarly, in our chapter on "Abuse in Gay/Lesbian/Bisexual/Transgender Relationships" (Chapter 9), we incorporated all we could find about aggression in these relationships. In nearly every state, gays and lesbians are denied the right to marry; consequently, we were forced to concentrate on violence in their nonmarital intimate relationships. Because the types of aggression discussed in these two chapters occur at rates at least equal to those of wife abuse, we considered it important to devote a chapter to each of them, even though systematic research is just beginning in these two areas and is certainly not as extensive as research on wife abuse.

In the first three chapters of this book, we put family violence in a cultural context. Chapter 1 begins with a discussion of the problems and controversies surrounding the process of defining family violence and abuse. It also discusses the different perspectives that can be found in the legal, medical, and social service professions, and the differences between these professional points of view and lay opinions. We then introduce a cognitive-affective-ecological conceptual approach, which provides a unifying framework for the book. Chapter 2 discusses the contemporary and ecological context of family violence in the U.S. today, including the role of culture in contributing to, maintaining, and providing rationalizations for violence in interpersonal relationships. Chapter 3 considers a specific cultural context, that of religious affiliations and values, in tolerance for violence in families.

In Chapters 4 through 11, we discuss specific types of aggression in family relationships. Chapter 4 is devoted to child physical abuse, Chapter 5 to child sexual abuse, and Chapter 6 to child neglect and psychological maltreatment. In Chapters 7 through 9, aggression within adult intimate relationships is discussed. Specifically, we provide a discussion of abuse toward wives, husbands, and gays and lesbians within adult intimate relationships. In Chapter 10, elder abuse is discussed, and in Chapter 11 we provide the limited information available on so-called "hidden" types of family violence; that is, violence against siblings, parents, and people with disabilities. In all these chapters, we begin with illustrative case studies, then address definitional issues and provide a discussion of the prevalence of the particular type of aggression. Next, predictors and correlates are discussed, structured according to our conceptual model, followed by information on the possible consequences of each type of aggression. Finally, efforts at preventing and intervening in family violence are presented. Within these chapters, we also provide information on the most extreme type of family violence, that of homicide of family members. Most chapters also have a "special issues" section addressing a specific issue related to that particular type of family violence. Our book concludes with a discussion of how we as a society are currently responding to the problem of family violence; the majority of this discussion is devoted to the criminal justice response.

Our decision to write this book was a result of writing our first book with Sage Publications. When writing that book, *Family Violence in a Cultural Perspective*, we became so involved in the research that we could not contain everything we wanted to write in just one book. Our editor at Sage, Jim Brace-Thompson, noticed the problem and suggested the possibility of a second book, a possibility we had already thought of but had not yet mentioned to him. We were thrilled that all of us were on the same page, and we would like to thank Jim for giving us the opportunity to write this second book. We would also like to thank Karen Ehrmann, Jim's editorial assistant, for the valuable help she provided in getting the manuscript ready for publication. Also, our thanks to the manuscript's reviewers: William Curcio, Montclair State University; Phil Davis, Georgia State University; Elwood W. Hamlin II, Florida Atlantic University; and Margie L. Kiter Edwards, University of Delaware.

Some of Kathie's undergraduate students also provided much-needed help with tracking down sources, doing Internet searches for important documents, and putting together the reference section. They include Kristine Burkman, Johnna Seaward, and David Oh. Members of the Family Research Laboratory and the Crimes Against Children Research Center at the University of New Hampshire, including Murray A. Straus, David Finkelhor, Theodore Cross, Kimberly Mitchell, Kathy Becker-Blease, Emily Douglas, Wendy Walsh, Janis Wolak, Lisa Jones, Melissa Holt, and Glenda Kaufman Kantor, provided valuable feedback on earlier versions of this manuscript. Denise would like to acknowledge the support of her many family members and friends through the years. Her parents especially have supported her and expressed much pride and love at every step along the way. Likewise, Kathie extends her appreciation to her family and to her friends David and Marilyn Glater and Eli and Carolyn Newberger.

1

Issues in the Definition of Family Violence and Abuse

Women can verbally abuse you. They can rip your clothes off without even touching you, the way women know how to talk, converse. But men . . . weren't brought up to talk as much as women do . . . So it was a resort to violence, if I couldn't get through to her by words . . . On some occasions she was the provoker. It didn't call for physical abuse . . . [but] it did call for something. You know, you're married for that long, if somebody gets antagonistic, you want to defend yourself. (Ptacek, 1998, p. 188)

What are your views on this case? If this man struck his wife when she antagonized him, has he been abusive? How about his wife? If she verbally attacked him, put him down, tried to antagonize him with her words, was she also guilty of spousal abuse? Is one form of abuse (e.g., physical aggression) more abusive than another form (e.g., verbal aggression)? What is the basis for your judgment? Does abusive behavior by one spouse justify retaliatory aggression?

He always wanted to have sex. He was jealous and if he didn't have sex with me every single day that means that I was with another guy and that was his theory. From the time I was 18, I had sex every single day for the first year we were married . . . [We] did it every day because he wanted to and I thought I had to. (Bergen, 1998, pp. 240–241)

Is the woman who reported this experience a victim of domestic violence, wife abuse, or maltreatment? Apparently she did not resist his advances even if she was often unhappy with his insistence on daily sex. If she did not resist, can his behavior be called "marital rape"?

> His hand on my throat, pressing me into the bed . . . I never called it . . . rape. I called it rough sex. Forcing himself on me. Being selfish and inconsiderate, a beast, a monster. He called it getting what he wanted. What he was entitled to. (Letellier, 1999, p. 10)

Does this appear to be a more obvious case of marital rape? A definite case of domestic assault? The victim in this case is a gay male. Does that affect your judgment of the case in any way? If so, how? How likely is it that the victim in this case will be able to find appropriate support services to help him deal with this relationship and its effects on him?

> When my son was a toddler . . . he would often attempt to squeeze past the front door where stone steps awaited his fall. Verbal reprimands and redirecting his attention elsewhere were fruitless, as he attempted time and again to get out . . . Rather than allow him to experience for himself the consequences of wandering too close to those steps, I swatted him smartly a couple of times on his diapered behind . . . ! It took two more swattings before he became convinced of the certain connection between trying to get out the front door and the painful consequences, but after that, he needed no more reminders! (Newberger, 1999, p. 79)

The woman who told this story did so proudly, pleased with her ability to discipline her son. What is your view? Is corporal punishment an appropriate response to self-endangering behaviors in a toddler? Is it appropriate in other circumstances? Or, is it possible that there are always better methods for dealing with child behavior that is considered undesirable?

> After being hospitalized with a broken hip, 80-year-old Mr. Jaffin began living with his daughter and son-in-law. His daughter repeatedly berated him for not being able to clean his room and began referring to him as their third child. Reminding him what an avid golfer he'd been before the accident, she ridiculed him for now being unable to go to the mailbox without his walker and assistance from a family member. Within a month, Mr. Jaffin began to feel worthless and withdrew from his family and friends. He spent the day in his room sleeping or watching television, no longer socializing with or even phoning old friends. (Humphries Lynch, 1997)

What are your views of Mr. Jaffin's situation? His daughter and son-in-law have taken him in, provided him with his own room and television, presumably keep him warm, safe, and well fed, and try to motivate him to be the active man he once was. Given these circumstances, would you say he is being maltreated in any way? Should the medical personnel who are overseeing his recovery from the broken hip intervene in the family situation? If so, how?

In the United States, one can expect considerable disagreement on the answers to these questions. Consider the first case. The problem of wife abuse has received a great deal of attention as an important social problem in recent years, but many people argue that husband abuse is not a social problem in need of a societal response, however painful it may be for its victims (e.g., Kurz, 1993; Walker, 1990). The Violence Against Women Act (VAWA) of 1994, expanded in 1998 and 2002, was designed to provide women with broad protections against violence not only in their communities but also in their homes, but society has been slower to provide protections for male victims of intimate abuse. As recently as July 2003, a Los Angeles man filed a sex discrimination lawsuit against ten local battered women's shelters because none of them would provide him with a bed (Zwerling, 2003). Some people assume that cases where women abuse their partners are so rare there is no need to provide shelters for their victims. Others assume that shelters for men are unnecessary because most women, like the one in Case 1, are only verbally abusive, or that verbal aggression such as name-calling is in poor taste but calling it abusive is a misuse of the term. What do you think of these arguments?

Consider the other cases. Does marriage—or cohabitation, for that matter—entitle individuals to engage in sexual behavior with their partners whenever and however they want? Although most readers would probably say "No," when the second author of this book submitted an article on dating aggression to a major interpersonal violence journal, one reviewer criticized the article for including the item, "pressured you to have sex in a way you didn't like or didn't want," in the (previously published) measure of interpersonal abuse completed by the participants. According to that reviewer, pressuring someone to have unwanted sex is abusive only in the context of prior physical assault. What is your view of this?

Battered women who have also experienced sexual abuse often find that shelters provide very little assistance with the aftermath of their sexually abusive experiences. Historically, many shelters did not consider cases of marital rape to fall within their domain because it is not life threatening, and many rape crisis centers did not want to deal with wife abuse victims;

consequently, battered women who had also been sexually assaulted were sometimes shuffled back and forth, unaided, between facilities (see Malley-Morrison & Hines, 2004, Chapter 15). This situation appears to be changing, with efforts being made to promote greater responsiveness to wife rape victims by domestic violence shelters (Wellesley Center for Women, 1998).

For many decades after the public recognition of child abuse and wife abuse as serious social problems, members of the gay/lesbian community, fearing additional stigmatization, were reluctant to admit that abuse took place within their relationships. Only recently have victims of intimate violence in those communities begun to speak out, although services are still limited. There are now some states that legally recognize gay/lesbian unions (e.g., Vermont, Massachusetts), and many gay/lesbian couples are raising children. In response to these developments, many researchers have begun studying violence in gay/lesbian relationships.

The extent to which corporal punishment should be considered abusive is very controversial. Although the percentages of adults who approve of corporal punishment may be declining, a majority of parents in the U.S. spank their children and consider spanking appropriate and necessary (Straus, 1994), despite the fact that many professional organizations, such as the American Psychological Association and the American Academy of Pediatrics, have issued statements recommending that children not be subjected to corporal punishment.

Finally, should we view Mr. Jaffin's feelings and behavior simply as inevitable outcomes of the medical and other problems associated with aging, or is he a victim of elder maltreatment? He is not being physically abused. Nobody is trying to hurt or exploit him. But is he being emotionally abused? If so, how should he be helped? When he went to his orthopedist for a follow-up on his hip, he was referred to a psychologist who diagnosed him as clinically depressed. This psychologist began seeing him and his overwhelmed daughter regularly and helped them both work out a better way of dealing with their circumstances (Humphries Lynch, 1997). Given that elder abuse is a reportable offense, should the hospital personnel have notified Adult Protective Services rather than privately initiating the counseling program?

Definitional Issues

At the heart of many of the debates concerning whether particular behaviors are abusive are inconsistencies in the definitions of terms. Definitions of abuse, for example, have varied in the extent to which they incorporate assumptions about causes (e.g., people who hurt the ones they love are "sick"); effects (e.g., abusive behaviors are those that cause harm); motivations (e.g., abusive behaviors are intended to hurt rather than discipline); frequency (e.g., slapping is abusive only if it is chronic); and intensity (e.g., hitting is abusive if it is hard enough to cause injury). Such definitions, which vary in their inclusiveness and differ within and across fields, influence the likelihood that individuals subjected to unwanted behaviors within domestic settings will receive interventions from the legal, medical, and/or social service communities.

Efforts to distinguish among terms such as violence, abuse, and maltreatment have not led to any consensus. Definitions continue to vary in their inclusiveness (how broadly the construct is defined) and their abstractness (the extent to which they focus on specific behaviors or define one abstract construct in terms of another). For example, Levesque (2001) held that "family violence includes family members' acts of omission or commission *resulting in* physical abuse, sexual abuse, emotional abuse, neglect, or other forms of maltreatment that hamper individuals' healthy development" (p. 13, italics added). Emery and Laumann-Billings (1998) distinguished between two levels of *abuse—maltreatment* (i.e., minimal or moderate forms of abuse, such as hitting, pushing, and name-calling) and *violence* (i.e., more violent abuse involving serious endangerment, physical injury, and sexual violation). Here, *abuse* is the broader term, and *maltreatment* and *violence* are considered subtypes of abuse, varying in level of intensity. According to the American Academy of Family Physicians (2004), "Family violence can be defined as the intentional intimidation or abuse of children, adults, or elders by a family member, intimate partner or caretaker to gain power and control over the victim. Abuse has many forms, including physical and sexual assault, emotional or psychological mistreatment, threats and intimidation, economic abuse and violation of individual rights" (paragraph 7). Thus, the Academy defined family violence as abuse, emphasized the intention of power and control, and included "mistreatment" as a form of abuse. Finally, Straus, in his early work (Straus, Gelles, & Steinmetz, 1980), distinguished between *socially accepted violence* (e.g., spanking) and *abusive violence,* defined as an "act which has a high potential for injuring the person being hit" (pp. 21–22).

One of the biggest debates in the field is whether corporal punishment should be considered inherently abusive. Straus and Yodanis (1996) defined corporal punishment as "the use of physical force with the intention of causing pain but not injury for purposes of coercion and control" (p. 826)—thus emphasizing both intent and expectations concerning outcomes. Straus and Runyan (1997) noted that most cases of physical abuse happen when corporal punishment gets out of control, and that ordinary corporal punishment of adolescents is associated with a heightened risk for many social and psychological problems. If child abuse is defined as behaviors that put children at risk for injury, and both psychological and physical injuries are considered, then there is a basis for considering corporal punishment abusive because of the demonstrated negative effects of corporal punishment. Straus (1994) has made this argument in his efforts to ban corporal punishment in the United States, as it is in many European countries. There has been considerable resistance to a ban among professionals as well as laypeople. Box 1.1 provides a sampling of major social science perspectives on the issue. Each of these authorities presents empirical data in support of his or her position, yet there are no signs that the differences will be resolved soon.

Box 1.1   Is Corporal Punishment Abusive?

Corporal punishment of children is essentially a legalized form of assault. Acts of "minor violence" that would be crimes if committed on an adult are legal when they occur as "discipline." (Berliner, 1988, p. 222)

In the toddler years, parents should first respond with the least aversive discipline they think will stop the misbehavior. If that does not elicit compliance, they should then turn to more aversive responses . . . If the child still fails to comply . . . , such noncompliance should result in a mild prescribed spanking. (Larzelere, 1994, p. 205)

Almost without exception, when harsh punishment is mentioned and has [negative] long-term consequences . . . what is being referred to is corporal punishment. (Carey, 1994, p. 1007)

Authoritative parents endorse the judicious use of aversive consequences, which may (but certainly need not) include spanking . . . The prudent use of punishment within the context of a responsive, supportive parent-child relationship is a necessary tool in the disciplinary encounter. (Baumrind, 1997, p. 330)

Regardless of the numerous factors . . . that might provoke the use of spanking in a given instance, even the most abusive parent believes he or she is spanking as a response to some child behavior that requires discipline. (Benjet & Kazdin, 2003, p. 220)

In our opinion, motivating parents to change from corporal punishment to alternative methods of discipline would seem to be the most productive public mental health program known. (Ontario Consultants on Religious Tolerance, 2003, paragraph 4)

Another perspective comes from Emery (1989), who holds that "calling an act 'abusive' or 'violent' is not an objective decision but a social judgment, a judgment that is outside of the realm of responsibility of social scientists" (p. 322). Similarly, Zuriff (1988) argued that "the definition of psychological maltreatment is not a task appropriate for psychologists as scientists or researchers . . . [The] problem of defining 'maltreatment' is one of determining a point on a set of continua at which the psychological effects of parental behavior are to be designated 'harmful.' I suggest this is not an empirical question . . . [Psychologists] should leave the determination of good and evil, benefit and harm, to the law, ethics, and religion" (p. 201).

While agreeing that terms like maltreatment represent social constructions and value judgments, we disagree that psychologists and other social scientists cannot aid in the definitional process by means of scientific data and scientific thinking. Social and medical scientists are in some ways uniquely qualified to provide evidence concerning the harmfulness of particular behaviors for the well-being of their recipients, others with whom those recipients interact, and even the larger community within which the recipients of those behaviors must function. Indeed, in considering the kinds of behaviors and interactions that may be harmful to members of families (broadly defined to include gay/lesbian relationships and cohabiting couples), we prefer the term *maltreatment* to the other commonly used terms, in part because of the explicit value judgment built into the prefix "mal."

Our term *maltreatment* embraces *corporal punishment* as well as *abuse, family violence, wife beating, domestic violence, spousal abuse,* and *elder abuse,* as these are commonly defined. We acknowledge that some forms of maltreatment are more serious than others. Children who receive a single slap on the hand or the buttocks during childhood are not being maltreated to the same degree as a child who is raped, or beaten every day, or constantly criticized and humiliated. However, we view all these behaviors as forms of maltreatment, and not beneficial ways for individuals to treat each other, inside or outside of families. As Straus has repeatedly pointed out, even acts that seem like relatively minor forms of maltreatment (e.g., spanking, name-calling) are risk factors for negative outcomes for individuals and society (Straus, 2001; Vissing, Straus, Gelles, & Harrop, 1991). Although our conceptual preference is for the term *maltreatment,* most researchers in family violence study the more extreme forms of maltreatment, and therefore, throughout this book, we generally use the term that the researchers used to describe the particular form of maltreatment of interest to them.

Definitions of terms such as maltreatment are embedded in broader perspectives on human beings, families, and intimate relationships. During the second half of the 20th century, new perspectives emerged within the international community, including the view that the more vulnerable members of the human race (particularly women, children, the elderly, and people with mental and physical disabilities) have an inherent right to freedom from exploitation and abuse. Concurrent with the evolution of that perspective, many countries criminalized forms of family aggression that had a long history of normative acceptance—for example, the beating and rape of wives and children. Accompanying the criminalization of such behaviors has been a *medicalization* of their effects (Newberger & Bourne, 1978). *Medicalization* refers to perceiving a behavior, such as child maltreatment, as a medical problem or illness, and expecting the medical profession to treat the problem. The medical communities in many countries, including the United States, have increasingly been given and/or have assumed the responsibility not just to heal intentional burns, set broken bones, and mend bruised and battered skin, but to alert legal and social service agencies about behaviors now deemed abusive.

Just as the concept of "family" has been broadened to include nonmarital cohabiting relationships and same-sex intimate relationships, legal protections against spousal abuse have increasingly been expanded to include nonmarital relationships. Also, because most definitions of abuse emphasize negative outcomes, the social science community has directed intensive efforts at providing a scientific basis for defining, studying, and intervening in situations of family violence and abuse. In the next sections, we provide a brief introduction to major perspectives on maltreatment in family settings. Many of these perspectives reflect assumptions held before individuals selected a profession or developed as part of their professional training and experience. These perspectives, which may guide important decisions concerning the current or future well-being of victims of family maltreatment, may or may not have a solid theoretical or empirical basis. This section is followed by an overview of several theories of familial maltreatment. During the past several decades, increasing work has been done to empirically test such theories in order to improve our understanding of the predictors and consequences of maltreatment and to provide a foundation for intervention and prevention efforts.

Perspectives on Maltreatment

The Human Rights Perspective

One major view on human rights is that they are privileges granted by people in power to those who are less powerful. For much of human history, women and children were seen as having no rights separate from those that men offered them—and such rights were generally extremely limited. A second major view is that human rights are *inherent* in being human. From this latter perspective, human rights are, in the words of Amartya Sen (1998), "entitlements of every human being" (paragraph 1). This second view that is embodied in the United States Declaration of Independence: "We hold these truths to be self-evident, that all men are created equal, that they are endowed by their Creator with certain unalienable Rights, that among these are Life, Liberty and the pursuit of Happiness." It is also embodied in international human rights agreements promulgated by the United Nations and other nongovernmental organizations (NGOs).

Emerging from the horrors of World War II, wherein "disregard and contempt for human rights have resulted in barbarous acts which have outraged the conscience of mankind," the newly born United Nations adopted the task of establishing a lasting peace. One of its first accomplishments (1948) was the Universal Declaration of Human Rights, which proclaimed "all members of the human family" have "equal and inalienable rights" and that recognition of these rights is "the foundation of freedom, justice and peace in the world" (Universal Declaration, Preamble, paragraph 1). Article 5, which is most relevant to family maltreatment, states, "No one shall be subjected to torture or to cruel, inhuman or degrading treatment or punishment."

Since the passage of the Universal Declaration, the United Nations has promulgated other international treaties addressing the rights of individuals to freedom from maltreatment even within their own families. The Convention on the Rights of the Child (1989) specifies that member states "shall take all appropriate legislative, administrative, social and educational measures to protect the child from all forms of physical or mental violence, injury or abuse, neglect or negligent treatment, maltreatment or exploitation, including sexual abuse, while in the care of parent(s), legal guardian(s) or any other person who has the care of the child" (Article 19). According to this Convention, assuring such rights to children is necessary in order to rear them "in the spirit of the ideals proclaimed in the Charter of the United Nations, and in particular in the spirit of peace, dignity, tolerance, freedom, equality and solidarity." Thus, the international promulgators of this document, like many social scientists in the United States, recognize a connection between eschewing violence in the home and promoting international peace.

Child advocates in many countries have argued that corporal punishment violates the United Nations Convention on the Rights of the Child. The European Network of Ombudsmen for Children (ENOC; 2001) urged the governments of all European countries as well as NGOs concerned with children to work to end all corporal punishment. In their view, "eliminating violent and humiliating forms of discipline is a vital strategy for improving children's status as people, and reducing child abuse and all other forms of violence in European societies." ENOC concurred that no level of corporal punishment is compatible with the Convention on the Rights of the Child and that legal and educational steps should be taken to eliminate it. The United States is one of only two countries (the other being Somalia, which has no central government) that have not ratified the Convention on the Rights of the Child, and parents in the United States appear to be very resistant to the notion that corporal punishment may violate a child's rights.

Another important declaration adopted by the United Nations General Assembly was the Declaration on the Elimination of Violence Against Women, endorsed by all member states of the United Nations. According to this Declaration, "violence against women means any act of gender-based violence that results in, or is likely to result in, physical, sexual or psychological harm or suffering to women, including threats of such acts, coercion or arbitrary deprivation of liberty, whether occurring in public or in private life" (Declaration on the Elimination of Violence against Women, 1993, Article 1). Other NGOs taking a stand against maltreatment in domestic settings include Amnesty International, which in 2001 released a statement asserting that violence against women is a human rights issue, and that if a government fails to "prevent, prosecute and punish" acts of violence, those acts should be considered forms of torture—and therefore a violation of the United Nations Declaration on Human Rights.

The international human rights perspective emphasizes the relationship between social justice and individual rights to freedom from abuse, and between peaceful resolution of conflict in the home and peaceful resolution of conflict in the international community. Proponents of a human rights perspective are often critical of *systemic* or *structural* abuse—that is, abuse of individuals by the very systems or structures responsible for protecting them (e.g., Hearn, 1988). One human rights advocate has argued that attention to international human rights principles can help Americans "move away from practices and assumptions that condone, encourage, and improperly respond to family violence" (Levesque, 2001, p. 17).

Legal/Criminal Justice Perspectives

Although the United Nations Convention on the Rights of the Child has some legal status in international law, it has functioned not so much to enforce children's rights judicially or to criminalize violation of those rights, as to establish a universal standard that the international community has agreed to adopt. To our knowledge, the World Court has not tried any cases of family maltreatment. However, the European Court of Human Rights, established by the European Convention On Human Rights and Its Five Protocols, has addressed cases of family violence originating in a number of different European countries.

In general, the legal approach to family maltreatment in the United States has been to criminalize it. The focus is on both punishment and deterrence. Criminalization has involved mandating members of medical and social service professions to report suspected cases of abuse and imposing criminal penalties on perpetrators of acts identified as abusive. Although the United States has not ratified the Convention on the Rights of the Child, it has ratified the Declaration on the Elimination of Violence Against Women and has criminalized abuse of children, domestic partners, and the elderly. According to the federal Child Abuse Prevention and Treatment Act (CAPTA), "child abuse and neglect is, at a minimum, any recent act or failure to act on the part of a parent or caretaker which results in death, serious physical or emotional harm, sexual abuse or exploitation of a child (individual under the age of 18) and any act or failure to act which presents an imminent risk of serious harm" (42 U.S.C. 5106g). However, each state has its own set of laws, and, in contrast to the stance taken in many European countries, corporal punishment by parents is legal in every state.

The Violence Against Women Act (VAWA), passed in 1994 as part of an Omnibus Crime Bill, and modified in 2002, was revolutionary in its provisions for addressing violence against women, including wife abuse. The federal Older Americans Act provides definitions of elder abuse and authorizes expenditure of federal funds for a National Center on Elder Abuse but does not fund adult protective services or shelters for abused older persons. Every state has its own set of statutes criminalizing abuse of women and elders and its own procedures for investigating complaints and prosecuting violators. Actual practices often fall far short of the intent of the law; however, there has been enormous change since the days when the criminal justice system saw itself as not concerned with any violence short of murder when it took place behind the closed doors of the family home (Iovanni & Miller, 2001). Nevertheless, although there are laws addressing family violence against children, wives, and the elderly, no legislation deals specifically with family violence against siblings or husbands, who are also frequent victims of maltreatment in the family.

Although physical assault of wives has received increasing attention over the years, *marital rape* has been a virtual oxymoron until very recently. The so-called "marital rape exemption," mandating that forced sex of a wife by a husband could not be considered a form of rape, had its basis in English common law, according to which wives, by virtue of the marital contract, gave themselves willingly and irrevocably to their husbands. The nature of the marital contract was interpreted as negating the possibility of marital rape and ensuring the husband's right to have his desires satisfied by his wife (Bergen, 1998). Consistent with this perspective, rape was traditionally defined as a male's "sexual intercourse with a female, not his wife, by force and against her will" (Finkelhor & Yllo, 1985, p. 1). It was not until July 5, 1993 that all states had enacted legislation to criminalize the rape of wives; however, many laypeople are unaware that wife rape is now considered a crime, and still others do not believe it can or should be a crime (Malley-Morrison & Hines, 2004). There are still no laws against sexual assault of husbands, although we know that this form of sexual aggression takes place.

Although the principal legislation relating to domestic maltreatment provides funding for educational and social service programs, the legal perspective emphasizes the criminal justice system response to violation of federal and state statutes. Studies using legal definitions of abuse typically report the number of cases of identified child, spouse, or elder abuse reported to protective service agencies. Such reports provide a vast underestimation of the actual frequency of maltreatment in families because many cases are never reported to any agency. Moreover, many statutes related to maltreatment have exemptions. For example, in every state, the child abuse statutes have exemptions allowing parents to use "reasonable force" for purposes of child discipline and control. Legal definitions of physical abuse reflect cultural norms concerning what exceeds reasonable force, but the boundaries between physical abuse and corporal punishment have been generally left to the discretion of the legal and criminal justice systems (Straus & Runyan, 1997). Moreover, many states still have exemptions from prosecution for a husband raping his wife (Rennison, 2002), such as when he does not have to use force to make her have sex (e.g., if she is physically or mentally impaired and unable to give consent).

Medical Perspectives

Maltreatment in families has been recognized not just as a human rights and a legal issue but also as a medical issue. On an international level, the World Health Organization (WHO, 1997) identifies violence in the home as a public health problem. Within the United States, professional organizations such as the American Academy of Family Physicians (2004) have also noted that family violence is a public health issue of epidemic proportions. The medical perspective on maltreatment tends to focus on recognizing symptoms, identifying causes, and providing treatment. Medical practitioners frequently emphasize the causes of maltreatment having a biological component (e.g., substance abuse, psychiatric disorders). Perpetrators are often viewed as victims themselves and more in need of treatment than of criminal prosecution. For this and many other reasons (including assumptions that the social welfare system does not always respond appropriately), medical personnel often do not report the cases of maltreatment they are mandated to report (Zellman & Fair, 2002) (see Chapter 12).

Social Service Perspectives

The social service system has generally had a much broader perspective on family violence than the medical or legal systems, traditionally viewing maltreatment within family settings as a symptom of family crisis and a need for services. The social service system has been more concerned with ameliorating conditions that give rise to maltreatment than with promoting the prosecution of offenders or providing medical treatment of victims. Much of the emphasis on acts of omission (neglect) in definitions of child and elder maltreatment is derived from social service perspectives. Workers within the field have often emphasized the role of external forces—for example, poverty and discrimination (Beckett, 2003)—in contributing to family maltreatment. Of all the relevant systems, the social service system probably has the greatest familiarity with social science research relevant to family maltreatment, its causes and outcomes.

Need for Multidisciplinary Cooperation

In many cases of family maltreatment, representatives of the legal, medical, and social service professions all become involved. A coordinated approach of these various services is often hard to achieve because of the differing definitions and perspectives within these professions. Members of the legal profession want to pursue prosecution of the perpetrator if they believe they can "win" their case. Medical practitioners are more concerned with providing treatment for victims and perhaps perpetrators, but typically see it as beyond their purview to address any problems of poverty, community violence, and despair besetting the family. Social service personnel may believe that any focus on helping, prosecuting, or changing individuals is shortsighted, and emphasize the need to find better housing and employment for family members and address substance abuse problems. Perhaps in part because of the very breadth of their perspective, social service systems have been overwhelmed by family violence cases in recent decades and are not always able to respond appropriately. A number of legal cases have been brought against local social service agencies for maltreating their clients.

Box 1.2   Brief of DeShaney v. Winnebago County Dept. of Social Services

---

Born in 1979, Joshua DeShaney was placed in his father's custody when his parents divorced a year later. Joshua's father's second wife and neighbors reported that Joshua was frequently abused by his father. Following a police report and hospital treatment of Joshua's bruises and abrasions in January 1983, the local Department of Social Services (DSS) obtained a court order to keep Joshua in the hospital's custody. However, a child services protective team returned Joshua to his father three days later when his father agreed in writing to enroll Joshua in Head Start and enter counseling himself. A few weeks later DSS was informed that Joshua had again been seen at the hospital but concluded there was no evidence of abuse.

During 1983, a social worker visited the father's home five times. Although she observed that Joshua had bumps and scrapes on several occasions and that the father was not adhering to the terms of his agreement, she took no action. There was also no action taken when the hospital reported in November 1983 that Joshua had again been treated for suspicious injuries. When the social worker visited the home in January 1984, she was told she could not see Joshua because he had the flu. When she tried again to see him in March, she was told that he had recently fainted, but she did not request to see him. The next day Joshua's father beat him severely, causing brain damage and permanent retardation. A medical examination revealed evidence of multiple previous injuries to Joshua's head and body.

SOURCE: Adapted from online news reports.

Consider the case of Joshua DeShaney in Box 1.2. How did the differing perspectives of the various relevant agencies play out in his case? At what points in the process did systems fail him? Do any of the systems seem particularly culpable? After the final critical beating, Joshua's father was arrested, indicted, tried, convicted, and sent to jail for child abuse. Subsequently, Joshua's mother and Joshua (represented by a guardian ad litem) sued the county Department of Social Services (DSS) for depriving Joshua of "his liberty interest in bodily integrity, in violation of his rights under the substantive component of the Fourteenth Amendment's Due Process Clause" (*DeShaney v. Winnebago County Department of Social*

Services, 1989, p. 1). Their argument was that by failing to intervene in a way that protected Joshua against his father's violence, the DSS violated the Fourteenth Amendment statement that "[n]o State shall . . . deprive any person of life, liberty, or property, without due process of law." Joshua and his mother lost their case and appealed ultimately all the way to the Supreme Court, where they again lost. In a dissenting opinion, Justices Brennan, Marshall, and Blackmun argued, "The most that can be said of the state functionaries in this case . . . is that they stood by and did nothing when suspicious circumstances dictated a more active role for them. [We] cannot agree that respondents had no constitutional duty to help Joshua DeShaney" (*DeShaney v. Winnebago County Department of Social Services*, 1989, p. 17). In recent times, more cases have been brought against social service agencies. Does that appear to be the best way to deal with the staggering problems of maltreatment in society today? What other approaches might be better?

Disrespect for each other's professions may often hamper cooperation among representatives from different agencies. For example, although several United States Supreme Court decisions in the post-World War II years (e.g., *Brown v. Board of Education*, 1954; *In re Gault*, 1967) provided some recognition that juveniles have rights protected by the Constitution, more recent decisions by a more conservative Supreme Court have eroded some of these rights, in part because of a decreased willingness to attend to social science data (Walker, Brooks, & Wrightsman, 1999). For example, "Justice Scalia consistently has considered social science studies to be irrelevant when deciding on constitutional law; for him, the only 'empirical' materials of relevance . . . are legislation and jury decisions" (Walker et al., 1999, p. 11).

Ecological Models of Maltreatment

Many theories have been formulated about various forms of family violence, and most of these theories reflect broader views (paradigms) about human nature. *Theories* are useful both to organize and integrate knowledge and to guide research (Baltes, Reese, & Nesselroade, 1977). *Paradigms* are more abstract and general. Recognizing the place of theories within paradigms can help reduce the confusion that can come from trying to understand and evaluate what may seem like an overwhelming morass of theories about family violence. The dozens of competing theories concerning the causes of child, intimate partner, and elder maltreatment can all be incorporated into an ecological paradigm—which we do in the sections that follow.

In general, the prevailing ecological paradigm within the field of family violence derives from the work of Bronfenbrenner (1979), who argued that human development and behavior should be analyzed within a nested set of environmental contexts or systems. The *microsystem* consists of the relations between developing individuals and their immediate settings (e.g., the home). The *mesosystem* consists of relations among the settings in which the developing individual is involved (e.g., between home and school). The *exosystem* includes the larger neighborhood, the mass media, state agencies, and transportation facilities. Finally, the *macrosystem* consists of broad cultural factors, including views about the role of children and their caretakers in society. Building on Bronfenbrenner's model, researchers have identified causes of child maltreatment (Belsky, 1993), spousal abuse (e.g., Dutton, 1985) and elder abuse (e.g., Schiamberg & Gans, 2000) at different ecological levels. In an important modification of the theory, Belsky (1993) argued that the ecological system includes an *ontogenetic* or *individual/developmental* level—that is, the unique biological/genetic characteristics that exist even before birth and that individuals bring to every interaction. These biological/genetic characteristics change during the process of development under the influence of both nature and nurture.

Inherent within an ecological perspective is the dictum that, to understand how so many people can maltreat family members or other intimates, we need to understand the genetic endowments of those individuals, the microsystem in which they grew up, the microsystem in which they are currently embedded, characteristics of the neighborhood within which their family functions (including the availability of social support and social services, and relationships between the community and the criminal justice system), and the larger society that embraces all the separate neighborhoods. From this ecological perspective, maltreatment is the product of the genetic endowments, behaviors, cognitions, and affects of the individual at the center of the nested set of ecological contexts, as well as the genetic endowments, behaviors, cognitions, and affects of the other actors at each ecological level.

There have been and continue to be single-factor or single-process theories of maltreatment that focus on causes at just one level of the ecological framework. Empirical research addressing hypotheses concerning causes of maltreatment has confirmed that there are identifiable risk factors at every ecological level. Table 1.1 provides a brief summary of the basic assumptions of several current theories of maltreatment in families. It also indicates the ecological level being addressed by each theory, and an example of at least one empirical study addressing hypotheses associated with that theory.

Table 1.1 Major Theories of Family Violence within an Ecological Paradigm

Theory	Supportive Study	Focus	Key Assumptions and/or Findings
BIOLOGICAL INDIVIDUAL/DEVELOPMENTAL THEORIES			
Behavioral Genetics	Hines & Saudino (2004a)	Intergenerational transmission of intimate aggression	Intergenerational transmission of intimate partner aggression is due to shared genes. Genetic influences account for approximately 20% of the variance in physical and psychological intimate aggression.
Other Biological Theories	Frodi & Lamb (1980)	Association of child abuse with physiological reactivity	Child abusers are more physiologically reactive to a child's cry and smile, suggesting that they find both of these aversive.
	Wolfe, Fairbank, Kelly, & Bradlyn (1983)	Association of child abuse with physiological reactivity	Child abusers have large increases in physiological reactivity, as measured by skin conductance and heart rate, in response to stressful and nonstressful mother-child interactions.
	Hucker et al. (1988); Langevin et al. (1988); Wright et al. (1990)	Temporal lobe dysfunction and incest	Incest offenders and pedophiles show abnormalities within the temporal lobes, abnormalities that do not characterize any other types of offenders studied.
	Rosenbaum (1991); Rosenbaum & Hodge (1989)	Association of male battering with head injuries	In comparison to nonbatterers, wife batterers have had significantly more head injuries in their histories, the overwhelming majority of which preceded the intimate partner aggression.
	Booth & Dabbs (1993)	Association between testosterone and wife abuse	29% of high-testosterone versus 23% of low-testosterone men said they had hit or thrown things at their wives.
	Soler, Vinayak, & Quadagno (2000)	Association between testosterone and wife abuse	Higher testosterone levels were related to higher physical and verbal aggression against wives, even after controlling for demographic characteristics and alcohol consumption.
NONBIOLOGICAL INDIVIDUAL/DEVELOPMENTAL THEORIES			
Attachment	Alexander (1992)	Theory re: child sexual abuse	Insecure attachment may either help set the stage for sexually abusive behavior or interfere with its termination.
	Dutton (1995)	Husband to wife physical violence	Fearful attachment associated with men's propensity for psychological aggression against partners.
	George (1996)	Harsh punishment mother to child	Mothers with disorganized attachment style felt helpless and out of control with children, viewed children as out of control, described harsh punishment.
	Moncher (1996)	Physical child abuse mother to child	Mothers with insecure attachment style were at significantly higher risk for child abuse than mothers with secure attachment style.

(Continued)

Table 1.1 (Continued)

Theory	Supportive Study	Focus	Key Assumptions and/or Findings
Social Information Processing (SIP)/ Cognitive Behavioral	Holtzworth-Munroe (1992)	Theory re: wife physical abuse	Social skills deficits related to decoding social information, making decisions in social situations, and responding competently lead to marital violence in men.
	Milner (1993)	Theory re: child physical abuse	Perceptual distortions, interpretations of noncompliant child behavior as hostile, evaluations of noncompliance as morally wrong, inattention to mitigating events, and power-assertive techniques lead to child abuse.
	Hastings (2000)	Verbal and physical abuse of women	Personality variables with a social information processing orientation (e.g., adversarial sexual beliefs) were more closely related with aggression reported on the Conflict Tactics Scales (CTS) than more general attitudinal variables.
	Montes et al. (2001)	Child physical abuse	Consistent with SIP theory, mothers at risk for child abuse reported more hostile intent, aversiveness, annoyance, and use of power-assertion discipline.
	Haskett et al. (2003)	Child physical abuse	Unrealistic parental expectations, misattributions re: child's behavior, and overreactions to misbehavior produce child abuse.
Social Learning	Howell & Pugliesi (1988)	Men's minor or severe violence against wives	Observation of aggression between parents during childhood positively associated with use of violence against wives in adulthood.
	Gelles & Cornell (1990)	Intergenerational transmission of family violence	Individuals learn to be violent from growing up in violent homes, where they observe violent models, see violence being reinforced, and learn justifications for violence.
	Simons et al. (1991)	Intergenerational parenting practices	Parents who behaved aggressively with their children had parents who had used similar aggression with them when they were growing up.
	Mihalic & Elliot (1997)	Men's and women's minor and severe marital violence	In women, witnessing parental violence in childhood positively associated with higher rates of child abuse and less marital satisfaction; in turn, less marital satisfaction was positively associated with marital violence perpetration and receipt.
	Burton et al. (2002)	Sexual offending by adolescent males	Nearly 80% of a sample of male offenders had themselves been sexually victimized, typically by a father or other male relative, but often by perpetrators of both genders.
Psychiatric (including substance dependence)	Beasley & Stoltenberg (1992)	Characteristics of wife-abusing men	Compared to distressed but nonabusive men, abusive men viewed their parental relationships more negatively, had witnessed more conflict between their parents, and showed elevated scores on several clinical symptom scales.
	Dutton (1994)	Characteristics of wife-abusing men	Borderline personality disorder significantly positively related to the frequency of anger in intimate relationships, to chronic trauma symptoms, and to abusiveness.

(Continued)

Theory	Supportive Study	Focus	Key Assumptions and/or Findings
	Hwalek et al. (1996)	Analysis of 552 cases of substantiated elder abuse in Wisconsin	When the perpetrator was a substance abuser, the case was more likely to involve physical and emotional abuse and less likely to involve neglect. Substance-abusing perpetrators were more likely to be men, children of their victims, and not primarily responsible for caretaking. Victims of substance-abusing perpetrators were also likely to be substance abusers.
	Gelles (1999)	Theory re: family violence	Offender's personality characteristics are chief determinants of violence against intimates. Personality and character disorders, mental illness, and substance abuse are linked to family violence.
	DeBellis et al. (2001)	Case-control study of child-maltreating and control families	Maltreating mothers exhibited significantly higher levels of several forms of psychopathology and substance abuse than control mothers, and the majority of maltreated children and adolescents also showed significantly more psychopathology.

MICROSYSTEM THEORIES

Theory	Supportive Study	Focus	Key Assumptions and/or Findings
Systems	Straus (1973)	Theory of family violence	Family violence is a product of the family system, not of individual behavioral pathology. The strain of everyday interactions among family members produces conflicts, including violence.
	Giles-Sims (1983)	Wife battering (shelter sample)	100% of men and 77% of women had used physical violence in the year before the women came to shelter. Major stressors upsetting family system: arguments over money, jealousy, man's use of alcohol/drugs, sex/affection; 1/3 of women went back to husbands.
	Howes et al. (2000)	Case-comparison study of families with maltreated preschoolers	Sexually abusive family systems characterized by difficulty managing anger, disorganized roles, chaotic interactions, and restricted ability to manage interactions.
	Marcus & Swett (2003)	Theory of couple violence	Risk-increasing emotions within the couple system increase the likelihood of couple violence.
Stress	Farrington (1986)	Theory of family violence	Stress (changes/disruptions requiring adjustment and exceeding one's adaptive resources) associated with instrumental (e.g., spanking children to stop crying, slapping a spouse to settle an argument) and expressive family violence (e.g., lashing out in anger).
	Whipple et al. (1991)	Case comparison study of families of abused and nonabused children	Physically abusive parents reported significantly lower income. Abusive mothers reported more stress due to frequent life events, more negative perceptions of these events, and higher rates of both depression and state anxiety.
	Tajima (2000)	Analysis of data from 1985 National Family Violence Resurvey	Presence of wife abuse in home significantly increased odds of child abuse. Wife abuse was strongest risk factor for physical punishment of children. Strongest predictor of verbal child abuse was nonviolent marital discord.

(Continued)

Table 1.1 (Continued)

Theory	Supportive Study	Focus	Key Assumptions and/or Findings
	Crouch & Behl (2001)	Case comparison study of generally population and high risk parents	Level of parenting stress was positively associated with physical child abuse potential in parents who believed strongly in the value of corporal punishment but not in parents low in this belief.
EXOSYSTEM THEORIES			
Ecological Theory	Garbarino & Sherman (1980)	Child maltreatment report data and census data	Areas at high risk for child maltreatment characterized as socially impoverished—fewer neighbors exchanging child care, lower involvement in community exchanges, fewer people can be counted on.
	Godkin et al. (1989)	Case comparison survey study of abused and nonabused elders	Abused elders and caregivers increasingly interdependent prior to abuse because of loss of family members and growing social isolation.
	Lynch & Cicchetti (1998)	Case comparison study of maltreated and non-maltreated children ages 7-12	Children from high-violence neighborhoods were significantly more likely to have been physically abused than children from low-violence neighborhoods.
	Miles-Doan (1998)	Examination of law enforcement and census data	Neighborhoods in a Florida county with the greatest resource deprivation also had the highest rates of violence against intimates.
	Korbin et al. (2000)	Interviews with parents from neighborhoods with differential risk for child maltreatment report rates.	Causes of child abuse identified by parents included poverty and family disruption; substance abuse and stress; lack of moral and family values; and individual pathology—all of which were related to neighborhood conditions and individual perceptions of neighborhood characteristics.
	Van Wyk et al. (2003)	Survey data from Wave 2 of the National Survey of Families and Households and from the 1990 Census	Male-to-female partner violence was lowest for couples residing in the least disadvantaged neighborhoods (3.5%) and highest (7.9%) in the most disadvantaged neighborhoods.
Sociocultural Theories	Jennings & Murphy (2000)	Theory re: male-female battering	Female battering derives in part from compulsive masculinity and rigid sex role stereotyping, and is influenced by fear of humiliation, the social form of male shame; humiliation is deeply rooted in the same-sex relations of boyhood groups, male rites of passage, and troubled relationships with father figures.
	Rosen et al. (2003)	Study of 713 married U.S. Army soldiers	Wife abuse associated with a culture of hypermasculinity (operationally defined as group disrespect for women) and lower support for wives.
MACROSYSTEM THEORIES			
Feminist Theories	Yllo (1993)	Feminist perspective on family violence	Violence is a form of coercive control used by men to maintain their power in a patriarchal society and by husbands to maintain their power in the family.
	Yllo & Straus (1990)	Analysis of survey data and census tract data	Patriarchal ideology was positively correlated with wife beating, especially in states with high levels of structural inequality between men and women.

Individual/Developmental Theories

Biological Theories

Behavioral Genetics. A defining goal of behavioral genetic research is to estimate the extent to which genetic and environmental factors contribute to variability in selected behaviors in the population under study. This involves decomposing phenotypic (observed) variance of a trait or behavior into genetic and environmental components. *Heritability* is the proportion of phenotypic variance attributable to genetic factors. The remaining variance is attributed to environmental factors including both *shared environmental* influences (common to all members of the family) and *nonshared environmental* influences (the ones unique to each individual that operate to make members of the same family different from one another). Nonshared environmental influences include microsystem factors such as differential parental treatment; differential extrafamilial relationships with friends, peers, and teachers; and nonsystematic factors such as accidents or illness (Hines & Saudino, 2002; Plomin, Chipuer, & Neiderhiser, 1994).

To date, only one empirical study has specifically addressed the issue of genetic influences on family violence (Hines & Saudino, 2004a). In this twin study, genetic and nonshared environmental influences were the only significant contributors to individual differences in the use and receipt of both physical and psychological aggression in romantic relationships. These findings, which supported the hypothesis that familial resemblance in intimate partner aggression is due to shared genes, not shared environments, are consistent with those of many other studies showing that genes and nonshared environments influence the use of aggression in general (e.g., Carey & Goldman, 1997; DiLalla & Gottesman, 1991). This study further showed that many of the same genes and environments influencing the receipt of physical aggression also influence its use. Thus, there appear to be genetic and nonshared environmental influences on a tendency to get involved in aggressive romantic relationships, and aggressive people tend to choose aggressive partners (Hines & Saudino, 2004a).

The statement that behaviors are genetically influenced means that certain people, due to their genotype, may be more likely to commit aggressive acts in their relationships than people who do not have that same genotype. In other words, genetic influences are *probabilistic,* not deterministic. Genetic influences on aggression in family relationships must be seen as a predisposition towards aggression, not as destiny (Gottesman, Goldsmith, & Carey, 1997; Raine, 1993). The environment and manipulations in the environment can be very successful in reducing aggressive behaviors and preventing the full expression of any genetic predisposition (Hines & Saudino, 2004a; Hutchings & Mednick, 1977; Raine, 1993).

Other Biological Theories. Behavioral geneticists are not the only theorists interested in the role of biology in family violence, although research on biological factors is sparse. A possible reason for this lack of research is the assumption that if aggression in the family has a biological component, nothing can be done to ameliorate it—an assumption that is incorrect. Although biological factors contribute to family violence, the environment also plays a large role and most likely interacts with biological traits. Identifying people who are biologically at risk for behaving aggressively, and altering their environments, may help reduce family violence.

Studies linking biology to different forms of family violence come in four areas: head injuries, physiological reactivity, testosterone, and temporal lobe dysfunctions. Studies of the link between head injuries and family violence have been limited to male batterers. This research shows that a majority of male batterers may have a history of head injuries severe enough to result in concussion or loss of consciousness (Rosenbaum & Hodge, 1989). Furthermore, in comparison to nonbatterers, batterers have suffered significantly more head injuries, the overwhelming majority of which preceded the intimate partner aggression (Rosenbaum, 1991). Head injuries are thought to lead to aggressive behaviors because they tend to damage the frontal lobes, an area of the brain related to aggression.

It has also been shown that child abusers have large increases in physiological reactivity, as measured by skin conductance and heart rate, in response to stressful and nonstressful mother-child interactions (Wolfe, Fairbank, Kelly, & Bradlyn, 1983). In addition, child abusers are more physiologically reactive to a child's cry and smile, suggesting that they find both of these aversive (Frodi & Lamb, 1980).

Testosterone has been linked to two different types of family aggression: wife abuse and incest. Among military servicemen and low-income, culturally diverse groups of men, higher levels of testosterone were related to both physical and verbal aggression against wives (Booth & Dabbs, 1993; Soler, Vinayak, & Quadagno, 2000). However, because these are correlational studies, we do not know which came first—the high testosterone or the aggression—and aggression has been shown to lead to increases in testosterone (e.g., Gladue, Boechler, & McCaul, 1989). Furthermore, most of the studies in humans linking aggression to testosterone levels show that the links are neither simple nor direct (Englander, 1997). In studies linking testosterone to incest, castration significantly reduces recidivism rates (Wille & Beier, 1989), suggesting that testosterone levels and sexual aggression are related. What may be the problem, though, of concluding from these studies that high testosterone *causes* sexual aggression?

Finally, there is evidence linking temporal lobe dysfunctions to sexual offenses, specifically incest and pedophilia. Sexual activity is related to activity in the temporal lobe, and changes in sexual functioning are related to damage to that lobe (see Hucker et al., 1988). Incest offenders and pedophiles show abnormalities within the temporal lobes, abnormalities that do not characterize any other types of offenders studied (e.g., Hucker et al., 1986, 1988; Langevin et al., 1988; Wright et al., 1990).

There is also a significant amount of research linking nonfamily violence to frontal lobe dysfunctions (see Raine, 1993, for a review). Because head injuries are related to both family violence and frontal lobe damage, this is an important area to investigate further. Also potentially fruitful is an association found between minor physical anomalies (e.g., having only one crease in the palm instead of three, having low-seated ears) and family violence. Minor physical anomalies are an indicator of minor brain damage and have been empirically linked to violent crimes in people environmentally at risk (Kandel et al., 1990). It may also be useful to investigate the link between family violence and the neurotransmitters serotonin and norepinephrine, which appear to be reduced in aggressive individuals (Scerbo & Raine, 1992). Finally, birth complications have been linked to violence, particularly in people genetically at risk (Brennan, Mednick, & Kandel, 1993). Thus, the link between biological traits and family violence needs to be explored further, and interactions between biological characteristics and environmental or genetic factors are particularly important avenues of research.

Attachment Theory

The basic assumption of attachment theories (e.g., Bowlby, 1969, 1973) is that early experiences with caregivers contribute to the development of internal prototypes of human beings and human relationships. According to this perspective, individuals who develop a secure attachment style have positive feelings toward self and others, whereas individuals who develop an insecure attachment style have negative feelings toward self and others. In relation to family violence, the basic propositions tested are that child abuse leads to an insecure attachment style and that an insecure attachment style leads to family violence. The representative studies presented in Table 1.1 are among those providing some support for these theoretical propositions; however, based on a review of the relevant literature, Bolen (2000) concluded that overall the evidence is mixed and does not provide full validation of attachment theory.

Social Information Processing Theories

Social information processing theories, and cognitive behavioral theories, place much greater emphasis than other theories on social cognitive processes in individuals who maltreat others. These theories emphasize that it is the perpetrator's judgments about the behaviors, thoughts, and feelings of family members and their limited response repertoire for dealing with frustrations, disappointments, and negative emotions that lead to a reliance on aggression. These theories are important in their emphasis on the role of cognitive processes in family maltreatment; indeed, it is likely that most abusers do not see their behavior as abusive but as an appropriate response to the inappropriate behavior of family members.

Social Learning Theory

One of the most popular explanations for family violence comes from social learning theory, which posits that individuals learn "appropriate" situations and targets for aggression the same way they learn everything else— that is, through the patterns of reinforcements and punishments that they experience and through observing both the behaviors of significant others and the consequences of those behaviors. Thus, social learning theorists predict children who observe interparental violence or experience violence at the hands of their parents are likely to repeat this behavior in their own family relationships as adults. Considerable support for such suppositions comes from extensive research on the intergenerational transmission of family violence, which clearly indicates that violence in the family of origin, either witnessing and/or experiencing it, is predictive of violence in later close relationships. However, there are two major caveats for this support: 1) there is also strong evidence that not everyone who grew up in an aggressive family will become aggressive (Kaufman & Zigler, 1987); and 2) intergenerational continuity in aggression can also be due to the genes, not just the environments, that family members share.

Microsystem Level Theories

Systems Theory

Systems theorists emphasize the importance of analyzing families as dynamic, adaptive social systems with feedback processes taking place among family members in ways that maintain the stability of the system (Kazak, 1989). Systems theorists typically view influences in families as bi- or multidirectional; for example, children influence parents as well as parents influence children;

interactions between husbands and wives can influence the interactions of both with elderly parents; how a husband treats a wife may be related to how she treats their children. From this perspective, maltreatment in families is not a simple matter of one disturbed family member harming an innocent victim; rather, it results from everyday stresses and strains on the family system that produce conflicts, accommodations, and various responses, sometimes including violence. In some family systems, wives may tolerate maltreatment from their husbands because the husbands are providing a home for their children, or may sacrifice a daughter to the incestuous behaviors of their husbands to protect themselves from his aggression; or husbands may stay with violent wives to try to protect their children. Systems conceptual frameworks have been valuable in highlighting the complexity of most forms of family violence but have been consistently challenged by feminists, who view family violence— particularly wife abuse—as a gendered problem; that is, a problem residing in men due to patriarchal norms.

Stress Theory

Stress theorists have identified stressors at many levels of the ecological context in which individuals develop and family interactions take place. Stress is typically defined as the experience individuals have when the demands of the situation exceed their ability to deal with it. Within the family micro-system, common stressors may be too many children, not enough income, absence of one parent, and marital conflict. In particular, poverty is a risk factor for various forms of family maltreatment (e.g., Caetano & Cunradi, 2003; Sedlak & Broadhurst, 1996), particularly child neglect (Drake & Pandey, 1996). Abuse of one parent by another may result from one stressor—for example, the victim's substance abuse—but also constitutes a stressor for other members of the family, leading, for example, to the victim's abuse of the children. There is considerable empirical evidence supporting a link between stressors and family violence, but stress does not necessarily lead to violence. Although Straus (e.g., 1980) found that the greater the number of stressors parents were experiencing, the higher their level of child abuse, he also found that stress did not lead to child abuse except within the context of several other variables—specifically, growing up in a violent family, low attachment to the marital partner, a dominant role for the husband, and isolation from social support.

Events or changes that overwhelm the resources of the individual and send reverberations into family life often take place outside the home. A family member may lose his or her job, face discrimination because of a handicap, be arrested for drunk driving, or seek affordable mental health services and be unable to find them. The impact of such stressors appears to be particularly destructive when the individual lacks social support. For example, Garbarino and Sherman (1980) found that neighborhoods characterized by greater social isolation and "social impoverishment"—that is, neighborhoods where isolated families compete for scarce resources rather than assisting each other—had higher rates of child maltreatment. The interaction of exosystem characteristics of the neighborhood with characteristics of the families can exacerbate or ameliorate the conditions leading to maltreatment.

Exosystem Level Theories

Ecological Theories

In this book, we follow the contemporary practice of using the term *ecological paradigm* to refer to a broadly integrative conceptual framework that encompasses theories addressing factors at different levels of the human ecological system. However, the label *ecological theory* has also been applied to theoretical frameworks (e.g., Garbarino & Kostelny, 1992) focusing on neighborhood variables such as social cohesion and social isolation. Proponents of this approach emphasize the importance of paying attention to neighborhood variables when addressing problems of family violence. In a study of greater Chicago communities, Garbarino and Kostelny (1992) found that neighborhoods with high rates of child abuse were characterized by social disorganization (e.g., criminal activity) and lack of social coherence (e.g., lack of availability and knowledge of social services and support networks).

Sociocultural Theories

Although social support has frequently been identified as a factor that can reduce the likelihood of violence as a response to stress (Milner, 1994), there is also evidence that norms within an individual's peer group and community can contribute to the likelihood that violence will be viewed as an acceptable solution to difficulties within the family (Straus, 1980). Men's peer groups may support rigid sex role norms designed to ensure a superior status for men and subordination in women and children. Religious groups may endorse corporal punishment of children and encourage women to stay within abusive marriages in order to keep the family together. Evidence of differences among religious groups and in different areas of the country in tolerance for and use of aggressive tactics within the family supports the view that local norms play a role in the sanctioning of some forms of maltreatment.

Macrosystem Level Theories

Feminist Theory

At the heart of a variety of feminist perspectives is the assumption that domestic violence, or violence within the family, is a *gendered* problem; moreover, feminists generally concur that characteristics of perpetrators, victims, and interactions among perpetrators and victims, as well as expectations about families and society, are all profoundly influenced by gender and power (Yllo, 1993). From these perspectives, use of terms like *domestic violence* and *intimate partner violence* is inappropriate because of their gender-neutral quality. This gender neutrality fails to place the responsibility for *family terrorism* where it belongs: on males operating within a global patriarchal system that denies equal rights to women and legitimizes violence against women, children, and the elderly (Hammer, 2003).

Patriarchy, defined by Kurz (1993) as "the system of male power in society" (p. 49), has been identified by feminists as one of the most powerful forces contributing to wife abuse in the United States and most other countries. In patriarchal societies, men have more social, economic, and political power and status than women. They consider themselves superior to women and children and feel entitled to use force if necessary to maintain dominance in family decision making. Men's sense of entitlement, gender inequality, and patriarchal values are seen as causes of wife abuse (Barnett, 2000), femicide (Smith, Moracco, & Butts, 1998), media portrayals of sex and rape (Bufkin & Eschholz, 2000), and sexual abuse of children (Candib, 1999).

Violence against women and children has been connected to patriarchal norms around the world (Levinson, 1989). Moreover, within the United States, there is evidence that the greater the social inequality between men and women, the higher the levels of wife assault (Straus, 1994). Male dominance within the family has been found to predict not just wife beating but also physical child abuse; moreover, the higher the level of husband dominance in the family, the stronger the likelihood of child abuse (Bowker, Arbitell, & McFerron, 1988). On the other hand, Sugarman and Frankel (1996) found, in a meta-analysis of studies addressing the link between wife assault and maintenance of a patriarchal ideology, that the only component of patriarchal ideology that consistently predicted wife assault was the perpetrator's attitude toward violence.

Summary

As was illustrated in Table 1.1, there is some empirical support for each of the major theories of family violence, but there are also limitations to the empirical support. None of the individual theories can account for all forms of family violence. Essentially, research designed to test the validity of the theories has been effective in identifying risk factors for maltreatment at each of several different levels of the ecological systems in which development takes place. Conversely, the reciprocal of many of these risk factors can serve as protective factors against maltreatment or the negative impact of maltreatment. For example, social support within the micro- and exosystem has long been recognized as an important protective factor, as have higher income and higher education.

Even what may seem like a simple and clear-cut case of maltreatment of one family member by another is likely to have multiple causes. For example, a father may commit incest on his prepubertal daughter because of his sexual inadequacies with adult women, *and* his wife's overt contempt for him, *and* his wife's unconscious denial that her husband is doing to their daughter what her father did to her, *and* the norms of his peer group that a man's home should be his castle, *and* his personal belief that his family should be obedient to him, *and* society's tolerance for pornography, *and* the lack of availability of appropriate services within the community, *and* inadequate funding for intervention programs for incest offenders.

In our previous book (Malley-Morrison & Hines, 2004), we provided an extensive analysis of family violence in ethnic minority communities in the United States. In the current book, we focus more on family violence in the majority White European American community. We begin in Chapter 2 by describing the historical and ecological context of family violence in the contemporary United States, including the role of culture in contributing to, maintaining, and providing rationalizations for violence in close relationships. In Chapter 3, we expand on the description of the contexts of family violence by considering religious influences on family violence. Then, in Chapters 4–11, we review the research on the major forms of family violence, including violence in gay/lesbian relationships, in the United States today. As you read these chapters, consider the different perspectives that individuals from the different relevant professions would bring to an interpretation of the findings and judgments about the implications of those findings for alleviating family maltreatment. Also consider how the findings would be interpreted within the different theoretical frameworks described here. Finally, consider your own experiences of maltreatment as well as your views and values concerning various forms of maltreatment. Which, if any, of the forms of maltreatment seem worse in their consequences? For example, is the abuse of children worse than the abuse of adults? Is physical abuse worse than psychological abuse? Consider your answers to these questions. Now, does it surprise you to learn that women staying in shelters because they have been beaten by their husbands often say that it was his psychological abuse that was more damaging (Follingstad, Rutledge, Berg, Hause, & Polek, 1990)? What do you see as the best approaches to ending maltreatment in families?

2

Cultural Contexts of Family Violence

Jesse is a White, six-year-old male from a two-parent family. In the current incident, the father hit the child with a belt, leaving deep bruises and welts covering the child's back. There have been three prior investigations: The parents denied hitting the child in the first two; the father admitted doing so in the third. The parents were then mandated to attend parenting classes and cooperate with periodic home visits. This time, the parents deny hitting the child, but the physical evidence is clear. Complaints have also been made by neighbors about the parents' fighting, but no assault charges have ever been filed. (Britner & Mossler, 2002, p. 323)

He came home drunk one night and started harassing and abusing me. I told him to get out. I was six months pregnant at the time. He sat on my stomach. He followed me into the kitchen. I grabbed a knife and told him to get out of my house. When he turned around, I stabbed him in the back. I was protecting myself and my unborn child. He had to go to the emergency room to get stitches. (Swan & Snow, 2002, p. 286)

In considering the role of cultural context in family violence, we focus on "culture-as-meaning"—for example, notions within a social group as to what it means to be abusive and attitudes as to the role of government in family affairs. Although the United States is a diversified, multicultural society, certain forms of violence, and certain sets of hierarchical relationships, have been considered by large segments of society as not just justifiable but morally correct. Moreover, the traditional valuing of freedom from government control has, in some ways, helped to maintain violence in the home.

Most Americans know it is against the law to abuse children and the elderly. Many also know that wives can take out restraining orders against husbands to protect themselves from abuse. It is probably less well known that it is possible for husbands to seek similar legal protections if they fear abuse from their wives. Yet despite all the recent violence prevention legislation, the United States has had a long tradition of tolerating violence within the home and by its own institutions. We begin this chapter by briefly placing America's culture of violence into an historical framework. We then provide an overview of the research literature concerning tolerance of violence and abuse in the majority culture as well as in several of the minority cultures. (See Malley-Morrison and Hines [2004] for a much fuller discussion of family violence within ethnic minority communities in the United States.) Finally, we identify selected norms within the macrosystem that are an important part of the cultural context of family violence.

The Contemporary Context

In the United States today, homicide, generally as a result of maltreatment, is the leading cause of injury deaths in infants under one year of age (Child Trends Data Bank, 2002). Homicide, typically through beatings or suffocation, has also been identified as the fourth leading cause of death of children ages 1–4. It is committed largely by family members—typically mothers in the first few months and often fathers or stepfathers in later months (Child Trends Data Bank, 2002). Moreover, these rates are likely to be underestimates. According to one analysis, from 1985 through 1996, the total number of homicides in children younger than 11 years due to abuse was more likely to be 9,467 rather than the 2,973 reported (Herman-Giddens et al., 1999).

The rates of child maltreatment deaths in the United States are among the highest in the industrialized world, with approximately 25 children dying every week from maltreatment. This rate is either the highest or third highest rate (following Mexico and Portugal) of child maltreatment deaths of the 30 industrialized member nations of the Organization of Economic Cooperation and Development (UNICEF, 2003). The main conclusion of this UNICEF study was that the "challenge of ending child abuse . . . is . . . breaking the link between adults' problems and children's pain. It ought not to be part of family culture, or of our societies' culture, for the psychological, social, or economic stresses of adults to be vented on children, or for problems and frustrations to be so easily translated into abuse of the defenseless. The task is therefore one of creating a culture of nonviolence towards children" (UNICEF, 2003, p. 22).

Violence against women also continues to be an epidemic in the United States. Women are more likely to be murdered by intimates than by any other assailant. Homicide by an intimate partner is the seventh leading cause of premature death for women in general in the U.S. and is the leading cause of death for African-American women between the ages of 15 and 45 years (Greenfeld et al., 1998). According to a study conducted by the U.S. Department of Justice on violence-related injuries treated in emergency departments, approximately 37% of injuries to women involve a spouse, ex-spouse, or boyfriend (Rand, 1997). American women are also more likely to be raped by men they know and often love than by strangers (Bachman, 2000). It appears that in approximately 70%–80% of all intimate partner homicides, no matter which partner was killed, the woman was involved in a physically abusive relationship (e.g., Greenfeld et al., 1998). Although homicide is only one index of violence, it is noteworthy that the U.S. has the highest female homicide rate of 25 high-income nations. Although the country contains only 32% of the female population living within these 25 nations, it accounts for 70% of all female homicides and 84% of all female firearm homicides. Approximately 4,000 women are murdered in the U.S. each year—at a rate that is five times that of all the other high-income countries combined. A woman in the U.S. is eight times more likely than a woman in England and Wales, three times more likely than a woman in Canada, and five times more likely than a woman in Germany to be murdered (Hemenway, Shinoda-Tagawa, & Miller, 2002).

These high rates of violence against women are consistent with the United States' relatively high rate of violent crime overall (Felson, 2002). Thus, violence against men is also an epidemic in the United States. The majority of this violence is perpetrated by other males—most typically unrelated males, but violence by female partners also occurs. In fact, the majority of victims of both male and female violence in this country are male (Felson, 2002). Between 1976 and 2000, over 500,000 males and females over the age of 12 have been murdered—the majority by somebody they know, and about 11% by an intimate. As perpetrators of murder, women were more likely to kill somebody as a result of an argument or by poison (Fox & Zawitz, 2002). In addition, American men may be more at risk of sustaining violence from their wives than men in other countries (Kumagai & Straus, 1983). Finally, every year, there may be at least two million older Americans who are maltreated by family members. According to the National Elder Abuse Incidence Study of 1996, two thirds of the perpetrators of elder abuse were adult children or spouses. These maltreated elders tend to die earlier than nonabused elders, even in the absence of chronic conditions or life-threatening disease (American Psychological Association, 2003).

In the following sections, we discuss majority and minority culture views on maltreatment in families. A more extensive discussion of family violence in minority communities can be found in our book, entitled *Family Violence in a Cultural Perspective* (Malley-Morrison & Hines, 2004). However, it is important to note that there are many reasons one might expect ethnic group differences in tolerance for family violence. As compared to the majority culture, ethnic minority groups often confront considerable discrimination and major economic and social disparities—circumstances that can produce high levels of stress, affect perceptions and feelings, and lead to greater tolerance of aggression in the family. Moreover, some immigrants bring with them patriarchal cultural values favoring men over women and children and an emphasis on family privacy that can prevent abused family members from seeking help. For these reasons and more, we sometimes find cultural differences in tolerance for family violence.

Perspectives on Corporal Punishment and Child Physical Abuse

Before proceeding, read the items in Box 2.1, and indicate the extent to which you agree with each statement. Now consider the first case at the beginning of this chapter. Is Jesse an abused child? What if he had been hit just as often as he was, but the hitting did not leave bruises? What factors influence your judgment of abusiveness? Suppose there is evidence that Jesse's parents hit each other. Does that affect the extent to which you view Jesse as being maltreated? Do you think assault charges should be filed? Should anything else be done?

Box 2.1   Attitudes Toward Corporal Punishment

Read each statement below and indicate the extent to which you agree with it by choosing a number from 1 (totally disagree) to 7 (totally agree).

1. Parents should have the right to hit children (Graziano & Namaste, 1990).

2. It is sometimes necessary to discipline a child with a good, hard spanking (Flynn, 1998).

3. Spanking is an effective form of discipline (Graziano & Namaste, 1990).

4. If a girl refuses to clean up her room after being asked several times, it is appropriate to spank her on the rear with the hand (Herzberger & Tennen, 1985).

5. If a boy breaks the neighbor's window with a baseball after being told never to play except in the park, it is appropriate to hit him with a leather strap on the bare skin (Herzberger & Tennen, 1985).

6. Hitting a child with a paddle is abusive.

SOURCE: Bower & Knutson, 1996

A major issue in the American macrosystem is the extent to which spanking should be considered a form of abuse. Some investigators have argued that spanking is at the root of societal violence (Straus, 1996). Whether the label of abuse is applied to spanking or not, various professional groups and international human rights organizations have argued that spanking is not a disciplinary tactic that promotes the well-being of the child and that it violates the innate human rights of the child (e.g., Walker, Brooks, & Wrightsman, 1999). Nevertheless, in the United States, there is strong cultural support for physically punishing children. In response to a 1988 National Opinion Research Center (NORC) survey, over 80% of the respondents agreed or strongly agreed with the statement, "It is sometimes necessary to discipline a child with a good, hard spanking" (Flynn, 1994). Of nearly 700 college students, 69% considered spanking to be an effective disciplinary technique, 45% believed children need to be spanked, 85% held that parents have the right to spank children, and 83% reported an intention to spank their own children (Graziano & Namaste, 1990). In another college sample, a majority of the participants considered physical punishment to be appropriate for three- and four-year-olds who took something that did not belong to them, misbehaved in public, and talked back to or hit one of their parents; indeed, nearly 90% of the students considered spanking to be appropriate discipline for a toddler in at least one of the hypothetical situations (Flynn, 1998). Although support for corporal punishment of older children was less strong, 60% viewed physical punishment as appropriate for seven- or eight-year-olds who stole or talked back (Flynn,

1998). What are your views concerning how such forms of misbehavior should be handled?

Current cultural values supportive of corporal punishment are strongly instilled in the nation's children, who generally accept the message that even very harsh physical discipline may be necessary and therefore not abusive. For example, among college students, (1) only 26.6% of students meeting criteria for having been abused (i.e., subjected to injurious or potentially injurious parental aggression) classified themselves as abused (Rausch & Knutson, 1991); (2) students who considered their own harsh disciplinary experiences to be deserved rated harsh punishment as more appropriate than other students (Kelder et al., 1991); and (3) in comparison to nonabused students, students abused as children were more likely to regard severe physical discipline as deserved, appropriate, and even beneficial (Herzberger & Tennen, 1985).

Although the majority of families in the United States view corporal punishment as appropriate and necessary, such support may vary somewhat across ethnic and socioeconomic groups. For example, in a study of Whites, Latinos, and Chinese Americans, the Chinese judged beating a 12-year-old girl with a cane and burning a mark into her arm for stealing, and beating a child for not doing homework, as less serious than the other groups did (Hong & Hong, 1991). Similarly, within many Black communities, spanking and other forms of physical punishment are considered "appropriate discipline" rather than abuse (e.g., Giovannoni & Becerra, 1979; Alvy, 1987). However, ethnicity is often confounded with income, and low-income parents generally approve of spanking more than middle-income parents do (e.g., Heffer & Kelly, 1987). Another major ethnic group in this country, Latinos, appear to have *more* stringent definitions of child abuse than do other cultural groups, including Blacks, Whites, and Asians (Malley-Morrison & Hines, 2004).

There is also a tendency for levels of tolerance of violence against children, and readiness to label violence as abusive, to differ between the lay public and relevant professionals and among relevant professional groups (e.g., social service workers, criminal justice workers). When asked to judge the seriousness of vignettes dealing with psychological abuse (publicly ridiculing a child), neglect (leaving a child outside in a car at a shopping center for several hours in the cold of winter), and physical abuse (slapping a child in the face with an open hand, knocking him down, and splitting his lip), lay respondents and juvenile detectives judged the situations as more abusive than did Child Protective Service (CPS) workers (Kean & Dukes, 1991).

Review your responses to the survey questions in Box 2.1. How do your views compare to the ones you have read about in this section? If you responded with any number other than 1 for Questions 1–6 and any number other than 7 for item 7, does that mean you are showing some tolerance for child abuse? Would you interpret scores other than 1 for Questions 1–6 and other than 7 for item 7 as reflecting some reasoning or process other than tolerance for aggression?

Perspectives on Child Sexual Abuse

In 1982, a judge in Wisconsin sentenced a man to 90 days in a work-release program following his trial on the charge of sexually assaulting the five-year-old daughter of the woman with whom he was living. The explanation provided by the judge for his lenient sentence was, "I am satisfied we have an unusually sexually promiscuous young lady. And [the defendant] did not know enough to refuse. No way do I believe he initiated sexual contact" (Nyhan, 1982, in Finkelhor & Redfield, 1984, p. 108). Do you think a five-year-old child can be sexually promiscuous? Should a child ever be held responsible for an adult's behavior? What kinds of cultural values might contribute to the judge's analysis of this case? This case occurred in 1982; do you think a judge today would have a similar interpretation?

Research subsequent to that judicial decision has revealed a number of variables that influence the extent to which people view specific types of child sexual abuse as serious and/or abusive. If the perpetrator is an adult, parents tend to view sexual intercourse or attempted intercourse as "definitely" abusive; however, if the perpetrator is another child, or involves sexually related *verbal* aggression (e.g., telling girl she's a whore), parents tend to rate the behavior as less abusive, even when they are told that the consequences of the behavior are quite negative (Finkelhor & Redfield, 1984). If children are portrayed as acting passively in response to sexual abuse, the situation is rated as less abusive, even when the child is very young (e.g., age 0 to 6) (Finkelhor & Redfield, 1984; Mellott, Wagner, & Broussard, 1997). What is your view? If an adult seeks sexual pleasure by fondling the genitals of a 2-year-old, 5-year-old, or 12-year-old, is the behavior any less abusive if the child does not resist strongly? And does it matter how old the child is? Some researchers have found that older victims (e.g., 13- or 15-year-olds) of child sexual abuse are blamed more for their abuse than younger victims (e.g., 6- or 7-year-olds) (Back & Lips, 1998; Maynard & Wiederman, 1997). Is this judgment appropriate?

Although the evidence is mixed, both the gender of the child and the gender of the perpetrator sometimes influence people's judgments about whether a particular sexual behavior is abusive. In one study, students rated scenarios where the abuser was of a different sex from the child as less abusive than same-sex scenarios (Maynard & Wiederman, 1997). When the victim was described as a 15-year-old, less blame was attributed to the perpetrator than when an adult of either sex was described as molesting a 7-year-old; the least blame was attributed to an adult involved with an opposite-sex adolescent. When the perpetrator was described as a male, the harmful effects were assumed to be worse if the victim was a boy than a girl (Mellott et al., 1997). Do you agree with the judgments of these students? Why or why not?

Other characteristics influencing the extent to which individuals judge sexual aggression against children as abusive include amount of parenting experience—the greater the experience, the more serious the abuse is judged to be (Portwood, 1998). Moreover, greater valuing of children is associated with lower tolerance for sexual maltreatment (Ferrari, 2002), and there is some evidence that the greater one's degree of conformity to traditional sex roles, the more one blames sexually abused children for their own abuse (Ford, Schindler, & Medway, 2001).

Although some ethnic groups (e.g., Latinos) appear to be less tolerant of child sexual abuse than other groups are (Fontes, Cruz, & Tabachnick, 2001), overall, Americans seem somewhat reluctant to put full blame on perpetrators of sexual abuse—or even to call their behavior abusive—except in the most egregious circumstances or when the perpetrator is the same sex as the victim. According to Daro (2002), "It is not uncommon for individuals to identify extenuating circumstances when things that are labeled 'abuse' reflect behaviors or attitudes commonplace among their peers or understandable given their social context. This type of definitional manipulation recently led a U.S. Catholic Cardinal to distinguish between a priest who repeatedly abuses young boys versus a priest who, under the possible influence of alcohol, returns the sexual affections of a 16-year-old girl" (p. 1132). This tendency to apportion degrees of blame in relation to sexually abusive behaviors makes it difficult to develop a coherent and effective public policy response (Daro, 2002).

Perspectives on Child Neglect

Research addressing cultural values and judgments related to child neglect has indicated that community samples consider scenarios such as, "There is trash in the child's home, and some corners of rooms have piles of junk or trash," to be more serious than social workers and child welfare workers do (Dubowitz, Klockner, Starr, & Black, 1998; Rose & Meezan, 1995). Do you have any speculations as to the causes of such differences? What should be the criteria for judging signs of neglect as serious enough to justify reporting a family to CPS? In one community sample, 25% of the respondents did not consider regularly sending a child to school in dirty clothes as a form of child maltreatment (Price et al., 2001). Do you agree?

The tendency of mothers to rate neglect as more serious or abusive than CPS workers has been found in several different ethnic groups. In a study comparing White, Latino, and Black mothers, and CPS workers, the mothers rated all the categories of child neglect (e.g., physical neglect, emotional neglect, and lack of supervision) as more serious than the CPS workers did (Rose & Meezan, 1995). Moreover, Rose (1999) found that there may be important differences in judgments of neglect between ethnic groups as well: Black and Latino mothers rated the types of neglect as more serious than the White mothers did. Another cross-cultural study showed that Chinese respondents judged leaving a 9-year-old boy alone to feed and take care of himself at night, and refusing to take a withdrawn 8-year-old girl to a counselor, as significantly less serious than Latinos judged them to be (Hong & Hong, 1991). What kinds of experiential or cultural factors might account for such differences?

Perspectives on Wife Abuse

Despite all the laws, programs, and shelters developed to address wife abuse, in our culture, excuses are often made for men who hit their wives, thereby presumably reframing their violence as nonabusive. Although, in the abstract, most people in the United States may claim to be against wife abuse, many think it is acceptable for men to hit their intimate partners under some circumstances (Arias & Johnson, 1989; Greenblat, 1985; Stark & McEvoy, 1970). In a sample of male college students, 75% indicated they would be likely to hit a wife if she had sex with another man, and nearly 40% said there was some likelihood they would hit a wife if she refused to have sex or told friends that her partner was sexually pathetic (Briere, 1987). Overall, more than 79% of these men identified at least one circumstance under which they would be likely to hit a wife. Nearly 40% of another sample of college students approved of a husband hitting his wife if he catches her in bed with another man, nearly 25% approved of hitting a wife if she hit him first, and nearly 20% approved of hitting a wife because she was sobbing hysterically (Greenblat, 1985). Similar portions of adult community and prison samples said hitting a partner who has been unfaithful is acceptable, as is hitting the partner in retaliation for a slap (Herzberger & Rueckert, 1997). Among both college students and adults, 30% or more of each group said that battered women had at least some responsibility for their own battering (Aubrey & Ewing, 1989). Not surprisingly, women tend to be less tolerant of a husband hitting his wife than men are (Greenblat, 1985; Koski & Mangold, 1988; Locke & Richman, 1999).

Findings concerning ethnic differences in rates of approval of wife maltreatment are fairly limited and indicate that sometimes gender plays a more important role in judgments than ethnicity. Studies comparing tolerance of wife maltreatment in Blacks versus Whites generally reveal no ethnic difference or indicate that White males are more tolerant of wife maltreatment than Black males. Women of both ethnic groups appear to be more opposed to violence against women than men in both groups are. For example, in one study of Black and White college students, men were more likely than women to endorse the statement, "Sometimes a husband must hit a wife so she'll respect him," but there were no ethnic differences in support of marital aggression (Finn, 1986). However, according to national survey data from 1968, 1992, and 1994, Blacks were *less* likely to approve of a husband slapping his wife than Whites were, and from 1968 to 1994, the percentage of respondents approving of a husband slapping his wife decreased for Blacks but not for Whites (Straus, Kaufman Kantor, & Moore, 1997).

By contrast, Latinos appear to be somewhat *more* tolerant of aggression against wives than members of other ethnic groups—in part because of cultural values of *machismo* (Malley-Morrison & Hines, 2004). Although White women found "pulling hair," "biting," "constraint against will," "slapping," "throwing objects," and "pushing, shoving, or grabbing" to be abusive, Mexican-American women did not. Moreover, White women judged acts such as burning with a cigarette, throwing things, slapping, and constraining against one's will as more serious than did Mexican-American women (Torres, 1991). Similarly, *Asian* Americans appear to be less likely than Whites to define a husband's shoving his wife or smacking her in the face as domestic violence (Family Violence Prevention Fund, 1993). What are the possible implications of these definitional differences when Latina or Asian women are faced with a situation that many White women would label "abusive"? Does it seem that perhaps Latinas and Asians may be less likely to see themselves as victims and less likely to seek help when assaulted by their husbands?

In addition, Asian Americans may be more tolerant of wife abuse than both Whites *and* Latinos (Gabler, Stern, & Miserandino, 1998). In many Asian languages, there is not even a term for "domestic violence" (Lemberg, 2002). Perspectives on the extent to which there are particular circumstances justifying wife abuse also vary among the many different Asian American communities. For example, about half of a sample of Chinese Americans thought that hitting a spouse was justifiable in cases of defense of self and defense of a child (Yick & Agbayani-Siewert, 1997). Moreover, the older Chinese American respondents showed tolerance of intimate aggression perpetrated in response to a wife's extramarital affair. By contrast, a sample of

Filipino American students tended to view physical violence as unjustifiable under any circumstances, although the Filipino males were more likely than the Filipino females to endorse physical violence if the woman was flirting or unfaithful (Agbayani-Siewert & Flanagan, 2001). Among respondents with Chinese, Vietnamese, Cambodian, and Korean heritages (Yoshioka, DiNoia, & Ullah, 2001), the two groups with Southeast Asian origins (Cambodians and Vietnamese) were more likely than the Chinese and Koreans to endorse male privilege (e.g., "A husband is entitled to have sex with his wife whenever he wants it.") and less likely to endorse alternatives to living with violence (e.g., "Wife beating is grounds for divorce."). Of all groups, the Korean Americans were the least likely to view violence as justified in particular situations (e.g., when the wife has sex with another man). Thus, Asian communities differ not only in cultural aspects such as language and religion, but also in their attitudes toward wife abuse. Consequently, all studies in which Asians with different heritages are combined in a single "Asian American" sample must be viewed with caution (Yoshioka, DiNoia, & Ullah, 2001). These examples also show the danger of "ethnic lumping" in general. That is, there could also be important variations in tolerance for abuse within the larger Latino, Black, Native American, and White communities.

Perspectives on Marital Rape

Consider the statements in Box 2.2 made by undergraduate students asked for their opinions of "marital rape." Their opinions were gathered in the 1980s. Do you think such views could still be found today? What is *your* view—should sexual assault of a wife be just as serious a crime as the sexual assault of a stranger? Should there be any extenuating circumstances making it acceptable for a husband to physically force his wife to have sex?

Box 2.2   Is Rape Something That Can Happen in Marriage?

"No. When you get married you are supposedly in love and you shouldn't even think of lovemaking as rape under any circumstances."

"[No.] Sexual relations are a part of marriage and both members realize this before they make a commitment."

(Continued)

(Continued)

> "[No.] If the wife did not want to have sex . . . after many months the husband may go crazy. [Rape] would be an alternative to seeking sexual pleasure with someone else."
>
> "[No.] If she doesn't want to have sex for a long amount of time and has no reason for it—let the old man go for it."

SOURCE: Finkelhor & Yllo, 1983, p. 128

One approach to the cultural context of marital rape is to consider the relevant laws. In the United States, rape was traditionally considered to be intercourse forced by a man on a woman who was not his wife (Russell, 1990). Until 1977, when the first marital rape law was passed, husbands could batter their wives to gain sexual access with no legal recriminations. Only gradually, through strong efforts of the feminist community, have sexual assault laws changed, but not without considerable resistance. In Maine, during the legislative battle to remove the marital rape exemption, one legislator argued, "Any woman who claims she has been raped by her spouse has not been properly bedded" (National Center for Victims of Crime, 2003, paragraph 2).

Spousal rape is now considered a crime in all 50 states and the District of Columbia; nevertheless, in many states, marital rape is still a lesser offense than other forms of rape. In West Virginia, spousal sexual assault is defined as "unconsented sexual penetration or sexual intrusion of the perpetrator's spouse involving forcible compulsion or deadly weapon or infliction of serious bodily injury" (National Center for Victims of Crime, 2003, paragraph 2)—an offense that is a felony, punishable by imprisonment for two to ten years. If the same acts are committed against a person not married to the perpetrator, the sentence is 10 to 35 years. In California, a person who commits nonspousal rape by means of force, violence, duress, menace, or fear of immediate and unlawful bodily injury may not be sentenced to probation or a suspended sentence—a prohibition that does not apply to individuals who use the same means to commit spousal rape. In Kansas, sexual battery consists of "the intentional touching of the person of another who is 16 or more years of age, who is not the spouse of the offender and who does not consent thereto, with the intent to arouse or satisfy the sexual desires of the offender or another" (National Center for Victims of Crime, 2003, paragraph 9). In Ohio, the offense of sexual battery does not apply to a spouse, and the offense of rape by the use of a drug or intoxicant impairing the victim's ability to resist applies only to a spouse living apart from the victim.

Despite differences among state laws, there is currently widespread recognition that marital rape occurs. In a national survey, 73% of the respondents agreed that husbands sometimes use force, such as hitting, holding down, or using a weapon to make their wives have sex, and a majority agreed that if a husband forces his wife to have sex because he thinks she is leading him on or because she continually refuses sexual relations, he has raped her (Basile, 2002). Minority participants were less likely than White respondents to think that wife rape occurs. Moreover, the older the respondents, the less likely they were to believe that marital rape can occur or that the behaviors described in the marital rape scenarios were really examples of rape.

To what extent do survey respondents have views consistent with laws—characterizing marital rape as a lesser offense than nonmarital rape? College students made the most strong rape-supportive attributions (e.g., not labeling the sexual assault as rape, not viewing it as violent, not viewing it as a violation of the woman's rights) when judging a rape that occurred within the context of marriage (Ewoldt, Monson, & Langhinrichsen-Rohling, 2000). However, the students were considerably less tolerant of a rape when the marriage was dissolving than when it was intact. In general (and not surprisingly), women's attitudes are less rape-supportive than men's (Ewoldt et al., 2000). Judgments are also less supportive of marital rape when the husband is described as having previously engaged in physical violence against his wife (Langhinrichsen-Rohling & Monson, 1998).

Perspectives on Husband Abuse

Consider the second quote at the beginning of this chapter. Would you agree that the husband is guilty of wife abuse if he has committed the acts she describes? How about the wife? Is she guilty of husband abuse? Jane, the woman who narrated the story, was a client in a domestic violence educational program. Do you think she needs help in handling her aggression?

There is, within the United States macrosystem and within some minority cultures, considerable tolerance for women's aggression against men. In general, intimate aggression is seen as more serious when performed by a man against a woman than by a woman against a man (Bethke & DeJoy, 1993). A series of four surveys with nationally representative samples (Straus et al., 1997) provided provocative data concerning approval of violence by women towards men. In 1968, approximately equal percentages of respondents (~20%; 26.4% of men and 18.4% of women) said they could imagine circumstances where they would approve of a husband slapping his wife in the face or a wife slapping her husband in the face. In subsequent decades, approval rates for slapping a wife declined—to 13% in 1985, 12% in 1992, and 10% in 1994. By contrast, rates of approval for a wife slapping a husband remained stable over time at a little over 20%. Interestingly, as recently as 1994, the percentage of men who approved of a wife slapping a husband (31%) was nearly twice that of women who approved of the same behavior (16%). On the other hand, among college students from the U.S., Latin America, and Asia, *women* showed significantly higher tolerance than men for a wife killing or mutilating her husband if he frequently abuses her—an approval rating that was significantly higher in the U.S.-born than in Asian or Latin American participants (Gabler et al., 1998).

It appears that one reason there is greater tolerance for physical violence by wives than by husbands is the assumption that women's violence is less likely to cause harm. Over 20% of a sample of college students agreed that women who hit their husbands are unlikely to do them any physical harm (Greenblat, 1985). Other relevant assumptions are that men are better able to escape an abusive situation if they want to and are more at fault if they get hit by their wives (Lehmann & Santilli, 1996).

Given considerable evidence that the best predictor of violence against a partner is the partner's violence (Billingham & Sack, 1986; Bookwala, Frieze, Smith, & Ryan, 1992; Vivian & Langhinrichsen-Rohling, 1994; also see Chapters 7–9), one extremely controversial issue is the extent to which intimate violence is mutual. College students appear to view mutual violence within marital conflicts as normative. In one study, students indicated that they expected violence between husbands and wives to increase in comparable increments during a conflict (Harris, Gergen, & Lannamann, 1986). Although these students viewed it as proper for both husbands and wives to become more aggressive and less conciliatory as their mate's aggressiveness becomes more intense, they also indicated that overall it was more desirable for the wives to use aggression than for the husbands to do so. Moreover, they considered it desirable for husbands to use conciliatory behaviors more than aggressive ones and for wives to use aggression more than conciliation.

Native Americans seem to have a somewhat different view of spousal aggression than found in the U.S. macrosystem, viewing it not as a gender issue but as a human or family issue (Malley-Morrison & Hines, 2004). This view probably stems from traditions of relative gender equality existing in many Native American communities prior to European contact. Currently, many Native American Indians acknowledge that both men and women use violence against their spouses. In their view, to eliminate family violence (or violence in general), both types of aggression must be addressed, and the common practice in the majority judicial system of blaming the male for the conflict and ignoring the woman's role is inappropriate and ineffective (Durst, 1991).

Perspectives on Elder Maltreatment

Before reading this section, consider the case in Box 2.3. Have you just read a case of elder abuse? If yes, who is the perpetrator, and who is the victim? What, if any, would be the appropriate response of a social service agency to this situation? (Note: Mr. Smith is too incapacitated to leave his home or be left alone. Mrs. Smith feels that after so many years of caring for her husband in the face of his abuse, she cannot now abandon him.)

Box 2.3   Mr. and Mrs. Smith

Mrs. Smith, age 65 years, is primary caregiver to her 75-year-old husband, who has been treated for manic depression for 20 years and was diagnosed a year ago as being in the early stages of Alzheimer's disease. Mrs. Smith reports to her caregivers support group that her husband's condition has been getting worse. In a recent incident, Mrs. Smith was babysitting her 9-year-old grandson when Mr. Smith became increasingly agitated, blocked the TV that Mrs. Smith and her grandson were watching, screamed profanities at her, called her names, lunged at her, and threatened her. She reported that she jumped up, pushed him back, sent her grandson out of the room, and gradually led Mr. Smith to their bedroom where she ordered him to stay. In a subsequent visit with a support group counselor, Mrs. Smith and her son revealed that Mr. Smith was a "mean alcoholic" who had been physically and emotionally abusive to his wife and son for many years and was becoming physically aggressive again.

SOURCE: Bergeron, 2001

In general, cultural values in the United States provide fewer rationalizations for physical aggression against the elderly than for physical aggression in parent-child and spousal relationships. Nevertheless, there is some tolerance under some circumstances for particular forms of elder maltreatment. For example, although White college students judge physical abuse of an elder as significantly more abusive than psychological abuse, financial abuse, and medical neglect, they tend to view abusive behaviors (e.g., tranquilizing an elderly mother to avoid embarrassment when there are guests in the home) as more justifiable when the elderly parent is agitated or senile. Indeed, students have rated the abusiveness of an agitated, unstable parent as higher than the abusiveness of the caretaker who sedated her (Mills, Vermette, & Malley-Morrison, 1998). Both college students and adult community members labeled physical aggression against the elderly as abusive and as having harmful results, but considered the physical aggression to be more abusive and more harmful when the perpetrator was middle-aged than when elderly—again implying some tolerance of *spousal* violence, independent of age (Childs, Hayslip, Radika, & Reinberg, 2000).

In general, it seems that the majority culture and experts in the field of elder abuse do not substantially differ in the behaviors they consider abusive (e.g., forcing an elder to do something he or she does not want to do) (Hudson, Armachain, Beasley, & Carlson, 1998); however, although most of the community members in this study said that even one incident of slapping or hitting was sufficient to warrant the label of abuse, the experts tended to reserve the term "abuse" for acts "of sufficient frequency and/or intensity" (p. 100). This finding presents an interesting contrast to the research on views on child abuse, where experts tend to see hitting and slapping of children as more abusive than community members do.

Much of the research on attitudes toward elder abuse has been done in ethnic minority communities or has compared views among different ethnic communities. For example, in judging hypothetical examples of elder abuse, Native American Indians gave higher ratings of severity to the abuse than did White Americans (Hudson & Carlson, 1999). Furthermore, Native American Indians were more likely than White experts to strongly agree that physically or verbally forcing elders to do something they did not want to do was abusive. Members of two different Native American Indian tribes generally agreed that elder abuse could be demonstrated in physical, psychological, social, or financial ways, and that abuse could have negative effects on elders in any or all of those ways. Moreover, some Native American respondents stated that elder abuse is "being disrespectful of or not honoring the elderly and their place in history;" "to treat an elder as less than human;" and "injuries to an elder's body, heart, mind, or spirit" (Hudson et al., 1998).

"You must don't hit your momma!" This injunction has been identified as a major theme in the Black community. According to Griffin (1999a), whatever other forms of abuse Blacks may commit, physical abuse of elderly mothers is typically not among them. Typically, Blacks are about as intolerant of elder abuse as Whites, and in many cases even more intolerant. For example, among older Black, White, and Korean American women recruited from churches and social service agencies, 73% of the Blacks rated an entire set of 13 elder mistreatment scenarios as abusive, as compared with 50% of Korean Americans and 67% of Whites (Moon & Williams, 1993). Interviews with Black, Korean American, and White elders reveal that Blacks are less tolerant than Whites of verbal and financial abuse but not of hitting (Moon & Benton, 2000). In focus groups composed of older Whites, Blacks, Puerto Ricans, and Japanese Americans, Black elders were distinguished by emphases on (1) love as the central responsibility of the family; (2) withholding love through ignoring elders or subjecting them to harsh or profane language as the worst thing family members could do to elders; and (3) the importance of context in determining whether an act is abusive (Anetzberger, Korbin, & Tomita, 1996). Black caregivers stressed the role of the oldest daughter in caring for elderly relatives and insisted that although certain behaviors, such as yelling at or ignoring a demanding elderly relative, when performed under stress might be understandable, they were not respectable or acceptable.

Although the popular image of the treatment of the elderly in Asian countries emphasizes honor and respect, these outward expressions may mask much tolerance of aggression towards the elderly (e.g., Koyano, 1989; Malley-Morrison, You, & Mills, 2000). When asked to judge scenarios portraying situations of elder maltreatment (e.g., a daughter-in-law drugging her mother-in-law when guests visited to avoid being embarrassed by her), fewer Korean American than White or Black women viewed the hypothetical situations as abusive (Moon & Williams, 1993). Compared to White and Black elderly, Korean American elderly were the most tolerant of elder abuse, the most likely to blame the victims for elder abuse, and the most negative towards outside help when elder abuse was taking place (Moon & Benton, 2000).

As was true in regard to wife abuse, Asian American groups appear to differ among themselves in judgments concerning the kinds of behavior that constitute elder maltreatment. Compared to Japanese Americans, Taiwanese Americans in one study were more likely to disapprove of a caregiver who ties a physically or mentally impaired adult in bed and more likely to disapprove of yelling at elderly parents (Moon, Tomita, & Jung-Kamei, 2001). Financial exploitation was most tolerated by the Korean Americans, who also were most likely to blame elderly victims for their abuse. Taiwanese Americans were least tolerant of leaving bedridden elderly parents alone occasionally for a few hours.

The Broader Ecological Context

Norms concerning the treatment of children, spouses, and elderly parents are embedded within a broader cultural framework. Although many characteristics of the American macrosystem play a role in family violence, there are four in particular that warrant special consideration: (1) protection of the right to own firearms; (2) media violence and pornography; (3) legalized capital punishment; and (4) tolerance for poverty and economic inequality.

Firearms

Personal ownership of guns appears to be more widespread in this country than in any other developed nation, ranging from a low of 29% in New England to a high of 60% in the East South Central region (Smith & Martos, 1999). Although a large majority of Americans support measures to keep guns away from criminals (e.g., by requiring a background check and a five-day waiting period before a handgun can be purchased), more than 60% oppose limiting gun ownership to the police and other authorized persons, and only 16% support a total ban on handguns (Smith & Martos, 1999). Over 90% of respondents to this NORC survey supported the prohibition of gun sales to criminals convicted of domestic violence, but fewer than 30% thought guns should be banned in homes with children under age 18. In general, women, residents of large cities and their suburbs, liberals, and Democrats are most likely to support gun safety measures, whereas men, residents of rural areas, conservatives, and Republicans are least likely to support them.

The resistance to increased gun control laws in the United States is important because of the significant role guns play in violence within the home. According to a report from the Centers for Disease Control and Prevention (1997), "The United States is unequaled in its rate of firearm injuries and fatalities. An international comparison of 26 industrialized countries found that the firearm death rate for U.S. children under 15 was 12 times higher than the rates for the other 25 countries combined" (paragraph 2). Firearms are also involved in a substantial percentage of domestic violence homicides (see discussion in Chapter 7), as well as in crippling injuries. In the year 2000, 95% of female firearm homicide victims were murdered by a male—usually an intimate partner (Violence Policy Center, 2000). Other research has indicated that when there are one or more guns in the home, the risk of suicide among women increases nearly five times and the risk of homicide more than triples—due largely to homicides at the hands of a spouse, intimate acquaintance, or close relative (Violence Policy Center, 2000).

Despite the historic resistance to gun control in the United States, some legislation has been passed to reduce the role of firearms in domestic homicide. In 1994, Congress passed the Protective Order Gun Ban, which prohibits gun possession by a person against whom there is a restraining or protective order for domestic violence, and in 1996, the Domestic Violence Misdemeanor Gun Ban, which prohibits anyone convicted of a misdemeanor crime of domestic violence or child abuse from purchasing or possessing a gun.

Media Violence and Pornography

On average, Americans, either intentionally or unintentionally, are exposed to large doses of violence in the media and pornography. Content analyses of over 8,000 hours of television programming revealed that approximately 60% of TV programs contain violence (National Television Violence Study, 1996, 1997, 1998, cited in Bushman & Anderson, 2001). On average, American children see more than 8,000 murders and more than 100,000 other violent acts (e.g., assaults, rapes) on network television by the end of elementary school (Bushman & Anderson, 2001). Moreover, the level of violence portrayed on television is not a simple reflection of reality. For example, approximately 50% of the crimes portrayed in reality-based TV programs are murders, yet only 2% of the crimes reported by the FBI are murders (Oliver, 1994, cited in Bushman & Anderson, 2001).

Outside of the media industry, there is widespread acceptance of the extensive scientific data indicating that observing violence increases violence. Six major professional societies—the American Psychological Association, American Academy of Pediatrics, American Academy of Child and Adolescent Psychiatry, American Medical Association, American Academy of Family Physicians, and American Psychiatric Association—signed a joint statement on the hazards of exposing children to media violence, noting that "at this time, well over 1,000 studies . . . point overwhelmingly to a causal connection between media violence and aggressive behavior in some children" (Joint Statement, 2000, p. 1, cited in Bushman & Anderson, 2001, p. 488).

The widespread viewing of violence on television, in movies, and in video games is evidence of this country's acceptance of violence (Harrington & Dubowitz, 1999) and can lead to violence by and against children. Indeed, the more violent programs children watch, the more aggressive their behavior toward their peers (Eron, 1987)—an association that holds even when controlling for many possible confounding variables (Eron & Huesmann, 1980, 1985). In a comprehensive review of media violence and children's behavior, Wood, Wong, and Chachere (1991) concluded that viewing violent media *causes* children to behave aggressively. If children learn to view the use of violence as acceptable, how are they going to behave when they become adults and their children and/or spouses upset or frustrate them?

Research has demonstrated specific links between viewing television violence in childhood and later aggression, including partner maltreatment. For example, a longitudinal study of the relationship between TV violence viewing at ages six to ten and adult aggressive behavior approximately 15 years later showed that men who were high TV violence viewers as children were significantly more likely to have pushed, grabbed, or shoved their spouses than men who had watched less violent TV as children (Huesmann, Moise-Titus, Podolski, & Eron, 2003). Similarly, women who were high TV violence viewers as children were more likely to have thrown something at their spouses, and had punched, beaten, or choked another adult at over four times the rate of other women.

Pornography is another multibillion-dollar industry in the United States that may influence tolerance of family aggression. In the late 1980s, 20% of American youngsters had seen an X-rated movie (Linz, Donnerstein, & Penrod, 1988). In addition, the explosion of the Internet business has made pornography readily available to children of all ages. A Kaiser Family Foundation survey revealed that 70% of adolescents aged 15 to 17 years had unintentionally been exposed to pornographic Web sites (Kanuga & Rosenfeld, 2004). Moreover, the Internet has been used for trafficking in child pornography, engaging children in inappropriate sexual interactions, and attempting to recruit them for abusive purposes (Kanuga & Rosenfeld, 2004). A study of over 200 private school students from grades 7 through 10 revealed that 21% had visited a pornographic Web site at least once for over three minutes, and a third had visited such sites more than four times (Stahl & Fritz, 2002). Several students said they had been invited to participate in sexual activity as a result of their online communications.

Although studies of the association between exposure to media pornography and later maltreatment are sparse, there is some research linking exposure to pornography with various forms of maltreatment of women and children. In a sample of juvenile sex offenders, over 30% of the victimizers reported viewing child pornography (Zolondek, Abel, Northey, & Jordan, 2001). In a study of victims of wife rape, one third of the victimized women reported that their partners viewed pornography, and the partner's use of pornography was associated with the most sadistic rapes (Bergen, 1998). Reports from women in a battered women's shelter revealed that the likelihood of sexual abuse as part of the violent relationship significantly increased when the batterers used pornography (Shope, 2004).

Capital Punishment

The United Nations, the European Union, Amnesty International, the Organization of American States, and other groups have all passed resolutions against capital punishment, which has declined in recent years in most parts of the world. The United States is the only democracy where the death penalty is still frequently put into practice (Harries, 1995). Between 1976 (when the Supreme Court reinstated capital punishment after a nine-year hiatus) and March 2004, approximately 907 people were put to death in this country for capital crimes. Of these, 34.1% were Black, 57.2% White, 6.5% Hispanic, and 2.4% of another ethnicity. Currently, despite substantial evidence that capital punishment is not an effective deterrent to crime (Barak, 2003; Milburn & Conrad, 1996), the death penalty is supported by statute in 38 states. Executions take place primarily in the South (especially Texas, Florida, and Virginia) and disproportionately involve Black men. Nearly 400 juveniles (youth under the age of 18) have also been executed in the United States, including 22 in the period 1973–2003, which is a violation of international law (Child Welfare League of America, 2002). Nearly two thirds of these recent executions of minors occurred in Texas, which in this period has put to death more juveniles than the rest of the world combined. In all other nations, the practice of executing juveniles is no longer accepted (Streib, 2003).

When asked whether they support capital punishment, people in the United States overwhelmingly indicate their approval, a fact that has been used by the Supreme Court as evidence that the death penalty is consistent with community standards. However, when asked more specific questions, a substantial portion of respondents indicate that they are unwilling to put certain groups of defendants—for example, juveniles and the mentally retarded—to death (Haney, 1997). Moreover, many respondents who indicated support of the death penalty in the abstract also indicated that there were a number of circumstances that would move them in the direction of favoring life in prison without the possibility of parole (for example, when the defendant had been seriously abused as a child).

What does the nation's tolerance of capital punishment have to do with family violence?

- Many violent prisoners, and particularly many prisoners on death row, have themselves been victims of family violence (American Academy of Child and Adolescent Psychiatry, 2000).
- Support for the corporal punishment of children in homes and schools is highest in the same states that show the greatest support for the death penalty (Cohen, 1996).

- Many death penalty supporters grew up in homes where corporal punishment was accepted as necessary; they internalized childhood messages concerning punishment and retribution, and unconsciously seek outlets for their suppressed anger (Milburn & Conrad, 1996).
- Tolerance for the death penalty, in violation of international laws and trends, is part of a social and cultural context in which punishment, retribution, and violence are legitimized and sanctioned as appropriate responses to undesired behavior.

Social and Economic Inequality

Although the United States is in many ways one of the wealthiest nations in the world, it also has a persistently high number of individuals and families living in poverty. Indeed, according to Rank (2001), "Impoverishment in the United States exceeds that of all similar countries" (p. 885). In support of his assertion, Rank cited the results of a cross-national analysis of poverty in 18 developed nations (Smeeding, Rainwater, & Burtless, 2000), which revealed that the U.S. poverty rate of 17.8% was substantially higher than the rate in the other 17 nations. The second highest rate of poverty was in Italy (13.9%), with the Scandinavian countries, as well as Belgium, the Netherlands, and Luxembourg, showing the lowest rates.

Throughout its history, both the citizenry and the power structure in the United States have varied enormously in their willingness to address such social problems as family violence and poverty (Pleck, 1989). In recent decades the percentage of children living below the poverty level (which was $18,244 for a family of four with two children in 2002) has averaged around 20%, but the number of children living in extreme poverty (i.e., below 50% of the poverty threshold) has increased. By 2002, there were 11.6 million children living below the poverty level (ChildStats, 2004). The percentage of American children living in poverty varies disproportionately by ethnicity. By 2001, the percentage of Black children growing up in circumstances of extreme poverty reached its highest level in 23 years (Children's Defense Fund, 2003). Poverty is also high in children in immigrant families. In 1970, poverty rates of immigrant children (~12%) were somewhat lower than those of native-born children (14%). By 2000, over 20% of immigrant children, as compared with approximately 15% of native children, were classified as poor (Van Hook, 2003). Thus, although poverty rates remained stable over this 30-year period for native-born children, they almost doubled for immigrant children.

The evidence is overwhelming that poverty is associated with a broad range of social problems, including various forms of violence. These problems include higher birth rates in unwed mothers, teen pregnancy, crime and delinquency, drug-related problems, child maltreatment, higher rates of infant mortality (Drake & Pandey, 1996), and rape (Smith & Bennett, 1985). One analysis of data from the 1985 National Family Violence Survey revealed that overall violence toward children was 4% higher; severe violence was 46% higher; and very severe violence was 100% higher in poor families than in families living above the poverty line (Gelles, 1992, cited in Burgess, Leone, & Kleinbaum, 2000). A 1992 study of Missouri child abuse and neglect cases revealed that the high poverty neighborhoods had the highest number of reported and substantiated incidents of child maltreatment (physical abuse, sexual abuse, and neglect) (Drake & Pandey, 1996). Data from the National Longitudinal Survey of Youth revealed that low income is associated with both frequent spanking and overall child maltreatment (Berger, 2004). In a study of family violence cases in a very high crime county in Florida, the highest rates of family violence were found in neighborhoods with the lowest levels of economic resources (Miles-Doan, 1998). There are also numerous studies demonstrating a link between poverty and intimate partner violence (Rank, 2001), and, although there is less research on the link between poverty and elder abuse, there is some evidence that poverty is a risk factor for elder maltreatment (Lachs, Williams, O'Brien, Hurst, & Horwitz, 1997).

Despite the high correlation of poverty with family violence and other social ills, it may not be poverty per se that is the causal agent of the problems as much as related factors such as social and/or economic inequality. Income inequality (the juxtaposition of extreme poverty with extreme wealth) is associated with both interpersonal and collective violence (Mercy, Krug, Dahlberg, & Zwi, 2003). There is also evidence that income inequality favoring women rather than overall poverty contributes to wife abuse (McCloskey, 1996) and that income inequality favoring Whites over Blacks may contribute to intimate partner violence in the Black community (Aborampah, 1989). Moreover, whereas social inequality declined in the Western industrialized nations in the decades immediately following World War II, it increased during the 1980s and 1990s—as did crime and punishment (Hagan, 1993). In a cross-national analysis that included the U.S., previous findings that income inequality is associated with violent crime, including homicide, were confirmed (Fajnzylber, Lederman, & Loayza, 2001). Given that immigrants and people of color are disproportionately represented at poverty levels and the primary victims of social inequality, it should be no surprise that rates of family violence sometimes are higher in those communities than in the majority community, although differential reporting practices may also play a role in these figures (Malley-Morrison & Hines, 2004).

Summary

Rates of family violence are relatively high in the United States as compared with other wealthy industrialized nations, as are the levels of violent crime, gun ownership, capital punishment, and impoverishment. Norms supporting violence under particular circumstances are pervasive. In the next chapter, we give further consideration to the role of cultural context in contributing to violence in families, with particular attention to religious cultures. As you read this chapter, consider how your own background has contributed to your views on such issues as corporal punishment, punishment of marital infidelity, and the legitimacy of violence as a response to presumed transgressions by family members.

3

Cultural Contexts

Religion

Our pastor asked my wife and me, "When do you plan to start spanking your son? He's not a baby anymore, you know. Don't you think he's ready for his first spanking?" We were stunned. We didn't know how to answer. We had decided before we were married that we would not raise our children the way we were raised, that we would not spank. Now, my wife and I don't feel comfortable at church. Every time the pastor greets us, we imagine he is silently asking, "Well, have you started spanking yet?" (Riak, 2002, paragraph 2)

You know that fatalistic attitude that a lot of Muslim women have, whatever is Allah's will, okay, I went through that period, but then it is at the point where now it is Allah's will, but it is also up to me to determine how I am going to proceed with my life and my kids' lives . . . And so in that respect I am very grateful that I had to go through [wife abuse] because it has reawakened my spirituality as to why I became a Muslim in the first place. (Hassouneh-Phillips, 2001a, p. 428)

For years I was married to an abusive man . . . When I finally could stand it no longer I . . . moved to a new church, where I was treated with gentleness and loving concern and accepted as a worthy individual. People cared about me as someone Christ loved. As I experienced this love, some of my self-esteem returned and the healing process began. (Sutherland, 1996, pp. 17–18)

An important component of the cultural context in which most Americans are reared is religion. In this chapter, we consider the complex role played by religious affiliation, religiosity, spirituality, and religious conservatism in family violence.

Religion in the United States

"What is your religion, if any?" When a national sample of Americans was asked this question, 81% provided some affiliation—with a total of over 100 different affiliations being listed (Mayer, Kosmin, & Keysar, 2001). Religious membership at the beginning of the 21st century was approximately as follows: 55% Protestant, 24% Catholic, less than 2% Jewish, 14% none, 2% unidentified Christian, and less than 1% for each of other religious affiliations (e.g., Buddhism, Hinduism, Islam) (American Religion Data Archive, 2002). Although the percentages of non-Christian religions appear relatively small, in absolute numbers there are approximately 6 million Jews, 2 million Muslims, 400,000 Buddhists, and 227,000 Hindus living in the United States (U.S. Census Bureau, 2000). It is also important to note that Muslims, Hindus, Buddhists, Protestants, Catholics, and Jews can belong to a number of different religious communities within the broad umbrella of their religion. There are, for example, well over 100 different Protestant denominations in this country (Smith, 1987).

Necessary Distinctions

In this chapter, *religion* refers to religious affiliation—that is, membership in a particular religious community. *Religiosity* refers to the behavioral, motivational, and cognitive aspects of individual participation in a religious community and can be differentiated into two types: intrinsic and extrinsic. *Extrinsic religiosity* refers to religious behavior motivated by external and often self-serving reinforcers. *Intrinsic religiosity* refers to religious behavior motivated by feelings of personal conviction (Allport & Ross, 1967). *Spirituality* is "a personal belief in and experience of a supreme being or an ultimate human condition, along with an internal set of values and active investment in those values, a sense of connection, a sense of meaning, and a sense of inner wholeness within or outside formal religious structures" (Wright, Watson, & Bell, 1996, p. 131). Spirituality is thus closely akin to intrinsic religiosity, although individuals can be spiritual without belonging to a religious community. As we show, some organized religions have been criticized, particularly by women's rights advocates, for supporting—or at least not adequately resisting—subjugation of women and children in ways that can result in their abuse. By contrast, spirituality and religiosity, particularly intrinsic religiosity, can sometimes serve as protective factors, reducing the likelihood of abuse or reducing its negative effects.

Other important constructs in debates over the role of religion in family violence are religious conservatism and religious fundamentalism. Ellison and Sherkat (1993) emphasized that the ideological orientations of the conservative Protestant denominations (the "religious right") include a belief in biblical literalism, a view of human nature as sinful and prone to egoism, and a heightened sensitivity to issues of sin and punishment. Conservative Protestants include two main groups—fundamentalists, who believe in biblical literalness and infallibility, and evangelicals, who may see the Bible more metaphorically (Ellison, Burr, & McCall, 2003). According to some experts, there are four major types of Protestantism: (1) liberals (Congregationalists, Episcopalians, Methodists); (2) moderates (Presbyterians, Disciples of Christ); (3) conservatives (American Lutherans, American Baptists); and (4) fundamentalists (Southern Baptists, Missouri Synod Lutherans) (Glock & Stark, 1965). Other researchers have classified Protestant denominations in somewhat different ways, but Lutherans and Southern Baptists are typically viewed as conservative and/or fundamentalist. Moreover, the other major religions also have religious communities varying from somewhat liberal to conservative and fundamentalist (Hertel & Hughes, 1987).

Fundamentalists of any major religion are likely to believe that their holy texts are inerrant (i.e., infallible) and literally true. However, fundamentalists are selective in the texts they proclaim as establishing the laws of conduct. When feminists and others are critical of the negative role of religion in family violence, it is often the religious culture of the fundamentalist branches of different religions that are the primary concern.

Religions as Cultural Systems

All religious sects and denominations encapsulate a set of cultural values. Even within Christianity, Judaism, and Islam, there are enormous differences in cultural values across different religious subcommunities. Each community has, for instance, its own subjective meanings for the concept of punishment—who should be punished, by whom, and how. It has been argued that in the New England Protestant tradition, which still has a powerful influence on American culture, there is no room for the fuzzy, authoritarian, but forgiving orientation of Roman Catholicism (Melossi, 2001). From the Puritanical Protestant perspective, whoever is perceived to be disobeying a law must be punished. Although Catholic traditions are also paternalistic, the authoritarianism is softer, with more room for indulgences and absolution.

Each religious community also has its own subjective meanings for the roles of "husband," "wife," and "children." Each has scripture with descriptions of family roles and relationships—descriptions that have often become prescriptions, especially with the modern global movement towards religious fundamentalism (e.g., Jaccoby, 2002; Levesque, 2001). According to fundamentalist interpretations of the scriptures of all the major religions, woman's place is in the home, where her duties include raising her husband's children, serving as his helpmate, and respecting his authority in all earthly matters. In the face of feminist and human rights fights for gender equality, such tenets can lead to frustration and conflict in families, which can in turn lead to violence and abuse.

Consider the excerpts from several major holy books provided in Box 3.1. How do you interpret them? What are their implications for relationships between husbands and wives, parents and children? Do they seem to provide a rationale for men's dominance over women and parental beating of children? Do they lend themselves to interpretations providing justification for violence within domestic relationships?

Box 3.1   Excerpts from Holy Books on Family Relationships

The Old and New Testaments

- To the woman he said, "I will greatly multiply your pain in childbearing; in pain, you shall bring forth children, yet your desire shall be for your husband, and he shall rule over you." (Genesis 3:16)
- If anyone sets his heart on being a bishop he desire a noble task . . . He must manage his own family well and see that his children obey him with proper respect. If anyone does not know how to manage his own family, how can he take care of God's church? (Timothy 3:1–5)
- Discipline your son while there is still hope; do not set your heart on his destruction. (Proverbs 19:18)
- He who spares the rod hates his son. (Proverbs 13:24)
- "I took this woman (in marriage) and slept with her and did not find proof of virginity in her," then the girl's father and mother shall take the proof of her virginity to the elders of the town, at the town gate . . . if, on the other hand, the accusation is true and no proof of the girl's virginity is found, then they shall bring her out to the door of her

father's house and the men of her town shall stone her to death . . . She has committed an outrage in Israel by playing the prostitute in her father's house; you shall rid yourselves of this wickedness. (Deuteronomy 22:14–21)

- To the woman he said: "I will intensify the pangs of your childbearing, in pain shall you bring forth children. Yet your urge shall be for your husband, and he shall be your master"; and "for man did not originally spring from woman, but woman was made out of man; and man was not created for woman's sake, but woman for the sake of man." (I Corinthians 11:8–9)

The Qur'an and Other Islamic Holy Writings

- Men are the protectors and maintainers of women, because God has given the one more (strength) than the other, and because they support them from their means. Therefore the righteous women are devoutly obedient, and guard in (the husband's) absence what God would have them guard. As to those women on whose part ye fear disloyalty and ill-conduct, admonish them (first), (Next), refuse to share their beds, (And last) beat them (lightly); but if they return to obedience, seek not against them Means (of annoyance): For God is Most High, great (above you all). (Ayah 34 of Surah 4)
- In inheritance, the son inherits as much as two daughters.
- As long as a child has not reached the age of maturity, or is not mature, the child is under the guardianship of the father, or someone from the paternal lineage (i.e., the grandfather or uncle). After attainment of maturity, the father or paternal lineage no longer has guardianship, unless it is in regards to the marriage of a virgin girl, in which case the permission of the father or paternal lineage is required.

Hindu Texts

- A virtuous woman should serve her husband like a god.
- A wife, a son, and a slave, these three are declared to have no property; the wealth which they earn is (acquired) for him to whom they belong. (Manu VIII.416)
- In childhood a female must be subject to her father, in youth to her husband, when her lord is dead to her sons; a woman must never be independent. (Manu V.148)

Obviously, these passages do not provide all the relevant messages in the holy writings of the religions represented in the box. Each religion also has scripture with very different messages from these. Moreover, within every major religion there are controversies over the interpretations of these and other texts (e.g., Bartkowski, 1996; Hassouneh-Phillips, 2003). Even among members of the faithful who consider their sacred texts to be the direct word of God, there are debates over how particular words should be translated and the extent to which particular interpretations of texts are correct (Bartkowski, 1996).

In general, feminists and proponents of international human rights have argued that writings from all the major religious texts have been used—and misused—within patriarchal systems to maintain the power and authority of men and to keep women and children in subordinate positions, by force if necessary (e.g., Dasgupta & Warrier, 1996; Dutcher-Walls, 1999). Organized religions are viewed as upholding coercion by men—for example, in promulgating the notion that men are "entitled" to sex with their wives, whether their wives want to engage in sex or not (Basile, 1999). From this perspective, the problem is not so much with the religions per se as in the misuse of the texts for political purposes, specifically for maintaining male dominance. One psychiatrist described the case of an abused woman whose husband used the family's conservative Christian faith not only to justify his abuse but also to disrupt his wife's relationship with the therapist trying to help her with her anxiety and depression. This husband declared that the therapist "was a liberal and a feminist sent by Satan to tempt the patient to repudiate her marital and maternal obligations" (Stotland, 2000, p. 696). Moreover, the husband threatened, the wife would go to Hell if she stayed in therapy.

It is probably not coincidental that the feminist movement, the international human rights movement, and the new conservatism (which is closely tied to religious conservatism) have acted as competing forces over the past few decades. As women around the globe have fought to gain greater educational, employment, and political opportunities, and as their struggles have become incorporated into the human rights movement, political forces directed at maintaining the status quo have united with fundamentalist religious leaders with a similar goal of maintaining a hierarchical social structure with clearly delineated gender roles and power resting firmly in the hands of men (e.g., Gül & Gül, 2000).

Any adequate consideration of the role of religion in family violence must recognize that within the world's major religions and spiritual orientations there are countless sects and denominations, and the belief systems held by different sects within a broader religious grouping may vary widely. Indeed, it is possible that the highly conservative or fundamentalist factions of different religions have more in common with each other than with the more liberal denominations within the same religion. Given the international religious fundamentalist surge, and its pressuring of women into subservient roles, much of the research on the role of religion in family violence has involved comparisons of fundamentalists with other religious groups.

Consider now the excerpts in Box 3.2. These excerpts have quite different messages from those in Box 3.1 and are emphasized by more liberal congregational cultures preferring to focus on messages of mutual respect in the Bible (Alsdurf & Alsdurf, 1988), the Qur'an (Hendricks, 1998; Rahim, 2000), and Hindu religious texts (Rahim, 2000). These passages are often cited by reformists arguing that despite the traditional mistreatment of women and children within presumably religious homes and communities, it is possible to seek equal human rights and still stay within a religious fold (e.g., Hendricks, 1998; Rahim, 2000).

Box 3.2   Different Messages from Holy Books

The New Testament

- Husbands love your wives, just as Christ loved the church and gave himself up for her to make her holy . . . In this same way husbands ought to love their wives as their own bodies. He who loves his wife loves himself. After all, no one ever hated his own body, but he feeds and cares for it, just as Christ does the church. (Paul, Ephesians 5:25–28)
- Fathers, do not exasperate (scold or punish too harshly) your children; instead, bring them up in the training and instruction of the Lord. (Paul, Ephesians 6:4)

The Qur'an

- And among His signs is that He created for you mates from among yourselves that you may live in tranquility with them, and He has put love and mercy between you; Verily, in that are signs for people who reflect. (30:21)
- None honours women except he who is honourable, and none despises them except he who is despicable. (Hadith)
- Whosoever has a daughter and he does not bury her alive, does not insult her, and does not favor his son over her, God will enter him into Paradise (Ibn Hanbal, No. 1957). Whosoever supports two daughters

---

(Continued)

(Continued)

till they mature, he and I will come in the day of judgment as this (and he pointed with his two fingers held together).

Prophet Muhammad

- The best of you is the best to his family, and I am the best among you to my family. The most perfect believers are the best in conduct and best of you are those who are best to their wives. (Ibn-Hanbal, No. 7396)

Hindu Sacred Texts

- Where women are honored, there the gods are pleased . . . Where the female relations live in grief, the family soon wholly perishes; but that family where they are not unhappy ever prospers. (Laws of Manu, as cited in Radhakrishnan & Moore, 1989, pp. 189–190)
- He only is a perfect man who consists (of three persons united): his wife, himself, and his offspring. (Laws of Manu, as cited in Radhakrishnan & Moore, 1989, p. 190)
- The wife and husband, being the equal halves of one substance, are equal in every aspect; therefore, both should join and take equal parts in all work, religious and secular. (The Vedas, as cited in Polisi, 2003)

If religious people simply adopted the values expressed in Box 3.2, would that reduce family violence? Or is increased secularization the only effective response to violence in families? Does the increasing political power of the religious right, both in the U.S. and abroad, endanger women and threaten to enshrine corporal punishment of children even as European countries ban it from homes and schools? Can women fight for their rights for freedom from violence and still be at ease with their religions?

Controversies Over the Interpretation of Texts

A good example of controversies between conservative practitioners of a religion and more liberal practitioners can be found in the debate concerning how children should be reared. There is ample evidence that child-rearing practices of conservative Protestants are influenced by a particular set of theologically related beliefs and values. One important question is whether these Protestants interpret the Bible correctly. For example, many conservative Protestants argue that the Bible says, "Spare the rod and spoil the child," but it has been widely pointed out that this statement does not come directly from the Bible. Also, nowhere does the Bible say that if parents do not hit their children with a rod, the children will be spoiled, and numerous authors have noted that "the rod" refers to a shepherd's staff, which is used to guide, not hit, sheep (think of "Thy rod and thy staff they comfort me") (e.g., Carey, 1994).

Controversies concerning the meaning of texts in the "inerrant" Bible abound even among conservative Protestants (Bartkowski, 1996). Ross Campbell (1989), a popular conservative evangelical parenting specialist, urges Christian parents to rear their children with unconditional love rather than corporal punishment and admonishes conservative Christian parents to avoid the "punishment trap." In stark contrast to Dobson (1978; 1987), another author of child-rearing manuals for conservative Protestants, Campbell argues that the frequent use of corporal punishment (1) diminishes beneficial guilt in the child and inhibits the development of conscience; (2) degrades, dehumanizes, and humiliates the child; (3) teaches the child to be aggressive and punitive; (4) can be physically damaging; and (5) equates morally correct action with avoiding parental detection of infractions (Campbell 1989).

Religious Affiliation and Religious Conservatism as Risk Factors for Family Violence

Child Maltreatment

If you were asked to identify "religion-related" cases of child maltreatment, what kinds of maltreatment would come to mind? Sexual abuse by priests? Ritualistic abuse? Withholding of medical care? All of these and others were identified by a sample of over 2,000 mental health professionals from across the United States (Bottoms, Shaver, Goodman, & Qin, 1995). Among the examples given were withholding medical care, abuse perpetrated by persons with religious authority, and attempts to rid a child of evil. One clinician reported, "The aunt truly believed she could beat the devil out of the children" (Bottoms et al., 1995, p. 93). In most of these abuse cases, the perpetrators were Protestant.

Corporal Punishment and Child Physical Abuse

Much of the limited research on cultural values concerning child rearing and child discipline within different religions has focused on the values and beliefs of conservative Protestants as compared with other groups. Conservative Protestants have a distinct set of cultural values supportive of corporal punishment encapsulated in many of their child-rearing manuals. An analysis of best-selling conservative and mainstream Protestant child-rearing manuals revealed that mainstream Protestant manuals emphasized democratic parent-child relationships and open communication between parents and children, while conservative Protestant manuals emphasized obedience and submission (Bartkowski & Ellison, 1995). Whereas mainstream manuals emphasized the parental use of reasoning tactics, conservative manuals emphasized corporal punishment. In particular, the conservative manuals recommended that parents use a rod to administer physical punishment to children whenever children show willful defiance to parental authority. What are your views on the child-rearing advice of conservative Protestants? If punishing a child with a rod is advised in a child-rearing manual, can it still be abusive? If people believe that "a good beating" is not just condoned, but demanded, by their religious texts, what arguments might nevertheless justify labeling the beating as abusive?

It is somewhat difficult to obtain a clear picture of differences between religious groups in both endorsement and use of corporal punishment and in frequency of child abuse reports for several reasons: (1) various investigators have classified religions differently and grouped them together in different ways; (2) lumping together different Protestant denominations in order to compare Protestants and Catholics, as has been done in some studies, ignores the wide range of values characterizing the different Protestant denominations; (3) measures of constructs such as religiosity and religious conservatism have varied considerably across studies; and (4) variables such as geography, ethnicity, and social class tend to be confounded with affiliation.

Despite these limitations, some important findings have emerged. National survey data support the propositions that there are differences between religions in support for corporal punishment, and the endorsement of particular religious beliefs (e.g., that the Bible is the literal word of God) is positively associated with convictions concerning corporal punishment. For example, the first wave of the National Survey of Families and Households (NSFH) revealed that Jewish parents were significantly less likely than Protestant or Catholic parents, and Protestant parents were more likely than parents with no religious affiliation, to use corporal punishment (Xu, Tung, & Dunaway, 2000). According to the 1990 National Longitudinal Study of Youth (Giles-Sims, Straus, & Sugarman, 1995), Catholics had the lowest use of spanking of all the groups studied. In a more localized study of couples in a southwestern city (Gershoff, Miller, & Holden, 1999), conservative Protestant parents spanked their children more often than mainline Protestant, Roman

Catholic, and unaffiliated parents. Moreover, 29% of the conservative Protestant parents said they spanked their children three or more times a week, as compared to only 3% of Roman Catholic and none of the unaffiliated parents. Whereas the conservative Protestants emphasized the instrumental benefits of spanking, the mainline Protestants and Roman Catholics anticipated more negative effects of spanking. Mainline Protestant and unaffiliated parents were more likely to indicate they would try to reason with their children about misbehavior than conservative Protestants.

Two smaller religious denominations of particular interest to researchers are Quakers (because of their advocacy of nonviolence) and Mormons (because of their strict patriarchal traditions). A survey of nearly 300 Quakers revealed that 75% of mothers and 69% of fathers reported violence toward their children (Brutz & Ingoldsby, 1984). As compared to national norms, Quaker mothers and fathers reported more kicking, biting, and punching of their children, all of which are typically considered to be abusive in mainstream society. Interestingly, Quaker parents reported less slapping and hitting with objects, which may be the most normative forms of corporal punishment among Protestants in general. In contrast to the national sample, where some fathers reported beating their children and threatening or using guns or knives on their children, no Quaker fathers reported these forms of violence against their children. Thus, even though Quakers advocate nonviolence, it does not necessarily mean that they do not physically punish their children; they may just use different forms of physical punishment than the majority culture. In a study of Mormons, rates of severe child physical abuse were slightly lower in a Utah sample (80% Mormon) than the rates in the U.S. as a whole (9.3% versus 10.7%) (Rollins & Oheneba-Sakyi, 1990). Thus, rates of child maltreatment in Mormons appear to be the same or slightly less than nationwide rates, even though the patriarchal structure of their religion would lead some to theorize that they would be more likely to physically punish their children.

Despite considerable differences among congregational communities within religions in their values relating to the use of aggression in families, there has been very little research comparing the smaller Christian and non-Christian denominations and sects on their values and practices. In a rare comparative study, Malley-Morrison and Hines (2003) found, in a convenience sample of nearly 500 people, that Christians, as a group, reported having experienced significantly more corporal punishment than Jews, Hindus, and atheists/agnostics. When comparisons were made among principal denominational groupings, they found that Presbyterians tended to be somewhat less tolerant of corporal punishment than both conservative Protestants and Catholics. Jews had experienced less corporal punishment as children, and were also somewhat less tolerant of it, than conservative Protestants. Conservative Jews were more tolerant of corporal punishment than Reformed Jews. Finally, the agnostic/atheist group had experienced less corporal punishment than the Catholics but were more supportive of parental use of corporal punishment than Buddhists, Reformed Jews, and Conservative Jews.

Although denominational differences have been found, the espousal of particular religious beliefs may be a more important predictor of aggression in families than simple religious affiliation. In the NSFH, degree of endorsement of items such as, "I regard myself as a religious fundamentalist," was significantly positively correlated with parents' likelihood of using corporal punishment (Xu et al., 2000). An analysis of parents' reported use of physical punishment to discipline preschool and elementary school-aged children showed that parents who believed the Bible is the inerrant Word of God and provides answers to all human problems used corporal punishment more frequently than parents with less conservative theological views (Ellison, Bartkowski, & Segal, 1996). The national General Social Survey of 1988 had similar findings: Conservative Protestants were significantly more supportive of corporal punishment than members of other religious groups. Endorsement of the items, "Human nature is fundamentally perverse and corrupt," and, "Those who violate God's rules must be punished," was the strongest predictor of tolerance of corporal punishment (Ellison & Sherkat, 1993).

Other studies confirm the finding that a conservative world view may contribute more to approval of corporal punishment than religion or religiosity. A 1995 national Gallup Poll revealed that parents who were more conservative in their social ideologies had more positive attitudes toward physical discipline (Jackson et al., 1999). In a random sample from Oklahoma City, even after accounting for socioeconomic and demographic variables, belief in the literalness of the Bible accounted for a large share of the differences between religious groups in degree of advocacy of corporal punishment (Grasmick, Bursik, & Kimpel, 1991). Similarly, among Protestants from five central and southern states, both males and females who believed in a literal interpretation of the Bible showed greater endorsement of statements such as, "Children should always be spanked when they misbehave," and "Parents spoil their children by picking them up and comforting them when they cry" (Wiehe, 1990, pp. 178–179).

Independent of the issue of whether corporal punishment is abusive, some studies have specifically addressed physical child abuse as a problem that may vary by religion. In the first National Family Violence Survey, the rate of child abuse was lowest among Jews and highest in families where one or both parents had a "minority religious affiliation" (Straus, Gelles, &

Steinmetz, 1980). The lower rates of family violence among Jews may be related to their generally higher levels of income, education, and employment. Conversely, members of minority religions may experience more discrimination, stress, and isolation from mainstream society, characteristics that may lead to greater violence within their homes (Straus et al., 1980).

Psychological Maltreatment

Consider a parent saying to a child, "God will punish you for your misbehavior." Is this statement a form of verbal abuse? What is the likely effect of such a statement on the child? Do you think regular threats of this sort are more or less abusive than corporal punishment? Why? In one study, such a statement was reported by 10% of Methodists, Presbyterians, and Lutherans; 20% of Catholics; and 25% of Baptists and Fundamentalists (Nelsen & Kroliczak, 1984). Consider also the views of a woman who described "religious abuse" as well as sexual abuse by her brother during childhood. She holds that the sexual abuse by her brother, combined with sexist attitudes toward women within her family and community, and her father's appeal to religious dogma to support his denigration of women, all had serious negative psychological effects on her: "It was like the messages I got all the time, you know, 'God says you're second best.' . . . your dad saying, 'God's right, you are [second best].' And then my brother's taking on the same thing. I had all these things, like, 'oh my God, where do I fit in?' " (Smith, 1997, p. 423). Would you agree with her characterization of her family's behavior as "religious abuse"?

Child Sexual Abuse

There is little research with representative samples to provide reliable estimates of the prevalence of intrafamilial sexual abuse in different religious groups. Of nearly 1,000 women living in the San Francisco area (Bolen, 1998), the following percentages of women reported having experienced childhood sexual abuse: approximately 43% of Protestants, 37% of Jews, 37% of Catholics, and 30% of those with other affiliations; there were no significant differences in prevalence, and no breakdown was given for intrafamilial versus extrafamilial abuse. A survey of nearly 3,000 professional women from across the United States presented a somewhat different picture: approximately 39% of women with agnostic/atheistic parents, 32% with families of some religion other than conservative Christianity, and 18% from conservative Christian families reported some form of childhood sexual abuse (either intra- or extrafamilial) (Elliott, 1994). More specifically, women raised in conservative

Christian homes where there was little emphasis on incorporating religious values into family life (i.e., low intrinsic religiosity) reported significantly higher rates of sexual abuse than women raised in Christian homes with higher integration of religious values (i.e., high intrinsic religiosity) or women raised in nonconservative Christian homes. Elliott speculated that conservative Christians who do not integrate their professed beliefs into their lifestyles may use those doctrines primarily to justify their own controlling tactics with family members and intimidate their victims into silence.

Although incest is prohibited by both Hebrew and Christian scripture, it is clear that it occurs in many homes (see Chapter 5)—including conservative Protestant homes. In one study of 35 women who had been sexually abused in childhood in their conservative Protestant homes (Gil, 1988), the mean age of the girls when the incest began was just under six years, with a range from 2 to 16 years; 37% of the women had been preschoolers when the sexual abuse started. The abusers were fathers in 66% of the cases and stepfathers in 34% of the cases. The natural fathers were evenly distributed across the denominational categories (including Presbyterians, Methodists, Lutherans, Baptists, and Fundamentalists who espoused conservative Christian values); however, 82% of the abusive stepfathers were Fundamentalists. The natural fathers were more likely to perpetrate the more serious forms of sexual abuse (e.g., involving penetration) than the stepfathers.

Box 3.3   Three Cases of Childhood Sexual Abuse

Joan's father was an alcoholic who had left home for six years, joined Alcoholics Anonymous, and returned to the family within the previous year. Joan's mother had turned to religion for support in dealing with her husband's infidelity and abandonment of the family. Joan, a teenager, abused drugs and alcohol and was sexually abused by her father when she was seven. Her mother became very angry when Joan entered a rehab program, and showed more concern about appearances than about her daughter's substance or sexual abuse. (Schmidt, 1995)

Sarah, a 40-year-old Jewish woman, grew up in a religiously observant kosher home with parents who had fled Eastern Europe during World War II and had lost many relatives in the concentration camps. Her father was psychologically abusive to his wife, and physically abusive to Sarah's brothers. Because of the Holocaust, he isolated his family from Jews and non-Jews alike, and preached that his family should never trust a non-Jew. When Sarah was 4 or 5 years old, she was sexually molested by one of her brothers. She was able to prevent further sexual abuse by telling her mother that her brother had hit her, but he was psychologically abusive towards her throughout her childhood, and physically attacked her when she was 13. The mother walked in on this attack, but left the room without saying anything. (Featherman, 1995)

Lacey, a Seventh Day Adventist, was repeatedly raped by her older brother in her pre-adolescent years. Her sister reported being sexually assaulted by the same brother as well as by their father. Both women were completely ostracized by other family members, who said neither of the men could have been capable of such behavior. Lacey's husband was an administrator of one of the Adventist institutions, and Lacey did not want to damage his career by reporting on her brother's abuse, but became very concerned when her brother was appointed to a position in a boarding academy for youth in South America. (Taylor & Fontes, 1995)

Although our current knowledge of prevalence rates of child sexual abuse within families as a function of religious affiliation is incomplete, there is some agreement that the effects of sexual molestation may vary depending on the family's religious beliefs. Box 3.3 presents the stories of three women sexually abused in childhood—an Anglo-Saxon Protestant, a Jew, and a Seventh Day Adventist. What is your view of the role of religion in each woman's experience? In what ways, if any, did the particular religious culture in which each woman was raised contribute either to the abuse itself or to the girls' response to the abuse? What aspects of the abusive situations seem independent of the particular religious heritage of each woman?

Child Neglect

A major form of religion-related child abuse identified by a national sample of mental health professionals is the withholding of medical care for religious reasons (Bottoms et al., 1995). Approximately 10% of neglect cases identified by professionals involved this form of neglect. Most of the parents in these cases were classified as "fundamentalist" (e.g., Mormon, Pentecostal, Seventh Day Adventist). Among fundamentalists, a number of different religious beliefs lead to resistance against certain forms of medical care. For example, Jehovah's Witnesses believe that the Bible prohibits blood transfusions and therefore any procedures necessitating a transfusion

(Flowers, 1984). Christian Scientists believe that sickness and pain are errors of the mortal mind, all diseases are mental conditions, and healing the sick must involve driving out misperceptions rather than using procedures like medications, which only relieve suffering temporarily but do not cure the problem. The principal exception to the Christian Science system is the use of medical care in childbirth and to set broken bones.

Consider the case of *In re Sampson,* in which the court was asked to intervene on behalf of a 15-year-old boy with a serious facial deformity resulting from a medical condition called neurofibromatosis. Because of this condition, the boy was extremely emotionally withdrawn, would not attend school, and was virtually illiterate. Although his condition was not life-threatening, his physicians indicated that they could improve his appearance surgically. Because the procedure would involve considerable medical risk, they were unwilling to undertake the procedure without authorization to provide blood transfusions if necessary. The mother, a Jehovah's Witness, refused to agree to the surgery if there was a chance that it would require a transfusion (Knepper, 1995). Is the mother's refusal in this case to allow her son to have the corrective surgery a type of neglect, some other form of maltreatment, or appropriate parenting within the context of her religion? If you were the judge considering the physicians' argument that surgery would benefit the boy, what would your conclusion be? In this case, the court decided the boy was a neglected minor and ordered his mother to permit the surgery and to consent to a transfusion if the doctors decided it was necessary during the course of treatment. Do you concur with this decision? Why or why not?

The debate over parental failure to obtain medical care for their children for religious reasons has generally been one of balancing the children's medical needs with parental rights to rear children in accordance with their religious beliefs. The issue becomes more complex when we consider childhood immunizations, which involve not just the well-being of the individual child but the well-being of the community. If fundamentalist families objecting to immunizations are concentrated in fairly large communities, any refusal to allow immunizations has the potential for creating a rather serious public health risk for the whole community (Ross & Aspinwall, 1997). The American Academy of Pediatrics has taken the position that, "Constitutional guarantees of freedom of religion do not permit children to be harmed through religious practices, nor do they allow religion to be a valid legal defense when an individual harms or neglects a child" (Ross & Aspinwall, 1997, pp. 202–203). Do you agree? If so, what steps should be taken to ensure that all children in the United States get needed medical care?

Outcomes of Religion-Related Child Maltreatment

Research on outcomes of religion-based child maltreatment has typically focused on mortality and psychological effects. One nationwide study of child fatalities between 1975 and 1995 (Asser & Swan, 1998) revealed 140 cases where children died because parents withheld medical care for conditions which survival rates with medical care would have exceeded 90%; in an additional 18 cases, survival rates with medical care would have exceeded 50%. These children had a wide range of treatable medical conditions at the time of their deaths—for example, dehydration, diabetes, burns, measles, meningitis, pneumonia, and appendicitis. When one two-year-old choked on a bite of banana, her parents frantically called members of their religious circle for prayer during the hour she still showed signs of life. The father of a five-month-old son reported that after four days of fever, his son started having apneic (cessation of breathing) spells. During each spell, the father "rebuked the spirit of death" and his son perked up and began to breathe again. The next day he died from bacterial meningitis (Asser & Swan, 1998). A teenage girl asked teachers for help getting medical care refused by her parents for fainting spells. She ran away from home but was returned to the custody of her father by law enforcement officials. She died three days later from a ruptured appendix. Of a total of 172 religion-based child mortality cases, the breakdown by religion was as follows: 23 Church of the First Born; 12 End Time Ministries; 64 Faith Assemblies; 16 Faith Tabernacles; 28 Christian Science; and 29 other denominations or unaffiliated.

Among the cases of religion-based child maltreatment reported by mental health professionals (Bottoms et al., 1995), one third of the victims who had been abused to rid them of evil admitted in adulthood that they had considered suicide. Dissociative disorders were also fairly high in both the ridding-evil and medical neglect groups. A boy from a Jehovah's Witness family that had engaged in punitive efforts to rid him of evil spirits was dissociative, could not concentrate, and was doing poorly in school. Reports from professionals also reveal that victims of multiple forms of religion-based family maltreatment have significantly more depression, insomnia, somatic problems, fearfulness and phobias, substance abuse, social withdrawal, and inappropriate aggressiveness than victims experiencing sexual abuse only (Goodman, Bottoms, Redlich, Shaver, & Diviak, 1998).

One woman explains the impact of her experiences of abuse: "I was raped by my uncle when I was 12 and my husband has beat me for years. For my whole life, when I have gone to a doctor, to my priest, or to a friend to have my wounds patched up, or for a shoulder to cry on, they dwell on my bruises, my cuts, my broken bones . . . but it's not my body that I really wish could get fixed. The abuse in my life has taken away my trust in people and in life . . . It's taken away my faith in God, my faith in goodness winning out in the end, and maybe worst of all, it's taken away my trust in myself" (DeKeseredy & Schwartz, 2001, p. 29).

Spousal Abuse

As is true of the research on child abuse and religion, there is a dearth of national population-based studies providing comparative prevalence rates for spousal abuse across different religions. More typical are small studies comparing spousal abuse in samples of adults from particular religious communities with national spousal abuse rates and studies providing rates for different religious groups within a relatively circumscribed population. Although two nationwide studies have been conducted on religious affiliation and spousal abuse, both have significant methodological shortcomings. For example, among battered women nationwide who responded to a survey published in *Woman's Day* magazine, 39% of their batterers were Protestant, 31% were Catholic, 15% were other religions, and 5% reported no religion (Bowker, 1988). In this sample, the Protestants and unaffiliated respondents seem underrepresented in relation to their proportions in the population. In a study with a similar methodology, public service announcements published in local newspapers throughout the United States advertised that spouses who were physically abused were being sought for a study (Horton, Wilkins, & Wright, 1988). Among the nearly 500 people who responded, 58% were Protestant, 20% were Catholic, 1% were Jewish, 10% were another religion, and 11% indicated no religious affiliation. In this case, it is the Jewish, Catholic, and no-religious-affiliation respondents who appear underrepresented in relation to their proportions in the population. However, the samples in both of these studies are self-selected and non-population-based, so generalization is not possible.

In a somewhat more representative community sample (Makepeace, 1997), individuals identifying with Christian denominations (Protestant, Catholic, Mormon) had "moderate" courtship violence rates (no percentage provided), whereas the rates were relatively high (nearly 29%) among respondents reporting no religion and exceedingly low for Jewish respondents (less than 6%). Within particular Christian denominations, some of the early clinical data suggest that spousal abuse may be a bit higher among Catholics than Protestants. For example, in one study, nonbattering men who had religious affiliations tended to be Protestant, whereas the batterers were typically Catholic (Star, 1978).

Like child abuse, spousal abuse can be found among Quakers, known for their antiviolence attitudes. In one sample of Quakers, 86% of the wives and 77% of the husbands reported using at least one act of verbal/psychological spousal aggression in the previous year (Brutz & Allen, 1986). Rates of *physical* aggression were much lower—0.5% for women and 0.2% for men. For both men and women, the frequency of both verbal/psychological and physical aggression were negatively associated with level of religious participation; however, Quaker wives showed higher levels of physical violence than Quaker husbands at both the high and low levels of religious involvement. In the wives, the greater their involvement in peace activism, the lower their violence against their husbands. By contrast, in the husbands, the greater their involvement in peace activism, the higher their level of both verbal/psychological and physical aggression against their wives. Why do you think this pattern occurs?

A study of nearly 1,500 Utah households (80% Mormon) revealed that Utah rates of spousal abuse per 1,000 households were slightly higher than the national average—a rate that contrasts with their child abuse rates, which were slightly below the national average (Rollins & Oheneba-Sakyi, 1990). In the Utah sample, the average rate of severe husband-to-wife violence was 3.4% (compared with 3% in the national sample), and the average rate of severe wife-to-husband violence was 5.3% (compared with 4.4% in the national sample). Among the religiously orthodox Mormons, the rates of husband-to-wife severe physical violence were substantially lower than the rates for the less orthodox. Thus, as was the case with child abuse, it seems that members of a congregation who espouse particular tenets of a religion and participate in the external aspects of the religion, but do not integrate the values deeply into their lives, are more likely to use aggression in the family. Rates of severe wife-to-husband physical violence did not vary in relation to orthodoxy.

As is true in relation to child abuse, only a few nonrepresentative studies provide some evidence concerning the prevalence of spousal abuse in religions other than Christianity. Estimates of spousal abuse in American Jewish families range from 15%–30%, depending on the nature of the study and sample and the year in which the study was conducted (Horsburgh, 1995). There do not appear to be reliable statistics for rates of wife abuse among the different Jewish communities; however, some authors have argued that wife abuse may be especially acute in Orthodox communities, particularly among the Hasidim (Horsburgh, 1995). There is also some evidence that Jewish women stay in abusive relationships longer than other women—seven to thirteen years as compared to three to five years (Horsburgh, 1995).

Despite pervasive American stereotypes about the maltreatment of Islamic women, rates of wife abuse appear to be relatively low among American Muslim women—perhaps in part because of underreporting. An estimate of the North American Council for Muslim Women, based on reports from Muslim leaders, social workers, and activists, is that approximately 10% of the Muslim women in this country are physically or sexually abused by their Muslim husbands (Memon, n.d.). If verbal and emotional abuse are included in the estimates, this figure "would rise considerably" (Alkhateeb, n.d., paragraph 5). By comparison, 7% of American women in general were physically abused and 37% were verbally or emotionally abused (Alkhateeb, n.d.).

Some Muslim American women may have an experience that is relatively unique as compared to other American women (other than Mormons) and which they sometimes consider abusive, specifically their husband's "misuse of polygamy" (Hassouneh-Phillips, 2001b). In one sample of 17 Muslim American women, more than 50% had had some experience with polygamy; of these nine, seven had been victims of physical and emotional abuse, and the remaining two were victims solely of emotional abuse. Although the majority of these women did not think of polygamy in general as abusive, they considered it abusive for their husbands to distort or stray from Islamic principles in order to pursue additional wives for their own pleasure. One woman said, "I divorced him when he tried to marry my sister who was my ward. And she was 13. And he told me that this would be really good because my sister liked him. She was a child and he had all the flash and dazzle. But when he asked about marrying her . . . I knew then that I would never allow him to marry my sister—to touch my sister" (p. 742).

According to Islamic principles, marriage is a matter of entering into a contract, and wives have a right to a divorce if their husbands do not live up to the contract—for example, through being abusive (Hassouneh-Phillips, 2001c). One Muslim American woman described:

> I was kicked, and he could go on for hours, kicking me and calling me bad names. Finally, we got kicked out of that apartment complex, and then I got an Islamic separation . . . Then he says, "Oh, I am going to be good. I am not going to hit you. I am not going to do anything." All the things he said, "You can go to the sisters meetings. I'm going to stop smoking. I'm going to put it all in a contract," because we had an Islamic divorce . . . And he wrote it. He promised me he is not going to punch me or hit me or kick me . . . He must keep the promises made in the first marriage contract. And one, I am allowed to work with women and/or children. Two, I am allowed to go to the sisters meetings and their other activities . . . Three, I can have my own business . . . Four, he will quit smoking . . . Five, he will go to anger management classes and the program for batterers . . . Six, he will not hit me, will not curse me, he will not accuse me of

bad things, will not threaten to kill me, will not kick me, spit on me, urinate on me, or do anything to insult me or harm me, and seven if he does not keep and follow these conditions, I am not responsible to give him his rights, and I have the right to divorce if I request it. (Hassouneh-Phillips, 2001c, pp. 937–938)

This woman's story tells a lot about her marriage and her husband's behavior towards her. What kinds of abuse can you identify? In what ways, if any, does this woman's experience seem related to her religion? Does her religion serve as either a positive or negative role in her situation? What, if any, universal features does her case have?

Positive Roles of Religion and Religiosity

Affection in Conservative Protestant Child Rearing

Some efforts have been made to identify the potentially positive aspects of conservative Protestant child-rearing techniques. National survey data have indicated that when dealing with their children, conservative Protestants spank more but yell less than nonconservative Protestant parents (Bartkowski & Wilcox, 2000). Moreover, conservative Protestants report hugging and praising their children more than other parents, and within conservative Protestant parents, conservative theological values are significantly positively correlated with the amount of hugging and praising provided to the children (Wilcox, 1998). What do you think might be the overall effect on children of a pattern of child rearing where physical discipline is swift and often harsh, but the children are rarely yelled at and often receive praise and hugs for desired behavior?

Protective Role of Religion and Religiosity

The majority of the relevant literature, perhaps because much of it is written from a feminist or human rights perspective, focuses on the negative role of religion in family violence, especially particular religious affiliations and the espousal of particular fundamentalist values. However, there are also important findings on the protective role that religion can play in human life. In general, both religiosity and spirituality have been identified as protective factors against a broad range of personal ills, many of which are risk factors for violence in families. For example, among adolescents, personal religiosity is a protective factor against substance abuse (Benda, 1995; Resnick et al., 1997). Moreover, mother's religiosity is negatively associated with age of onset for adolescent drinking—the more religious the mother, the less likely her adolescent was to start drinking across a two-year time span. Among adults, religiosity appears to be a protective factor in regard to mental health (Strawbridge, Shema, Cohen, & Kaplan, 2001). Among both adolescents and adults, religiosity has shown an inhibitory effect on engagement in delinquent and criminal activities (Benda & Toombs, 2000) and on engagement in risky sexual activities (Kendall-Tackett, 2003).

Not only may religiosity operate against engagement in negative behaviors associated with family abuse, it may also contribute to the development of other characteristics that mediate against the use of interpersonal violence. For example, religiosity in parents is associated with prosocial attributes in their adolescent children (Lindner Gunnoe, Hetherington, & Reiss, 1999). Among adolescents, there are significant correlations between belief in the importance of religion and involvement in community service (Youniss, McLellan, & Yates, 1999) and between religiosity and altruistic attitudes and behaviors (Donahue & Benson, 1995).

There are also findings directly supportive of a protective role for religiosity in regard to family conflict, violence, and abuse. Religious beliefs directly related to perceived sacred qualities in a marriage and beliefs that God is manifested in marriage have been associated with overall marital adjustment, less marital conflict, and less verbal aggression in marriage (Mahoney et al., 1999). Moreover, the reciprocal of a risk factor can be a protective factor; thus, whereas religious conservatism is a risk factor for family violence, more moderate religious philosophies can serve as a protective factor.

Research concerning a protective role for religiosity in relation to family violence has tended to provide two types of support: (1) evidence that religiosity is an inhibitory factor that reduces the likelihood of adult family members aggressing against each other or their children; and (2) evidence that religiosity in victims of family violence and abuse serves as a protective factor reducing some of the negative effects associated with victimization.

Religiosity as a Factor Reducing Spousal Abuse

We have already shown that there is some evidence that spousal abuse is lower when there is greater rather than lesser integration of religious values into the family life (Elliot, 1994). A related question is whether frequency of attendance at religious services reduces the likelihood of marital aggression. Ellison, Bartkowski, and Anderson (1999) found a negative correlation between the frequency with which religious services were attended (a measure of extrinsic religiosity) and level of partner violence. For men, fewer than 2% of men who attended services weekly reported committing acts of partner violence, as compared with the men who attended once a year or less (6%). Among women, even those who attended only once a month were significantly less likely to perpetrate violence than those attending once a year or less. These negative associations between religious attendance and spousal abuse remain even after controlling for social support, substance abuse, and psychological problems (Ellison & Anderson, 2001).

It also appears that within homes in which violence has taken place, religiosity can have an ameliorating effect on some of the aggression. In a nationally based sample of battered women (the *Woman's Day* sample; Bowker, 1988), the women's reports revealed that husbands who attended church relatively frequently were less likely to drink alcohol before assaulting their wives, were less likely to use weapons in their assaults, and showed less involvement in a male subculture of violence (Bowker, 1988). In another national sample of abused women (Horton, Wilkins, & Wright, 1988), higher levels of religiosity appeared to increase the chances that abused wives could achieve an abuse-free, happy future, even when they remained in their marriage.

Religiosity Protecting Against Negative Effects of Maltreatment

In some cases, religiosity appears to help both adult and child survivors of maltreatment cope better with their abusive experiences. In a community sample of formerly abused women from around the United States, approximately 25% said that their religious beliefs gave them "hope," "strength," and "courage" (Horton et al., 1988). These positive emotions may help interrupt the intergenerational transmission of aggression. In one sample of high school students from public schools on the East Coast and in the South, a measure of religiosity that included both intrinsic and extrinsic components was inversely related to the use of violence in abused adolescents (Benda & Corwyn, 2002).

For sexually abused adolescents, being religious or spiritual appears to be a protective factor against negative psychological outcomes in females but not males (Chandy, Blum, & Resnick, 1996). Moreover, among sexually abused adolescent girls, religiosity may be a protective factor against purging as a psychiatric response to abuse (Perkins & Luster, 1999). It is possible that sexually abused girls who are highly religious may be more likely to view themselves and their bodies in a more positive light than girls low in religiosity. Their participation in religious activities and strong belief in the importance of God appear to act as buffer against the risk factors associated with engagement in purging.

The Response of Religious Institutions to Spousal Abuse

Many battered women may turn to religious leaders to try to find relief from their suffering. In one sample of battered women, 54% of the religious victims and 38% of the nonreligious victims sought help and/or guidance from clergy (Horton et al., 1988). Among these respondents, 30% viewed the contact as satisfactory or very satisfactory, 29% said it was dissatisfactory, and 42% said it was very unsatisfactory. The dissatisfied respondents indicated that their clerics had told them things such as, "Try harder not to provoke him," and, "Hope for the best. God will change him" (p. 242). Overall, the victims' criticisms of the clerical responses revealed five major themes: (1) making the victims feel trapped within a dangerous relationship; (2) showing a lack of understanding and validation of the women's experience and minimizing the abuse; (3) providing no alternatives or practical suggestions; (4) blaming the victims for their victimization or making them feel responsible for the abuse; and (5) leaving them feeling helpless with no way to escape. By contrast, the victims who reported positive experiences with their clergy commented that their advisors provided "validation" and "approval." Clergy who agreed that safety, even if it meant separation or divorce, was necessary or at least acceptable received the highest praise. Also appreciated within this sample of women were a few clergy who recommended, and in some cases even helped pay for, outside counseling or education.

Historically, across religions, a typical response of clergy has been to tell women to stay with their husbands, keep the family together, and try harder not to anger their husbands. Box 3.4 provides several of these women's stories. What similarities and differences do you see in these stories? What do you think may be the sources underlying the clerics' efforts to preserve the marriages of these women from diverse religious backgrounds?

Box 3.4   Religious Counsel for Abused Wives

I would never in my wildest nightmares have dreamed that my husband would ever abuse me but he did. My husband is a Christian, but his rage at things is unreal. I took our two-month-old son and fled after the fourth time he struck me, but I had received counsel that it was my duty to stay and suffer for Jesus' sake. (Alsdurf & Alsdurf, 1988, p. 221)

I have seen it presented that if you suffer at the hands of your husband that you will receive *ajars* [rewards] from Allah for doing so, so do not help him reform, do not get counseling, do not try to seek a better way. Just accept his abuse and *Allah* will bless you for it . . . you know, accept the oppression because when you die you will go to heaven. (Hassouneh-Phillips, 2003, p. 688)

One woman, beaten for years, was 60 years old when she left her husband and entered a shelter. The rabbi called her and convinced her to return home. When she again entered the shelter after another beating, the rabbi convinced her to give the husband a second chance. The third time she sought refuge (she suffered from a broken rib), the rabbi again advised her to come home. He told her that her husband wanted to repent by taking her to visit the Holy Land. (Horsburgh, 1995)

I talked with a group of priests who told me that they had no one in their church who needed "this kind" of help. They told me that women who were beaten could stop it if they really wanted to. They said such women do not honor their husbands and should expect some punishment. (Zambrano, 1985, p. 213)

Personal accounts of the response of religious authorities to women's experience of abuse often indicate that the message to battered wives is that suffering is good, that it helps ensure a place in heaven. "A number of Muslim leaders, religious counselors, and even parents . . . counsel abused women with these two notions (determinism and endurance). Their suffering at the hands of tyrannical husbands is a result of the decrees of Allah and therefore has to be born with the patience expected of pious and obedient women. To add insult to injury they are often told that their decreed misfortune is a result of their laxity in executing the tenets of the Shariah. The question I have to ask is simple: How much more perversion are we as the ummah of Allah and his prophet Muhammad (SAW)[1] expected to tolerate?" (Hendricks, 1998, paragraph 10).

Although there is little information in the literature about the response of clergy to sexual maltreatment in marriage, some research has indicated that more conservative clergymen, like many other members of society, tend to blame the victims when women are raped. In a scenario study with Protestant clergymen (Sheldon & Parent, 2002), the more sexist the clergymen were, the more unfavorable their attitudes toward rape victims. Moreover, the more fundamentalist their religious values, the more sexist their attitudes, and the more they blamed the victim. Even in response to the marital rape scenario, over 20% of the clergymen assigned some responsibility for the rape to the wife. One clergyman, in explaining his views, said that the wife "said 'no' but didn't resist as the Bible refers to resisting, as crying out" (p. 242).

However, clergymen are not necessarily unresponsive to spousal abuse. Among 143 Christian and Jewish clergy in Maryland, most indicated that they had been confronted with a case of spousal abuse within the past six months (Martin, 1989). A little over half of the respondents had counseled one or more abused wives. The amount of counseling varied by denomination: Whereas 36% of the Catholic priests had counseled one or more victims in the past six months, 82% of the Lutheran ministers did. Among the types of response provided by the clergy, the most common were information about treatment programs, counseling of the victim, and suggestions that the victim get professional help. A few victims were advised to get legal help or separate from their abusers. However, more than half the members of the congregations polled indicated that their congregation's response to spousal abuse needed improvement: Catholics, Methodists, and Presbyterians were the most dissatisfied with their congregation's activities.

In a more recent study, Cwik (1997) polled nearly 200 rabbis from across the country, divided fairly evenly among Orthodox, Conservative, and Reform rabbis, and found that they had seen 403 abused women over a five-year period. Most of the rabbis, including Orthodox rabbis, indicated that they did not tell abused wives to change their behavior or to trust in God and pray as methods of stopping abuse; rather, their recommendation was that abused women should contact the police or an attorney. Although rabbis from all denominations said they would intervene to stop abuse, only Orthodox rabbis said they would personally confront the abusive husband. Many rabbis indicated that they would like more information concerning domestic abuse, but few actually sent away for materials prepared by Jewish Women International.

Because so many family members turn to their clergy for help in dealing with issues of family violence, many religious organizations have responded with declarations and position statements designed to improve the response of their clergy to victims of violence in families. For example, the General Assembly commissioners of the Presbyterian Church, following two years of work by a domestic violence task force, issued a statement identifying three goals: (1) protect victims from further abuse; (2) stop the abuser's violence and hold the abuser accountable; and (3) restore the family relationship, if possible, or mourn the loss of the relationship (Silverstein, 2001). The Islamic Society of North America (2004) issued the following statement: "As Muslims we understand that violence and coercion used, as a tool of control in the home, is oppression and not accepted in Islam. Marriage in the Islamic context is a means of tranquility, protection, peace and comfort. Abuse of any kind is in conflict to the principles of marriage. Any justification of abuse is in opposition to what Allah (swt) has revealed and the example of Prophet Muhammad" (paragraphs 1–2). In addition, many programs have been developed in recent years to provide training, consciousness raising, and insight to religious leaders from many communities (e.g., Alkhateeb, n.d.; Friedrich, 1988; Hoffman-Mason & Bingham, 1988).

Summary

Family violence appears to occur in all major religions, although it is found in some congregational cultures more than others. Child and spousal abuse seem to be associated most strongly with fundamentalist religious values across religions. Many victims of family violence turn to their religious authorities for help but are often dissatisfied with the response they receive. In some cases, the experience of family violence turns individuals away from their religion. In other cases, victims find their religion a source of solace. Efforts are being made to help clergy be more responsive to family violence. As you read subsequent chapters, consider how the predictors, correlates, and outcomes of the various forms of family violence may either vary by or be consistent across religions. Do you think, for example, that being sexually abused by a parent has the same causes and/or consequences in a Catholic family as in a Methodist or Muslim family? Are the effects of battering likely to be the same for a wife who turns to a fundamentalist Christian or Muslim clergyman as for an unaffiliated wife who turns to her local shelter? How about abused men? If they belong to a conservative and patriarchal religious community, might being abused by their wives have a different effect on them than it would on men belonging to more liberal congregations?

Note

1. "SAW" and "PBUH" represent the Arabic phrase *salla Allah alaihi wa sallam,* meaning "may Allah's peace and blessings be upon him." It is a standard Muslim expression of love and respect for the Prophet.

4

Child Physical Abuse

Kayla, a painfully shy six-year-old, lived with her biological father and step-mother. One day, a family friend noticed that Kayla had two black eyes and several other bruises. This friend insisted they take Kayla to the local emergency room, where both the father and Kayla said she had sustained the bruises from a bicycle fall. After discovering that Kayla also had internal bleeding and a broken nose, and doubting the bicycle story, the doctor asked a nurse to call the Florida Department of Children and Families (DCF). A DCF worker interviewed everyone involved. At a court hearing, the judge found no reason to doubt the bicycle story, and Kayla was sent home. Although her case was supposed to be monitored by a private agency hired by DCF, neither the agency nor DCF ever followed up on it. Over two weeks later, Kayla appeared in school with heavy makeup covering facial bruising. A few weeks later, she had even more bruises and said she had fallen off her bicycle again. Her teacher called DCF, who sent someone to visit Kayla's father. The worker insisted he take Kayla to a doctor, who found large bruises on her back, stomach, and head, and said she exhibited symptoms of chronic physical abuse. Kayla's father admitted that on occasion he spanked her with a paddle because she was a difficult child. Kayla returned home with her father and was monitored by DCF. She came to school two more times with bruises on her face and body, which she said were the result of being pulled down the stairs by the family dog. Because of the huge lump on her forehead, Kayla's teacher called DCF again and discovered the case had been closed. (Russ, 2003)

K ayla's story is a notorious case from the DCF in Florida. It exhibits well the signs and symptoms of children who are physically abused. What types of physical abuse did you identify in this case? What factors may have caused the abuse? Kayla's father, Richard, had a history of violent criminal behavior and had served jail time for assault and other crimes (Associated Press, 1998). How might this history have contributed to his behavior towards his daughter? What were the consequences of the abuse? The ultimate consequence for Kayla, unfortunately, was death. Within several weeks of the last report to DCF, Kayla's body was found buried in a shallow grave in a national forest. Her father beat her to death in a fit of rage because she soiled her underwear. When her body was found, it was discovered that her skull was fractured, four ribs were broken, her lung was punctured, her liver was split in three places, and her back and buttocks bled internally, so much so that her muscles were saturated with blood (Bickel, 2000). Richard's punishment was life imprisonment with no chance of parole. What might the system have done to prevent such a tragedy from happening?

Systematic research into child maltreatment in the United States began four decades ago with the publication of Kempe and colleagues' 1962 article on the battered child syndrome. Since that time, researchers have become increasingly concerned with the ways in which child maltreatment can manifest itself; the extent to which children suffer at the hands of their parents; the possible causes and correlates of child maltreatment; the consequences for children of victimization; and the ways in which we can intervene in abusive families or prevent maltreatment from happening. Although we have learned much in all of these areas, we have not succeeded in eliminating or arguably even ameliorating the problem of child maltreatment in the United States. In this chapter, we discuss studies in the most well-researched area of child maltreatment: physical abuse. In the following chapters, we discuss sexual abuse, neglect, and psychological maltreatment. Although the major types of child maltreatment tend to overlap, with victims of one type of abuse likely to be victims of at least one other type as well, we follow the most common pattern of discussing each type separately. As you read the current chapter, keep Kayla's case in mind and consider what factors may have contributed to her death.

Scope of the Problem

Exact rates of physical abuse of children are hard to pin down, in part because studies vary in their definitions of the problem. Even within the same study, data may be reported from several sources using different definitions. For instance, states have different statutes for the definition of physical abuse, and child protective services (CPS) workers have their own biases and interpretations of the cases they investigate. Nevertheless, each CPS agency and each state provide data to the government for physical abuse rates, data that are used by both the National Child Abuse and Neglect Data System (NCANDS) and the National Incidence Studies (NIS).

One aspect of the definitional problem is that the lines between physically abusive versus sub-abusive versus nonabusive behaviors are blurred, and there is no consensus on the exact severity, frequency, or duration of physical aggression that is necessary for a behavior to be deemed abusive. Consider Kayla's case again. Although several people assumed that she was physically abused, her physical abuse may never have been substantiated by the state, and, therefore, she would not be counted in national statistics on physically abused children. Consider also the case in Box 4.1. Even this case might be hard to substantiate, especially if Jacinta's mother did not admit that her husband physically abused the child. Also, must injury occur in order for a behavior to be labeled physically abusive? Is risk of injury enough? What if no physical injury ever occurs but the child is slapped on a daily basis for no apparent reason? What if the child was punched "only" once with no lasting injury? Is a precise and universal definition of physical abuse possible?

Box 4.1   Case of Physical Abuse by a Caregiver

Jacinta was five weeks old when she presented at a local hospital. She had many unexplained injuries, including a torn frenulum (tissue that connects the tongue to the mouth), lacerated and infected gums, a fractured nose and upper jaw, and burns to the buttocks and thigh. The hospital staff decided to do a full skeletal x-ray to investigate whether there were any other injuries, and they discovered a fractured forearm that was approximately two weeks old.

Sally, Jacinta's mother, explained that the burn to the buttocks occurred when her 17-year-old niece gave Jacinta a bath and allowed her to roll over and come into contact with an electric heater. The thigh burn occurred when Peter, Jacinta's father, accidentally spilled hot milk on Jacinta. Sally explained that the facial injuries occurred when Sally was carrying Jacinta over her shoulder, tripped over the family dog, and fell such that she landed on Jacinta's face. Finally, she stated that the fractured arm must have occurred when the hospital staff held Jacinta's arm too tightly to collect blood.

(Continued)

(Continued)

> Sally herself presented with a large bruise on her cheek, but she denied any violence by her husband toward either her or the child and stated that the bruise was actually a birthmark that had always been there. A child welfare team investigated Jacinta's case and decided to place Jacinta in foster care. They informed Sally that court papers had been filed to make Jacinta a ward of the state. Sally then confessed that Peter abused both Jacinta and Sally and that Jacinta's facial injuries occurred when Peter punched her.

SOURCE: Oates, Ryan, & Booth, 2000.

According to the U.S. Department of Health and Human Services (DHHS, 2001), rates of substantiated physical abuse from child protective services (CPS) agencies reveal that physical abuse is the second most prevalent form of child maltreatment, second only to neglect and comprising 18.6% of all maltreatment cases. This DHHS report also indicated that rates of substantiated physical abuse declined steadily between 1993 and 1999 but then remained constant between 1999 and 2001. In 2001, approximately 168,000 children were physically abused by their parents, which is comparable to a rate of 2.3 per 1,000. Physical abuse was the least *recurring* type of maltreatment, probably because its indications are the most concrete; moreover, it was not the form of maltreatment most likely to lead to death (neglect resulted in more fatalities). In all, 26.3% of maltreatment fatalities were due to physical abuse, and an additional 21.9% of fatality cases were due to physical abuse in combination with neglect.

According to the Third National Incidence Study of Child Abuse and Neglect (NIS-3; Sedlak & Broadhurst, 1996), the NCANDS data reported by DHHS underestimate the number of children physically abused in this country. The rates of physical abuse found in the NIS-3 analysis are higher than the NCANDS rates, probably because the investigators had access to more sources of data than just CPS agencies. The three NIS studies (1980, 1988, and 1993) solicited information from community professionals mandated to report child abuse cases, specifically professionals from schools, hospitals, law enforcement agencies, and welfare offices. These mandated professionals were asked to report all cases of child maltreatment with which they had come in contact and to indicate which ones they had reported to CPS. The NIS investigators also solicited reports from CPS agencies, so that they could compare the cases reported by the professionals with the reports received by CPS from either professionals or nonprofessionals, such as neighbors, family members, and friends. In identifying maltreatment cases as physical, sexual, and emotional abuse, and physical, emotional, medical, and educational neglect, the NIS respondents were instructed to use a definition of maltreatment emphasizing deliberate acts or marked inattention to the child's basic needs that could result in foreseeable and avoidable injury or impairment to a child, or worsen an existing condition.

The most recent NIS survey (NIS-3) contains a nationally representative sample of over 5,600 professionals in 842 agencies serving 42 counties. Like the NIS-2, the NIS-3 used two sets of standardized definitions of abuse and neglect. Under the *Harm Standard,* the only standard in NIS-1, children were identified as maltreated only if they had already experienced harm from abuse or neglect. Under the *Endangerment Standard,* added to NIS-2 and NIS-3, children were included in the maltreatment group if they had experienced abuse or neglect that put them at risk of harm (Sedlak & Broadhurst, 1996). According to the Harm Standard, 381,700 children were physically abused in 1993, a 42% increase from the previous survey in 1986. According to the Endangerment Standard, 614,000 children were physically abused, a 97% increase.

Another source of data suggests that actual rates of physical child abuse are even higher than indicated in reports on identified cases. The 1985 National Family Violence Re-Survey (NFVR) was a national population-based survey that asked a representative sample of parents to self-report how often they used specific physical acts (e.g., hitting, beating up) to discipline their children. According to this survey, almost two thirds of parents hit their children at least once in the previous year (Straus & Gelles, 1990a). According to the Very Severe Violence Index (all serious forms of violence, except hitting with an object), 23/1,000 children were victimized by their parents in 1985, a rate that projects to 1.5 million children nationwide. According to the Severe Violence Index (very severe violence plus hitting with an object), which the researchers consider to be the best indicator of physical abuse, 110/1,000 children were victimized by their parents in 1985; that is, *6.9 million* children were physically abused by their parents during that year alone (Straus & Gelles, 1990a). Moreover, even these large numbers are likely to be underestimates because many parents may be unwilling to disclose the number of times they beat up, choked, kicked, or punched their children. In addition, although the rates were highest for the youngest children and steadily declined as the children got older (Wauchope & Straus, 1990), teenagers were not immune from physical aggression: In fact, 3.8 million (340/1,000) teenagers experienced some form of parental aggression; 800,000 teens (70/1,000) experienced severe violence, and 235,000 (21/1,000) experienced very severe violence in 1985 (Straus & Gelles, 1990a). Reports of overall violence did not decline between 1975 (the year in which the original survey was conducted) and 1985; however, three-year-olds were 97% more likely to be hit, whereas 15- to 17-year-olds were 33% less likely to be hit in 1985 than in 1975. In addition, both severe violence and very severe violence declined, with very severe violence down from 36 to 19/1,000, a 47% decline (Straus & Gelles, 1990b). The researchers argued that the dramatic decline in child physical abuse rates were most likely due to the increased media attention to the problem, which in turn created a social stigma against parents physically abusing their children.

Predictors and Correlates

Since, according to the NFVR, at least 6.9 million children are physically abused by their parents each year (Straus & Gelles, 1990a), it is important to identify the risk factors influencing parents to abuse their children. The identified risk factors reflect every level of the ecological theory discussed in Chapter 1. Furthermore, many researchers (e.g., Newberger, Hampton, Marx, & White, 1986; Straus & Smith, 1990) have found that it is the *combination* of risk factors, not any single risk factor alone, that puts children at risk for physical abuse.

Macrosystem

Economic Indicators

Families suffering from unemployment and/or low income seem to be at a higher risk for physically abusing their children than their more economically successful counterparts (Gillham et al., 1998; Sedlak & Broadhurst, 1996; Straus & Smith, 1990), especially if the parents are dissatisfied with their standard of living (e.g., Straus & Smith, 1990). Also, less-educated parents are more abusive than more highly educated parents (Jackson et al., 1999).

Cultural Values

The occurrence and perpetuation of physical child abuse seem to be supported by a culture in the United States that is accepting of violence and reluctant to get involved in family matters. Although the general cultural context of violence in the United States was discussed at some length in

Chapter 2, there are aspects of the community and broader society that have particular relevance to child maltreatment in the majority culture.

Parenting has traditionally been viewed as a private matter in this country, and the government has been reluctant to interfere in parents' rights to raise their children as they see fit (Harrington & Dubowitz, 1999). Although several laws have been passed since the early 1960s prohibiting excessive parental violence against children, there continues to be, in actuality, considerable tolerance for aggression against children. Such aggression may not be characterized as abusive by the average person but is considered maltreatment by many researchers (e.g., Straus, 1994). For example, the laws in this country support the use of physical punishment for children. Although physical abuse is illegal in all 50 states, all states allow parents to physically discipline their children (Davidson, 1997). Furthermore, as of 2001, teachers in 23 states are allowed to use corporal punishment on their students (National Coalition to Abolish Corporal Punishment in the Schools, 2001), and the Supreme Court upheld the right of teachers to use corporal punishment on children, even those with disabilities (*Ingraham v. Wright,* 1977; Lohrmann-O'Rourke & Zirkel, 1998). Moreover, while most professionals having contact with children are required to report suspected cases of child abuse (Davidson, 1988), some of these regulations are being undermined by the welfare reform act. The version of this act passed in 1996 weakens the definition of child abuse and may exclude many less severe cases; in addition, this act indicates that parents are not required to get necessary medical care for their children if it interferes with their religious beliefs (Harrington & Dubowitz, 1999).

This acceptance of violence can perpetuate the physical abuse of children. Cultural norms allow parents and teachers to aggress against children, and most parents accept physical punishment as an acceptable and even desirable method of child rearing (Harrington & Dubowitz, 1999; Straus & Stewart, 1999). The cumulative impact of ordinary physical punishment on the social and psychological health of this nation may be even greater than the cumulative impact of acknowledged forms of child physical abuse; thus, our goal should be eliminating all forms of physical punishment (M. Straus, personal communication, March 4, 2004). Studies have shown that both physical child abuse and corporal punishment are associated with criminal behavior (e.g., Gershoff, 2002). Straus argues that although the association with criminal behavior is larger for physical abuse than for corporal punishment, many more people experience corporal punishment than experience physical abuse. Thus, eliminating all forms of physical discipline would result in a drastic reduction of criminal behavior in this country.

Exosystem

Families characterized by physical abuse seem to be more socially isolated than nonabusive families (e.g., Coohey, 2000; Straus & Smith, 1990). Furthermore, they tend to have fewer sources of social support than nonabusive families. For example, in one study, mothers who were physically abusive reported fewer friends and less contact with their friends than their demographically matched controls (Bishop & Leadbeater, 1999). Moreover, these mothers rated their friendships as lower quality than the nonabusive mothers, and both relationship quality and social support were independent predictors of maltreatment.

Several studies have assessed neighborhood quality as a predictor of child physical abuse. For instance, neighborhoods with a high percentage of single-parent families (Drake & Pandey, 1996), evidence of social deterioration (Garbarino & Kostelny, 1992), inadequate social resources (Zuravin, 1989), impoverishment, child care burden, and instability (Korbin, Coulton, Chard, Platt-Houston, & Su, 1998), have all been shown to have higher levels of child physical abuse than their counterparts. However, one study showed that many of these neighborhood influences (e.g., concentrated disadvantage and neighborhood crime) no longer influenced child physical abuse after controlling for family composition and family socioeconomic status (microsystem factors) (Molnar, Buka, Brennan, Holton, & Earls, 2003). The only neighborhood-level variable that still predicted child physical abuse after controlling for microsystem level influences was immigrant concentration, in that higher immigrant concentration was a protective influence on child physical abuse rates (Molnar et al., 2003).

Microsystem

Characteristics of the Victim

Age is probably the most well-researched child-level risk factor for physical abuse; younger children are more likely to be abused than older children, and as children age their risk for being physical abused continually declines (e.g., Jackson et al., 1999; Straus & Smith, 1990). Other child-level risk factors are difficult temperament and behavior problems, and medical, developmental, or intellectual abnormalities (e.g., prematurity, low birth weight, low IQ or mental retardation, physical disability) (Belsky & Vondra, 1989).

If such characteristics of children put them at risk for physical abuse, does that mean the children are somehow responsible for their own abuse? Or, must responsibility for child abuse always lie in the parents and/or the ecological systems in which the parent-child interactions take place? A reasonable argument is that if the parents' dispositions are such that their child's traits lead them to behave in an irritable and punitive manner, it is still ultimately the parents' fault for the physical abuse (e.g., Harrington & Dubowitz, 1999; Kolko, 2002). Indeed, children's traits do not predict their own physical abuse above and beyond the parent's traits (Ammerman, 1991).

Characteristics of the Family

Several characteristics of the family seem to put a child at risk for physical abuse. For example, if there is only one parent or if one of the parents is nonbiological (i.e., a stepparent), the children in that family are at a greater risk for being physically abused (e.g., Gillham et al., 1998; Sedlak & Broadhurst, 1996; Straus & Smith, 1990). Abusive families also seem to be quite chaotic. They often appear to have limited cohesion, limited communication, and high conflict (Mollerstrom, Patchner, & Milner, 1992). In addition, the extent of the family violence may be quite pervasive. That is, mothers and fathers who physically and/or verbally abuse their spouses are more at risk for physically abusing their children, and parents who verbally abuse their children are more at risk for physically abusing them, too (e.g., Appel & Holden, 1998; Jackson et al., 1999; Straus & Smith, 1990).

Individual/Developmental

Family History of Aggression

One of the most consistent parental risk factors for the physical abuse of children is the parents' own childhood history of family violence—through either being victimized by child abuse (even just repeated corporal punishment) or witnessing interparental violence in childhood. Although not all children who grow up in violent homes go on to maltreat their own children, it is estimated that about 30% will, and those who were exposed to violence in their families of origin are at significantly greater risk for physically maltreating their own children than those who were not (e.g., Jackson et al., 1999; Kaufman & Zigler, 1987; Straus & Smith, 1990).

Psychopathology and Alcohol/Drug Abuse

Parental psychopathological problems have also been implicated in parents' tendencies to physically abuse their children. Mothers who are depressed and/or anxious (e.g., Whipple & Webster-Stratton, 1991) or sociopathic (e.g., Brown et al., 1998) are at higher risk for physically abusing their children. Parents who abuse alcohol or drugs are also at greater risk for abusing their children (e.g., Besinger, Garland, Litrownik, & Landsverk, 1999). Indeed, it has been estimated that substance abuse accounts for approximately two thirds of child maltreatment fatalities in this country (Reid, Macchetto, & Foster, 1999). Although parental psychopathology and drug/alcohol abuse are problems of an individual/developmental level, how might they be related to characteristics of the exo- and macrosystem levels?

Emotions and Cognitions

Physically abusive parents also differ from nonabusive parents in the way they think about, feel about, and interact with their children. For instance, physically abusive mothers, in comparison to nonabusive mothers, believe that using physical discipline is an appropriate way of rearing children and have higher expectations for their children's behavior (Simons et al., 1991). Perhaps because their children cannot possibly live up to these high expectations, abusive mothers perceive their children more negatively than nonabusive mothers (Azar & Siegel, 1990). Moreover, attributions for their children's behaviors differ. That is, in comparison to nonabusive mothers, physically abusive mothers tend to perceive a noncompliant child's behavior as being more intentional, stable, and global, more difficult to control, and more stressful (Dopke & Milner, 2000). In addition, they attribute their child's negative behavior to characteristics of the child but attribute the child's positive behaviors to their own successful parenting (Bugental, Mantyla, & Lewis, 1989).

Negative cognitions about their children seem to translate into negative interactions between abusive mothers and their children. Abusive mothers have few positive interactions and many negative interactions with their children (Caliso & Milner, 1992; Dolz et al., 1997) and may even respond aversively to any prosocial approaches by their children (Reid, Kavenaugh, & Baldwin, 1987). In addition, their interactions with their children seem less mutual than those of nonabusive mothers (Alessdandri, 1992). When faced with the inevitable misbehavior of their children, physically abusive mothers tend to use emotion-focused rather than problem-focused strategies to resolve the situation (Cantos, Neale, & O'Leary, 1997). Perhaps these coping strategies are why abusive mothers' child-rearing styles seem to be inconsistent, critical, hostile, and eventually aggressive (Whipple & Webster-Stratton, 1991).

Other Demographic Predictors

Younger, single, and/or nonbiological parents are more likely to physically abuse their children than their counterparts (Brown et al., 1998; Milner, 1998). Furthermore, female parents are more physically abusive than male parents (e.g., Sedlak & Broadhurst, 1996). Consider these findings. Why would younger parents be more likely to abuse their children? What kinds of situations and stresses would single parents face that might increase their risk for abuse? Why would nonbiological parents be more likely to abuse a child than a biological parent? And finally, why would mothers be more likely to abuse their children than fathers? Could it be because mothers spend more time with their children on average than fathers do and therefore have more opportunity to? What other variables might be at work here?

Application of the Ecological Model

Box 4.2 provides more information about the case of Jacinta, introduced in Box 4.1. As you read about the characteristics of her parents, consider how many of the risk factors you have just read about apply to her case. If you were a child protective worker involved in her case, what would your recommendations be? Would you allow Jacinta to return to the care of her mother? Would you allow her to live with her paternal grandmother? What do you think would be the best approach to keeping Jacinta safe from further abuse? The decision made by the professionals in this case was to place her in foster care. Do you agree with that decision?

Box 4.2   Jacinta's Parents

The investigation into Jacinta's case revealed several important facts about her parents, Sally and Peter. In high school, Sally had attended a special class for slow learners. She had been very unhappy in school, feeling teased and picked on by other children, and had become quite wild as a teenager, unable to keep herself from getting involved with abusive males. While Jacinta was in the hospital, the nursing staff observed that Sally did not seem to have the skills or intellectual competence to care for Jacinta adequately. Her parents were supportive of her, but were retired and in ill health, and unable to provide the necessary care for their granddaughter.

As a child, Peter had been removed from his parents' care because of abuse and spent many years in juvenile correctional institutions. He had a long history of increasingly violent crime. Although he did not have a formal psychiatric evaluation during the case review process, professional

(Continued)

(Continued)

> observers believed he had all the characteristics of a psychopathic personality. At the time Sally married him, he was an unemployed truck driver. His mother insisted that he was just a victim of circumstances and was willing to take Sally and Jacinta into her home.

SOURCE: Oates et al., 2000.

Consequences

Both short- and long-term consequences of child physical abuse have been found in almost every area of functioning. However, the link between child physical abuse and its consequences is not straightforward or direct. In fact, many aspects of the abusive situation itself and other areas of the child's life can mediate the impact of physical abuse on a child's adjustment. For instance, the perpetrator's gender and relationship to the victim can be important: One study showed that physical abuse by mothers may be associated with worse outcomes than physical abuse by someone else (Herrenkohl et al., 1983), whereas another showed that among women, father violence, but not mother violence, was predictive of low self-esteem (Downs & Miller, 1998). Furthermore, more than one abusive caretaker and more abuse types (e.g., physical abuse in combination with neglect and/or sexual abuse) can result in more severe outcomes (e.g., Brown & Anderson, 1991; Herrenkohl et al., 1983). Other aspects of the child's abusive situation, such as the timing, frequency, severity, and duration of the abuse may also aggravate negative outcomes (Brown & Kolko, 1999; Keiley, Howe, Dodge, Bates, & Pettit, 2001; Manly, Kim, Rogosch, & Cicchetti, 2001). The more chronic, frequent, and severe the abuse is, the more severe the outcome. Physical abuse during the preschool period is particularly risky for negative outcomes such as internalizing and withdrawn behavior (Manly et al., 2001), and physical abuse prior to the age of five has been associated with more negative outcomes than later physical abuse (Keiley et al., 2001). Overall family environment, such as level of poverty, may also make outcomes worse (Augoustinos, 1987; Malinosky-Rummell & Hansen, 1993).

Protective Factors

Although different studies show differing results, certain child characteristics, such as gender and age, may have a moderating effect on the extent or types of negative outcomes that a child experiences as a result of physical abuse (Malinosky-Rummell & Hansen, 1993). Several researchers have found that such child characteristics as relationship with a supportive nonabusing caregiver, high IQ, and some genetic factors can lessen the impact of child physical abuse (e.g., Kashani et al., 1992). In one study looking specifically at the role of support networks on children's adjustment (Ezzell, Swenson, & Brondino, 2000), peer support was particularly helpful in reducing negative outcomes among physically abused adolescents. Specifically, greater peer support was associated with less depression and anxiety in a sample of physically abused adolescents whose cases were substantiated by CPS. Family support was associated with less depression, although it was not clear which family member provided the support or whether that family member also abused the adolescent. Teacher support was not associated with any adolescent outcomes, nor were any of the types of support (family, peer, or teacher) related to a decrease in externalizing or aggressive behaviors (Ezzell et al., 2000). This study addressed teacher, peer, and family support. Can you think of any other sources of support that may serve as a protective factor against the effects of child abuse?

Short-Term Consequences

Physical Injuries

In the short term, the most obvious outcome of child physical abuse is physical injury. Skin lesions are the most common type of injury seen by medical personnel in cases of child physical abuse, followed by fractures (Kocher & Kasser, 2000). Belts or straps (23%) and open hands (22%) were responsible for a substantial percentage of the injuries in one emergency department (Johnson & Showers, 1985), with fists (8%); being thrown, pulled, dragged, dropped, or pushed (8%); being hit with another hard object (8%), a switch or stick (6%), and a paddle or board (6%) constituting some of the other major causes of injuries.

Recently, one of the most widely discussed types of injuries in infants has been shaken baby syndrome. Babies have large and poorly supported heads, and when they are shaken back and forth the brain moves in the opposite direction. Consequently, brain tissue and blood vessels are sheared. Infants under one year of age who present to medical personnel with shaken baby syndrome suffer from central nervous system damage and may present with seizures, failure to thrive, vomiting, hypothermia, bradycardia, hypotension, respiratory irregularities, or coma (Johnson, 2002). Unfortunately, these injuries can also lead to death. Further issues of child fatalities due to child maltreatment are discussed in Box 4.3.

Box 4.3   Child Maltreatment Fatalities

According to the most recent statistics released by the U.S. Department of Health and Human Services (DHHS, 2001), 1,300 child deaths in the year 2001 were a result of child maltreatment, which is a rate of 1.81 per every 100,000 children in the population. The majority of fatalities were due to either physical abuse or neglect (~93%), and more than one third of these fatalities were due to neglect alone. In over 80% of the cases, the perpetrator of the maltreatment was a parent, and in the majority of cases (greater than 90%), these families had no contact with child protective services (CPS) in the five years prior to the death of the child. In other words, the families in which children were killed because of maltreatment were not currently under any maltreatment investigation.

Perhaps one reason for the lack of involvement of CPS was that the majority (~85%) of the child victims were under the age of six years, and over 40% were under the age of one (DHHS, 2001). Death due to maltreatment during infancy is particularly problematic. In one review of infant injury deaths in the United States between the years 1983 and 1991, the leading causes of death were homicide, suffocation, motor vehicle accidents, and inhalation of foreign bodies, in that order. Moreover, between those years, infant homicides increased 6.4% per year (Brenner, Overpeck, Trumbel, DerSimonlan, & Berendes, 1999).

Therefore, being an infant puts one at a high risk for homicide; however, even more risky is being a newborn. According to the Centers for Disease Control and Prevention (2002), the risk for being murdered on the first day of life is ten times greater than at any other time throughout the life cycle. Although theories as to why parents murder their newborns abound (e.g., an unmarried teenage mother is ashamed of her sexual activity and is afraid to tell her parents that she is pregnant), there exists only one population-based study of neonaticide to date, and this study seems to refute the stereotype of the person who kills her newborn. Herman-Giddens and colleagues (2003) reviewed all death cases of infants under four days of age in North Carolina from 1985 through 2000. They found that 34 newborns were killed or discarded by their parents, which projects into 2.1 per 100,000 newborns per year. Males (58.8%) were significantly more likely than females (35.3%) to be murdered, and Black newborns were overrepresented (52.9%) as cases in comparison to their representation in the overall population of newborns (28%). The mothers' average age was

19.1 years and ranged from 14 years to 35 years. Approximately one fifth of the mothers were married, and, in all cases in which the perpetrator could be determined, the perpetrator was the mother. Mothers received prenatal care in just under one half of the cases, although eight mothers denied having been pregnant (Herman-Giddens, Smith, Mittal, Carlson, & Butts, 2003).

Particularly poignant in this analysis was the description of the methods that the mothers used to kill their newborns. In just over 50% of the cases, the mothers either left the newborns in the trash or other waste disposal area (i.e., landfill, dumpster) or in the toilet in which they had given birth. In addition, in just over 50% of the cases, the newborn was strangled or drowned or died from exposure to the elements. Although not included in the analysis because the babies had never been born, these researchers also found that three mothers had shot their fetuses in utero, and two mothers' fetuses died because of self-induced abortions (Herman-Giddens et al., 2003).

Although all of these statistics on child fatalities due to maltreatment are disheartening, it is important to note that these numbers are widely believed to underestimate the rate at which children die because of maltreatment. For example, the numbers that the DHHS collects are based on widely divergent measurement criteria among individual states, because states vary widely in their definitions of abuse and neglect; they vary widely in terms of which child fatalities they accept for further investigation for maltreatment; and they vary widely in the extent to which they classify, report, and track child maltreatment fatalities (Chance, 2003). Therefore, the numbers presented here are conservative estimates of the extent to which our children, infants, and newborns are dying because of maltreatment. What is also uncertain is how many of these child fatality cases were intentionally perpetrated by the parents. That is, how many parents intentionally killed their children versus how many child fatalities were the result of physical abuse or neglect getting out of hand? It is impossible to know for sure.

What is being done to prevent child maltreatment fatalities in this country? One step that has been taken in at least 42 states is the Safe Haven laws, which are supposed to prevent neonaticide. According to these laws, parents can drop off an unwanted newborn anonymously to certain authorities, such as hospitals, without being charged with abandonment (Herman-Giddens et al., 2003). To improve our ability

(Continued)

(Continued)

> to detect child maltreatment fatalities, child death review teams (CDRTs) exist in all 50 states. It is their job to obtain comprehensive information from multiple sources when a child fatality is referred to them from CPS, law enforcement, or medical examiners (Chance, 2003). However, even though these teams exist in all 50 states, there is still no uniform system for reporting child fatalities to the CDRTs or for deeming a child fatality as due to maltreatment (Block, 2003).

Psychological Injuries

In addition to sometimes experiencing physical injury, physically abused children also suffer socially, emotionally, and cognitively. For instance, physically abused children tend to develop an insecure attachment to their parents that has been labeled disorganized/disoriented attachment (e.g., Barnett, Ganiban, & Cicchetti, 1999). This insecurity in relating to parents seems to carry over into other social relationships. As toddlers and preschoolers, physically abused children notice distress in their peers but are less likely than their nonabused counterparts to show empathy or concern. Furthermore, they are more likely to show fear or anger at another's distress but rarely sadness, and they may even strike or withdraw from distressed peers, a behavior that replicates their parents' way of dealing with a distressed child (Main & George, 1985; Klimes-Dougan & Kistner, 1990). As schoolchildren, physically abused children, in comparison to nonabused children, tend to display less friendly positive peer interactions (Kaufman & Cicchetti, 1989). They have limited social competence (Feldman et al., 1989) and are less cooperative and more disturbed in their interactions—they tend to display fewer prosocial and more antisocial behaviors (Salzinger, Feldman, & Hammer, 1993). Indeed, their play is often marked by fights, war, and conflict (Harper, 1991).

These negative ways of interacting with peers may account in part for why physically abused children are often rated as the most disliked children in their classes (e.g., Dodge, Pettit, & Bates, 1994) and why they have problems making friends (Gelles & Straus, 1990). Dodge, Bates, and Pettit (1990) theorized that physically abused children's problems in interacting with their peers may stem from inappropriate cognitions about social behaviors resulting from their own experiences of abuse. In their study, abused children tended to attribute hostile intent to their peers' ambiguous behaviors; for example, if a peer accidentally bumped into them in a hallway, abused children would be likely to think the behavior was intentional. Moreover, the abused children had limited ability to generate alternative solutions to social problems, focused only on negative solutions, believed in the appropriateness of violence as a problem-solving technique, and had little ability to identify nonaggressive solutions to social problems.

Traditionally, family violence researchers have assumed that family members, including maltreated children, behave aggressively because they view aggressive behavior as morally justifiable, normal, and acceptable (Astor, 1994). However, research comparing nonabused to abused children on moral judgments of hypothetical and actual transgressions of: (1) aggression against another child; (2) psychologically harming another child; and (3) conflicts over objects with another child reveals no differences in the children's judgments of the severity or wrongness of these acts (Smetana, Toth, Cicchetti, Bruce, Kane, & Daddis, 1999). All children viewed these acts as deserving of punishment and as wrong whether an authority figure witnessed the acts or not. The only differences between abused and nonabused children's responses to these transgressions were in their affective responses. Physically abused boys were less angry at being victimized than non-physically abused boys, and the normally observed gender differences in children's responses to others' distress (i.e., girls feel greater distress and concern for another's hurt feelings) were not observed in physically abused children.

Difficulties in other areas of cognitive functioning identified in physically abused children include problems in motivation and task initiation (Allen & Tarnowski, 1989), lower intelligence (Erickson et al., 1989), and receptive and expressive language problems (McFayden & Kitson, 1996). Some cognitive problems have been shown to be due to head injuries caused by the physical abuse (Lewis, Shanok, & Balla, 1979), but others have been attributed to lack of communication and stimulation in the home environments (Augoustinos, 1987).

In addition to social and cognitive problems, physically abused children also tend to have higher rates of emotional difficulties. In a large community sample of children, physically abused children were more likely than nonabused children to suffer from major depression, conduct disorder, oppositional defiant disorder, agoraphobia, overanxious disorder, and generalized anxiety disorder, even after controlling for potential confounds, such as income, family history of psychiatric disorder, and perinatal problems (Flisher et al., 1997). Other studies confirm that physically abused children and adolescents tend to have higher rates of internalizing symptoms, such as depression, hopelessness, and low self-esteem (e.g., Keiley et al., 2001; Mahoney, Donnelly, Boxer, & Lewis, 2003). Clinical studies have shown that physically abused children display suicidal behaviors (Fantuzzo et al., 1998), although this outcome was not found in community samples (Flisher et al., 1997).

Perhaps the most well-researched short-term consequence of physical child abuse is in the area of externalizing symptoms. Physical abuse during the preschool period is predictive of externalizing and aggressive behaviors (Keiley et al., 2001; Manly et al., 2001); the more severe the abuse, the more severe the symptoms (Brown & Kolko, 1999). In a study of clinic-referred abused adolescents, severe parental aggression towards the adolescents predicted greater externalizing problems, even after controlling for witnessing interparental aggression (Mahoney et al., 2003). Externalizing problems tend to display themselves as oppositional defiant disorder or conduct disorder when they are severe enough to be diagnosable (Famularo et al., 1992; Flisher et al., 1997), but at the subclinical level, physically abused children tend to have problems with aggression and other forms of antisocial behavior. That is, physically abused children are more at risk than nonabused children for poor anger modulation (Beeghly & Cicchetti, 1994), increased rule violations, oppositionalism, and delinquency (Flisher et al., 1997; Malinosky-Rummell & Hansen, 1993; Walker, Downey, & Bergman, 1989). They are more likely to commit property offenses and be arrested and more likely to drink, use drugs, and smoke cigarettes (Gelles & Straus, 1990; Kaplan et al., 1998). Moreover, they are more likely than nonabused children to commit a serious antisocial act (Lewis et al., 1989; Widom, 1989).

Long-Term Consequences

In the long term, people with histories of childhood physical abuse have been shown to have problems in their financial, social, emotional, marital, and behavioral functioning. Again, the most researched forms of long-term effects have been aggressive and other antisocial acts. These studies indicate that people with histories of childhood physical abuse are at increased risk for being arrested for a violent crime and for being chronically involved in criminal behavior (Widom, 1989). They are also more likely to enter prostitution (Widom & Kuhns, 1996) and engage in other sexual risk-taking behaviors (Herrenkohl et al., 1998). Both women and men with histories of childhood physical abuse may also abuse drugs and alcohol, although the association is much stronger in women than in men (Langeland & Hartgers, 1998). Both men and women are at higher risk for being diagnosed with antisocial personality disorder (Luntz & Widom, 1994). Antisocial behavior tends to spill over into family life: People with histories of childhood physical abuse are at increased risk for physically abusing their children and significant others (e.g., Kalmuss, 1984; Kaufman & Zigler, 1987; Marshall & Rose, 1990; Straus & Smith, 1990; Widom, 1989). In addition, they are at increased risk for becoming the victim of spousal abuse (e.g., Cappell &

Heiner, 1990). Which of these outcomes are descriptive of Peter, Jacinta's father?

People with histories of childhood physical abuse may have other problems in their lives that have nothing to do with the perpetration of aggressive behavior. For instance, adult survivors of child physical abuse may have lower intelligence and reading ability than their nonabused counterparts (Perez & Widom, 1994). They may also be at increased risk for health problems (Lesserman et al., 1997), including chronic pain (Goldberg, Pachas, & Keith, 1999). Finally, in a 17-year longitudinal study of a community sample, people who had been physically abused were at an increased risk for depressive and anxious symptoms, emotional-behavior problems, and suicidal ideation and attempts (Silverman, Reinherz, & Giaconia, 1996). Given findings such as these, would you want to see Jacinta receive mental health services in foster care?

Special Issue: Corporal Punishment

Although all of the above studies have been concerned with the consequences of physical abuse, several researchers have pointed out that physical disciplinary practices need not be labeled abusive in order for them to have negative consequences (e.g., Straus, 1994). Corporal punishment is viewed by some researchers (e.g., Baumrind, 1997) as an effective and desirable method of raising children. The public tends to agree with this assertion; close to 95% of parents have used physical discipline on their children at some point in time. Although parents usually use corporal punishment when their children are under the age of three, nearly 25% of 17-year-olds were found to be subject to corporal punishment by their parents (Straus, 1994).

Straus (1994) has argued that even this socially approved form of discipline can have long-lasting negative effects because it teaches children that violence is an appropriate way of dealing with problems. In a recent meta-analysis of corporal punishment and its associated child behaviors, Gershoff (2002) concluded that corporal punishment is strongly associated with many negative behaviors, including: less moral internalization; more aggression; more delinquent, criminal, and antisocial behavior; poorer quality of parent-child relationship; poorer mental health; greater likelihood of abusing one's own spouse or child; and greater likelihood of being a victim of physical abuse by the parent. These effects were greatest for middle school children. Moreover, the association between corporal punishment and aggressive or antisocial behavior was stronger for boys than for girls. Finally, the association between corporal punishment and aggression became increasingly stronger as the age of the children increased.

Although the results of Gershoff's (2002) meta-analysis are strong and striking, she pointed out several limitations inherent in her analysis. First, all of the studies reviewed were correlational; thus, corporal punishment may not *cause* the negative outcomes. The direction of causation could be reversed, or some third variables, such as inconsistent parenting style, may be causing both the corporal punishment and the poor outcomes. Furthermore, her study does not give any indication as to which conditions may make these negative outcomes more likely. Not all children will display these negative behaviors, but the exact combination of factors that produce each outcome still needs to be discovered. Do these results on the possible consequences of corporal punishment influence your views on whether it should be considered a form of maltreatment? Do you think that, as discussed in the section on macrosystem influences, the cumulative impact of corporal punishment on the social and psychological health of this country could be worse than that of physical child abuse? Why or why not? Could eliminating corporal punishment and other forms of "minor" aggression against children result in reduced social ills?

Prevention and Intervention

Over the past 30 years, hundreds of child abuse prevention programs have been implemented with the goals of improving parenting skills and knowledge, increasing social supports, and altering child-rearing norms (Daro & Connelly, 2002). Generally, these prevention and intervention programs fall under three types: primary, secondary, and tertiary. *Primary prevention programs* are programs aimed at the general population, usually in the form of media campaigns, with the goal of educating the public about the nature of the problem and ways to prevent it. *Secondary prevention programs* are more focused: They tend to target populations at high risk for abusing their children. These programs are designed to increase parents' child-rearing skills and knowledge so that abuse may be prevented *before* it begins. *Tertiary prevention programs,* which in comparison to primary and secondary prevention receive the lion's share of funding in this country, are aimed at families that have come to the attention of CPS because they have already experienced abuse. These families are either referred for services by CPS or ordered by the courts to get help to ensure the safety of the child. The tertiary programs generally assess the families' strengths, weaknesses, and needs so that an appropriate intervention can be set up to address the child's individual needs and safety (Harrington & Dubowitz, 1999; DHHS, 2001). These programs are usually offered not only for the problem of physical abuse but also for all the other types of child maltreatment. In the following sections, we discuss each type of program and evidence regarding the extent to which they actually achieve the goal of preventing child physical abuse.

Primary Prevention

Primary prevention programs for child physical abuse were popular in the 1970s and 1980s and came in the form of public service announcements on television and radio, and news coverage of the most egregious cases of child abuse (Daro & Connelly, 2002). Although such programs are often viewed as a "soft" approach to preventing child abuse, research has shown that many parents who view child abuse prevention advertisements have changed their behavior, attitudes, and beliefs about child rearing (Daro, 1991). In addition, even though these programs may not alter the behavior, attitudes, and beliefs of the most abusive parents, they have been found to be valuable in changing the behavior of mildly to moderately abusive parents and in preventing the negative psychological outcomes these practices may cause in their children. Furthermore, by providing information on how to get help, they educate adults as to how they can assist children who are severely abused in their neighborhood (Daro, 1991).

Secondary Prevention

Secondary prevention programs are those aimed at parents who are deemed to be at greatest risk for abusing their children. These programs often target pregnant women or women with young children whose circumstances match those of parents who have abused their children. The most popular type of secondary prevention programs comes in the form of home visiting. These programs began in 1993, and the six most common ones serve 550,000 children each year (Daro & Connelly, 2002; Gomby et al., 1999).

The most widely cited home visitation project was conducted in Elmira, New York (Olds et al., 1986; Olds et al., 1997). The purpose of this study was to investigate whether intensive nurse home visitation was an effective way of preventing abuse by mothers considered to be at high risk (i.e., poor, unmarried teenagers). The mothers were recruited before the birth of their first child. One group was offered the intensive nurse home visitation program during pregnancy and the first two years of the child's life, in addition to the developmental screenings and transportation to prenatal and well-child care that the other groups were given. The early reports showed that only 4% of the high-risk mothers in the intensive home visitation program abused their children, compared to 19% of the other high-risk mothers.

These short-term benefits seemed to be sustained; 15 years later, the mothers who had received the home visitation still had lower rates of child abuse (Olds et al., 1986; Olds et al., 1997).

Home visitation programs are thought to work because they prevent the problem of maltreatment before it has the chance to occur. Mothers who are pregnant or just starting to parent are less likely to be offended by someone providing guidance in child care than mothers who have been reported for abuse. Moreover, by coming into the home, the service provider is able to evaluate and make changes in the home environment and work with the parents in the privacy of their homes at attaining healthy interactions with their children. In addition, they are able to educate the parents about appropriate child-rearing techniques and child development on a flexible schedule (Daro & Connelly, 2002).

Other home visitation programs have shown similar successes. For instance, in the Healthy Start Program in Hawaii, which also recruited high-risk mothers, no child abuse was reported in 99% of the program families. This rate was significantly different from that of the high-risk families not enrolled in the program. Moreover, at two years of age, the children in the program had more positive developmental outcomes and improved health status in comparison to the controls (Wallach & Lister, 1995). The success of this program led to the development of a national program disseminated by the National Committee to Prevent Child Abuse and Neglect entitled Healthy Families America (Wallach & Lister, 1995). As Daro and Harding (1999) found in their analysis of 29 evaluations of this program: "Healthy Families America (HFA) home visitation programs documented notable change among participant families, particularly in the area of parent-child interaction and parental capacity. Most families receiving these services appear better able to care for their children; access and effectively use health care services; resolve many of the personal and familial problems common among low-income, single-parent families; and avoid the most intrusive intervention into their parenting, namely, being reported for child abuse and neglect" (Daro & Connelly, 2002, p. 436).

However, one point that many researchers make (e.g., Gomby et al., 1999; Daro & Connelly, 2002) is that the positive outcomes of these projects are not always consistent; that is, not every participating parent improves in parenting skills. Indeed, some program "graduates" are reported for abusive behaviors, and some children do not fare well psychologically. As a matter of fact, researchers of the Good Start Project, which is a home- and clinic-based program designed to improve parenting skills, found that it was the families at moderate risk for abuse who were most responsive to the services offered. These services included developmental assessments, counseling, medical care, social advocacy, and parent-child enrichment sessions (Ayoub & Jacewitz, 1982; Kowal et al., 1989; Willett, Ayoub, & Robinson, 1991).

Tertiary Prevention

Tertiary services, aimed at preventing re-abuse in families already substantiated for abuse, are the most common type of prevention/intervention service. According to the DHHS (2001), approximately 528,000 children received tertiary prevention services in 2001, a count that is probably an underestimate. This count represents 58% of child victims substantiated by CPS, which means that a sizable minority of child abuse victims do not receive tertiary services, even though their cases were substantiated. Consider the case of Kayla at the beginning of this chapter. She was one of the children who never received any tertiary prevention. What were the ultimate consequences for her lack of services? Do you think service intervention could have prevented her death?

It is not quite clear how policies within each state's social service agencies influence which children get the needed care (Kolko, Seleyo, & Brown, 1999), but several factors have been identified by DHHS (2001). Consider which of these factors may have been present in Kayla's case. Victims who (1) had multiple forms of maltreatment; (2) were physically abused or neglected (i.e., not sexually abused); (3) had a prior history of maltreatment; (4) were non-White or non-Hispanic; (5) were reported by medical personnel rather than social or mental health professionals; (6) were under the age of four; and (7) were maltreated by their mothers were more likely to receive services than their counterparts. Many of these families have long histories of multiple forms of abuse, and dysfunctional methods of dealing with stress have been firmly established within the family relationships (Browne & Herbert, 1997). Moreover, because these are the most severe cases, many of these families are also characterized by severe depression, substance abuse, and interparental violence (Daro & Cohn, 1988). What are the implications of these findings? How might their situations affect their willingness and ability to receive help?

Because substantiated cases are generally from the most violent families with the most dysfunction, these families are the least likely to show improvements in treatment. Many times there is no change in their behavior, no matter how much treatment they receive (Ayoub et al., 1992; Willett et al., 1991). Approximately one third of families receiving tertiary services and half of service completers have a recurrence of abuse (Malinosky-Rummell et al., 1991). In addition, tertiary programs are characterized by problems with initial engagement and compliance and high dropout rates (Cohn & Daro, 1987). Up to 66% of those families receiving these services drop out before completion (Warner et al., 1990).

Although most tertiary programs are neither comprehensive nor built on an integrative conceptual framework (Daro & Connelly, 2002), there are programs offering hope for the abusive family. Kolko's (1996) Abuse-Focused Family Treatment program has three phases. In the first phase, *engagement,* the family's roles and interactions are assessed and reframed to enhance cooperation; the negative effects of physical force are discussed; and a no-violence agreement is signed. In the second phase, *skill building,* families are trained in problem-solving techniques and communication skills. Finally, in the third phase, *application/termination,* families establish their new problem-solving routines. In an assessment of the program, Kolko found that although 20%–30% of parents still used high levels of physical force at the end stages of treatment, many families displayed lower levels of anger and physical abuse. At a one-year follow-up, there was still evidence of improvements in the parent-child relationship—for example, there was more family cohesion and less conflict, physical abuse, and parental distress.

Two other programs have also helped some abusive families improve in functioning. For instance, Homebuilders, which offers intensive counseling, casework, and concrete services to abusive families, and Project SafeCare, which addresses the behavioral deficiencies of abusive parents, have both shown improvements in family functioning and parental skills. In the absence of comparison data, it is difficult to assess exactly how well these programs are doing (Kolko, 2002; Lutzker et al., 1998; Whittaker et al., 1990); however, the apparent successes of these programs have prompted some to argue that perhaps the best way of treating abusive parents is through a combination of therapeutic techniques and parenting instruction. That way, they can receive the support that they need from the group setting but still benefit from the individualized attention that addresses their psychological needs (Daro & Connelly, 2002; Gambrill, 1983).

Many of these programs focus on the parents; their aim is to teach abusive parents alternative techniques for raising their children. Only recently have researchers and clinicians begun to address the needs of the abused child. Although the ultimate policy goal is to keep the family together, an abused child will most likely have therapeutic needs following abusive episodes. Two types of programs have been shown to help abused children. First, therapeutic day programs, which offer developmentally appropriate and therapeutic activities, have led to improved intellectual and language abilities in abused children at a one-year follow-up (Oates & Bross, 1995). Also, an intensive, group-based treatment program for abused children, designed to offer supportive peer relationships to abused children and help them recognize their feelings through play, speech, and physical therapy, led to increased cognitive functioning, peer acceptance, maternal acceptance, and developmental skills (Culp et al., 1991).

Although such programs have proved valuable, not all treatment approaches are appropriate for all children. Bartholet (1999) suggests that treatments need to be individualized, and more treatment alternatives are necessary. Finally, as discussed in the final chapter of this book, not all suspected cases of maltreatment are reported to or even investigated by CPS, and over 40% of substantiated cases of child abuse are not referred for services (Zellman & Fair, 2002). What does this mean? These two facts taken together mean that even though we have programs available to help abused children and their families, the majority of abused children and their families will not receive any services at all (Kaplan et al., 1999). What is the impact of the lack of services on the future of these children and the future of child abuse in this country? What are the costs of child abuse and insufficient services to society as a whole?

Summary

In sum, it is difficult to ascertain exactly how many children in this country are physically abused each year. Self-report statistics suggest that nearly seven million are, whereas official statistics from CPS agencies suggest that only a minority of these children are referred for state intervention. In addition, only a fraction of the families who are reported for child physical abuse receive any services from the state. Thus, millions of children each year are being physically abused by their parents and yet receive no help, as was the case with Kayla, the young girl who died at the hands of her father even though her case was reported to the state. The short- and long-term consequences for these children can be debilitating, and some researchers suggest that even minor physical abuse, in the form of corporal punishment, can also lead to maladjustment. It is imperative that we encourage parents to use nonphysical means to discipline their children; that we encourage those who are physically abusing their children to get help; and that we also ensure that the children who are physically abused get the help they need. Research into the various predictors of child physical abuse shows that we have substantial knowledge of which families and children are most at risk for abuse. These risk factors come at every level of the ecological model, and, thus, we can target our interventions to those who need it most.

5

Child Sexual Abuse

A good man leans his weight upon my back, breathes in my ear, wraps his arm in affection high on my chest, near my neck. He means well, and whispers words of healing neither he [n]or I have ever heard said to us. "I'm glad you're here, son. I'm glad you're a boy." Memory floods and overwhelms me, rushing me back to a bedroom of a tenement on the South Side of Chicago . . . My father's arm is tight across my throat, arching my back. He is exhaling hard, in rhythm, as he shows me why he is glad I am his son. Twenty years since I found his body, dead by his hand. I rage that he is not alive today, so I could strangle him with my own hands. I hunger for a dish of vengeance, a repast of revenge . . . My predatory, molesting grandmother died in her sleep. The neighbor lady, at 32, seduced me at 14. Two months later, she ate a .22. The coach who groomed me drowned in alcohol. The apostate priest who raped me now lives down south, troubled only by the knowledge that his secret is broken. My mother . . . lives a mile from me. She shares a dilapidated apartment with my brother. He is forty years old, and still lives with her. He has never married. She has never remarried. No need to. She raised her sons to be her husbands. None of these baby-raping monsters will ever spend a day behind bars. None will pay a penny of recompense. None will know the thump of my fist, the crack of my boot . . . I have not been avenged. I shall never be avenged. Vengeance is the Lord's . . . I hope. But it is my experience that He leaves justice to mere mortals, and His servants have done a lousy job. (Abraham, 1997, pp. 1–3)

T his extreme example of child sexual abuse (CSA) described by the victim, Scott, portrays the myriad perpetrators who have been shown to victimize children sexually. Scott was sexually abused by his father, mother, grandmother, a neighbor, a priest, and a coach. In addition to these perpetrators, grandfathers, uncles, aunts, and siblings have all been implicated in the sexual abuse of both girls and boys. The current chapter deals solely with the sexual abuse of children by adult family members. The bulk of the chapter deals primarily with the most-researched type of CSA: father-daughter incest. However, we also give consideration to other types of incest, including incest perpetrated by mothers and sustained by boys.

Consider Scott's case. What consequences, both physical and psychological, do you think he suffers from as an adult? He was probably first sexually victimized by his parents and grandmother, and later by a neighbor, priest, and coach. Do you think this pattern of revictimization might be common among survivors of CSA? Why? Scott is very angry because his perpetrators were never held responsible for what they did to him. Sexual abuse, in some instances, can leave physical evidence (e.g., sexually transmitted diseases, pregnancy, and physical injuries such as vaginal tears), but it is often less detectable than physical abuse, in part because the child victims may not reveal the evidence. How often do you think children are willing to reveal their abuse, and how often do you think they are believed? What particular circumstances would dictate both the disclosure of sexual abuse and whether or not the child victims are believed? With the possible exception of the priest, Scott's abusers were never held accountable, and in the case of his mother, it seems as if the incest is ongoing. Perhaps one of the reasons his perpetrators were never prosecuted was because the CSA occurred at a time when people did not discuss or even acknowledge that the problem existed. How often do you think child sexual abusers are held accountable these days? What circumstances do you think mitigate whether they are held accountable? The questions posed here are among the issues we discuss in the current chapter. Think about your answers to these questions as you read the sections on the prevalence, predictors, and consequences of CSA, and approaches to intervention.

Scope of the Problem

Before we discuss the incidence and rates of CSA by family members, it is important to define exactly what CSA is. This is not easy, as researchers have been unable to formulate a consistent definition. Most studies have, as one criterion for judging sexual acts as abusive, an arbitrary age difference of at least five years between the victim and perpetrator (e.g., Finkelhor, 1979); however, this criterion does not take into consideration that siblings with less of an age discrepancy can be perpetrators of sexual abuse. Furthermore, there is no consensus on what acts constitute abuse. Sometimes *noncontact* acts such as voyeurism, exhibitionism, and exposure to pornography are included in operational definitions (i.e., measures) of sexual abuse, which leads to very high incidence rates. If only *contact* sexual abuse is included (e.g., fondling; oral, anal, vaginal intercourse), rates are lower. Most studies do not specify the frequency or duration of sexual contact necessary for it to be considered abuse, although many workers indicate that just one instance of sexual contact (or noncontact) is enough to label an act abusive. When we consider the causes and consequences of CSA, a child intentionally exposed once to pornography is going to differ dramatically from the children who experienced the abuse described in Box 5.1.

Box 5.1   Case of Child Sexual Abuse

The sexual abuse of Ann and Marie by their father began around the same time, when Ann was five and Marie was four. At the time of their evaluation by a psychologist, both were adolescents, and Marie was the one who had reported the abuse. Ann had mixed feelings about the fact that Marie had disclosed the abuse—she was happy that at least now the abuse would stop because her father would get help, but she was also upset that her father was in trouble.

"Ann reported that, most commonly, her father would fondle her in the breast area and between the legs. At different stages of her life, he would perform different sexual acts on her, which she and Marie would discuss on occasion. The most extensive abusive act Ann experienced was what the girls called the 'full treatment.' This consisted of the father's undressing her, fondling her, and lying in bed with her back to his chest. He would roll her around his genital area and touch her on her breasts and vaginal area. During the course of the interview, Ann related many incidents of this type being perpetrated against her, almost on a weekly basis during certain periods of her life . . . Marie described the same type of fondling as Ann . . . As Marie got older, the father would perform oral sex on her as well. In describing this, Marie became quite tearful and felt guilty and 'dirty' because of this particular abuse. She wondered if it was her fault and if she should be blamed for not stopping her father sooner."

SOURCE: Cohen & Mannarino, 2000, pp. 212–214.

The first source of data on rates of sexual abuse is the official NCANDS statistics provided annually to the U.S. Department of Health and Human

Services (DHHS). All states have laws prohibiting the sexual abuse of children (Myers, 1998), but each state has its own definition of what is considered sexual abuse and therefore what is prohibited. Because sexual abuse statutes vary from state to state, the official statistics need to be interpreted with caution. Generally, though, adult sexual contact with a child under the age of 14 is illegal, as is any kind of incest. According to the 2001 NCANDS statistics, 9.6% of all substantiated maltreatment cases were CSA cases, and nationwide 86,830 children, or 1.2 per 1,000 children, were victimized by sexual abuse. Parents were the perpetrators in approximately 40%, and other relatives in approximately 60% of these cases. Although these rates of substantiated cases of sexual abuse are high, they reflect a decrease from 1997, when the abuse rate was 1.7 for every 1,000 children (DHHS, 2001). All abusers were relatives of the children because CPS generally does not get involved in cases of extrafamilial child maltreatment (Finkelhor, 1994).

The Third National Incidence Study (NIS-3; Sedlak & Broadhurst, 1996; see Chapter 4), based on child maltreatment cases reported to CPS as well as cases known to community professionals, shows a similar picture of CSA in the United States. According to the Harm Standard, the number of sexual abuse cases rose from 119,200 in 1986 to 217,700 in 1993, an increase of 83%. Similarly, according to the Endangerment Standard, the number of cases rose from 133,600 to 300,200, an increase of 125%. Overall, the data indicate that the reporting of CSA was on the rise during the latter half of the 1980s and then declined in the mid- to late 1990s. According to NIS-3, reported cases of sexual abuse were three times more likely to be girls than boys. The lowest rate of CSA was in children in the zero- to two-year age range; the rates were constant for children over age three. Children in the poorest income category (less than $15,000/year) were 18 times more likely to be reported for sexual abuse than children whose parents earned more than $30,000 per year. In contrast to the official CPS statistics reported by the DHHS, only one fourth of children were abused by their birth parents (Sedlak & Broadhurst, 1996).

Although official statistics are very informative, Finkelhor et al. (1990) point out that most sexual abuse is not reported at the time it occurs; in fact, sexual abuse is often never reported. In addition, there is reason to believe that sexual abuse perpetrated by a family member is even less likely to be reported to officials than extrafamilial sexual abuse. Consequently, better sources for an estimate of the prevalence of CSA may be retrospective reports of CSA in nonclinical populations (Berliner & Elliott, 2002). In a review of these studies, a sexual abuse rate of 20%–25% for women and 5%–15% for men was determined (Finkelhor, 1994). This sexual abuse included both intra- and extrafamilial contact sexual abuse. The age range of victimization was from infancy through 17 years, with a peak age range of 7 to 13 years, and a mean of 9 years. In the general population, parents are the perpetrators 6%–16% of the time, and any relative is the perpetrator in one third of the cases. Strangers account for only 5%–15% of the cases; thus, 85%–95% of the perpetrators are known to the child victim (Finkelhor, 1994). Half of the victims seem to experience multiple episodes of abuse, and 25% experience completed or attempted oral, anal, and/or vaginal intercourse (Saunders et al., 1999).

These self-reported rates of CSA in nonclinical samples are dramatically lower than rates found in clinical samples. People who seek mental health help because of a history of CSA probably have had more severe abuse experiences, and the data confirm that supposition. Participants in clinical samples have identified 25%–33% of sexual abuse perpetrators as parents, and any relative as the perpetrator in half of the cases. Three fourths of the clinical victims experienced multiple episodes of abuse, and, in the great majority of the cases, the multiple episodes consisted of completed or attempted oral, anal, and/or vaginal intercourse (Elliott & Briere, 1994; Ruggiero, McLeer, & Dixon, 2000).

In the recent research literature on sexual abuse, more attention is given to girl than to boy victims—possibly because, given the likely social stigma and social mores, boys are less likely to report abuse and less likely to label an abusive sexual experience as such. The reported cases do provide some evidence of differences in sexual abuse experiences, based on the gender of the victim. Boys are more likely to be older when their abuse begins, to have a nonfamily member as the perpetrator, and to have a woman as their perpetrator (Holmes & Slap, 1998).

Predictors and Correlates

Macrosystem

The sexualization of children in the media and elsewhere, the view of children as property, and male entitlement have all been implicated as providing cultural sanctions for the sexual abuse of children (e.g., Russell, 1995), although an empirical link has yet to be established. The availability of child and adult pornography may lead to adult sexual interest in children (Russell, 1995); however, the link between pornography and CSA appears to be indirect (Knudsen, 1988). One study showing this indirect link looked at incarcerated sexual offenders: The offenders who reported experiencing sexual abuse as children and early exposure to pornography displayed less empathy toward abused children and had more child victims (Simons, Wurtele, & Heil, 2002).

Furthermore, although empirical evidence is lacking, the proliferation of child pornography since the introduction of the Internet cannot be helping to end CSA. In fact, the percentage of pornographic images displaying children and adolescents on the Internet seems to be on the rise (Mehta, 2001). A recent national study of law enforcement agencies showed that between July of 2000 and June of 2001, over 1,700 arrests were made in this country for Internet-related possession of child pornography (Wolak, Finkelhor, & Mitchell, 2003).

Microsystem

The majority of research on predictors of CSA has focused on the microsystem level, specifically on problems and dynamics within the family and the specific relationship between the perpetrator and the victim. Possibly the most researched family structure implicated in CSA is the stepfamily. Russell (1986) found that a stepdaughter is seven times more likely to be a victim of CSA than a natal daughter. Similarly, Finkelhor (1984) found that a stepfather was five times more likely than a natal father to abuse a daughter. In addition, girls with stepfathers were more likely to be sexually abused by other men than girls who lived with both natural parents or only their mother. It has been speculated that stepfathers may be more likely to abuse their daughters because they do not have an incest taboo to overcome. However, this leaves some questions unanswered. For example, why would these stepfathers find children sexually attractive? Possible reasons are discussed in the section on individual/developmental predictors.

Relationships with mothers may also play a role in risk for CSA. According to Finkelhor (1984), if a girl ever lived without her natural mother, she is three times more likely to be sexually abused than girls who never lived without their mothers. Moreover, if the mothers are perceived as unavailable (e.g., emotionally distant, often ill, or unaffectionate), girls are also at increased risk (Finkelhor, 1984; Finkelhor et al., 1990). Finally, in families in which both biological parents are present, CSA is most likely to occur if the father is a nondrinker but the mother drinks (Vogeltanz, Wilsnack, Harris, Wilsnack, Wonderlich, & Kristjanson, 1999); the mother's drinking behavior may lead her to be unavailable to her daughter, which may leave the girl emotionally needy and more vulnerable to the advances of a sexual predator.

In addition to the relationship between the child victim and her parents, other dynamics within the family are also predictive of CSA. Less cohesive families that are disorganized and generally dysfunctional, with perhaps other forms of child maltreatment, are more vulnerable to having a child victimized by sexual abuse (Elliott, 1994; Madonna, VanScoyk, & Jones, 1991; Mannarino & Cohen, 1996; Mullen, Martin, Anderson, Romans, & Herbison, 1994). These families may also be subject to much disruption, such as parental separation, absence, and conflict, and frequent moves (Mullen et al., 1994). Moreover, the parents may be characterized as uncaring, over-controlling, and rejecting (Fleming, Mullen, & Bammer, 1997; McLaughlin et al., 2000; Mullen et al., 1993). These dysfunctional families have been profiled as headed by a young mother who has had many negative life events and an unwanted pregnancy (Brown et al., 1998); the mother may also have mental health problems (Brown et al., 1998; Fleming et al., 1997). Further-more, the families may lack emotional closeness and flexibility, have problems with communication, and be socially isolated (Dadds et al., 1991).

Much research has been done on the types of incest described in Box 5.1: father-daughter incest. As in the case described here, families with father-daughter incest tend to be characterized by very traditional family values. Now read Box 5.2, a clinical description of the family from Box 5.1. What do you notice about this family? What characteristics do they have that may distinguish them from more functional families?

Box 5.2   Ann and Marie's Family

Ann and Marie's family did not interact or communicate in a healthy manner, which allowed the father's sexual abuse of the girls to commence and continue for many years. The family members were overdependent on each other and emotionally enmeshed, such that if something threat-ened the family's stability, each person's survival was at stake. The family was also characterized by an unhealthy marriage and the parentification of the children (i.e., the children were given adult responsibilities). The family members did not want to confront issues that were threatening to family stability; and therefore, their communication was vague and served to placate issues rather than to constructively resolve problems. Therefore, it is not surprising that Ann and Marie stated that they were afraid of reporting their father's sexual abuse because they feared the family would disintegrate, and they thus also felt an overwhelming sense of responsibility to hold together this dysfunctional family unit. Their mother's attitudes and behaviors contributed as well: She told the girls that she would fully support them during this time but also stated that she could not choose her daughters over her husband. In other words, she could not support her daughters over her husband. Ann and Marie subsequently began displaying signs of empathy towards their father.

SOURCE: Cohen & Mannarino, 2000.

Consistent with the family described in Boxes 5.1 and 5.2, families characterized by father-daughter incest tend to be patriarchal in structure, with the children subservient to the adults (Alexander & Lupfer, 1987; Wealin, Davies, Shaffer, Jackson, & Love, 2002). The most at-risk situation for a daughter is when her father is well educated and her mother is not (Finkelhor, 1984). Role reversal between the mother and daughter is also characteristic of father-daughter incest families (Ray et al., 1991), probably because there is a lack of intimacy between the parents (Hubbard, 1989) and marital problems or instability (Lang et al., 1990). Examples of some marital problems that incest offenders seem to have include mistrustfulness, lack of mutual friends and time together, low mutual give-and-take in disagreements, tendency to not confide in each other, poor quality of sexual relations, and being lonely within the marriage (Lang et al., 1990). Alcohol problems may be characteristic of both parents (Nelson et al., 2002).

As is probably self-evident, in father-daughter incest families, there is role and boundary confusion, affective enmeshment, and poor adaptation and problem-solving skills (Hoagwood & Stewart, 1989). When natal fathers are the abusers, they tend to have been out of the home or infrequently in the home during the daughter's first three years. If the abusing natal fathers were in the home during those years, they tended to be less involved in caregiving. Possibly because there was no opportunity for early bonding, there was no insulation from sexual desires for the daughter (Parker & Parker, 1986). In addition, sexual problems between the father and his wife may lead him to seek sex from his daughter as a surrogate source of affection (Ganzarain, 1992).

Individual/Developmental

Even though the culture may in some ways sanction CSA through the sexualization of children and child pornography, and even though some families may be characterized by dysfunctions, there are still many questions left unanswered as to why a person would sexually abuse a child, especially his/her own child. Finkelhor (1984) posed four questions that need to be answered in order for CSA to be understood more fully: "1) Why does a person find relating sexually to a child emotionally gratifying and congruent? 2) Why is a person capable of being sexually aroused by a child? 3) Why is a person blocked in efforts to obtain sexual and emotional gratification from more normatively approved sources? 4) Why is a person not deterred by conventional social inhibitions from having sexual relationships with a child?" (p. 17).

These questions need to be answered at the individual/developmental level. Motivations to sexually abuse children are vastly divergent (Ferrara, 2002). Moreover, much research is needed concerning gender preferences among sexual abusers—that is, why do some pedophiles prefer boys and others prefer girls—and on any differences between male and female sexual abusers. The research that follows has been done solely on male sexual abusers.

In answer to the first question concerning why a person would find relating sexually to a child emotionally gratifying and congruent, it has been shown that incarcerated child sex abusers have higher rates of psychopathology than other incarcerated individuals and that increased psychopathology is associated with increased sexual deviancy (Herkov, Gynther, Thomas, & Myers, 1996). However, incarcerated child sex abusers represent a very small minority of child sex abusers and probably are not representative of the offenders who are never reported or caught; this latter group probably has fewer detectable psychological abnormalities (Finkelhor, 1984).

It has been suggested that child sexual abusers have arrested psychological development, such that they experience themselves emotionally as children and can therefore relate better to children (Groth, Hobson, & Gary, 1982). However, arrested psychological development does not explain why adults would be sexually aroused by a child. In answer to Question 2, it has been shown that there is more childhood sexual victimization in the backgrounds of child molesters than in several comparison groups (e.g., Bard et al., 1983; Langevin et al., 1983) and that the families of child sexual abusers may be characterized by sexual abuse through the generations (New, Stevenson, & Skuse, 1999). This history of childhood sexual abuse is theorized to lead to a sexual interest in children. However, only about 28% of child sexual abusers report being sexually abused as children, and not all sexually abused children will become child sexual abusers; therefore, the relationship between childhood victimization and adult perpetration does not appear to be straightforward (Hanson & Slater, 1988).

Exposure to child pornography, especially when child sex is mixed with adult sex, may also lead to sexual interest in children (Russell, 1982). In partial support of this conjecture, Briere and Runtz (1989) found that masturbation to pornography was predictive of college males' sexual interests in children. Among those male students, 21% had experienced some sexual attraction to small children; 9% at some time had fantasies about sex with a child; 5% had masturbated during fantasies about sex with a child; and 7% stated it was at least somewhat likely that they would have sex with a child if there was no likelihood of detection or punishment. Along with masturbation to pornography, variables predicting attraction to children included negative early sexual experiences and self-reported likelihood of raping a woman.

Even if there is a relatively high rate of sexual attraction to children, the variables associated with this attraction do not explain why some men act upon it while others do not. In answer to Question 3 as to why some men are unable to get their sexual and emotional needs met in adult relationships, it has been found that child sex abusers tend to be "timid, unassertive, inadequate, awkward, even moralistic types with poor social skills who have an impossible time developing adult social and sexual relationships" (Finkelhor, 1984, p. 43). It has also been suggested that sexual abusers suffer from narcissistic personality disorder, as they display low self-esteem, lack of empathy for their victims, and self-centeredness (Ganzarain, 1992). In addition to these personality traits, some family-level characteristics, such as marital instability, may contribute to the men's inability to gain emotional and sexual satisfaction in adult relationships (Finkelhor, 1984).

Finally, even though all of these characteristics may exist, they do not explain why some men overcome the societal taboos against sex with children and/or incest. Several individual-level factors have been found to contribute to this disinhibition, including poor impulse control (e.g., Cohen et al., 2002), multiple major life stressors (e.g., Hermin, 1981), alcohol and drug abuse (e.g., Gordon, 1989), and psychosis (e.g., Marshall & Norgard, 1983). Although poor impulse control and psychosis have been implicated in some cases, alcohol and drug abuse seems to be the disinhibitor in a majority of cases (e.g., Greenfield, 1996).

Consequences

In addition to being correlates of sexual abuse, many of the variables already discussed also mediate the consequences of sexual abuse. Factors such as poverty, family dysfunction, parental psychopathology, and parental alcohol or drug abuse can serve to worsen the consequences of CSA. Other mediating factors that serve to worsen outcomes are longer duration of abuse, greater intensity of abuse, closer relationship to the abuser, greater frequency of abuse, and the use of force (e.g., Elwell & Ephross, 1987; Kendall-Tackett et al., 1993; Ruggiero et al., 2000).

In addition to mediating factors that can make outcomes worse, other factors, such as age and gender, can contribute to the outcome. Abused younger children tend to have more sexual and nonsexual behavior problems than abused older children, who have more internalizing problems (e.g., Ruggiero et al., 2000). Abused boys tend to have more externalizing problems than abused girls (e.g., Holmes & Slap, 1998). Other factors such as attributional style, coping strategies, and level of cognitive functioning have been implicated as mediating factors for outcome (e.g., Mannarino & Cohen, 1996; Shapiro et al., 1992; Spaccarelli, 1994). One of the most important mediators seems to be maternal belief in the child's disclosure of sexual abuse and her subsequent support of the child (Elliott & Briere, 1994).

This last point brings up a very important question: Why wouldn't a mother believe her own child's disclosure of sexual abuse? Several factors appear to have an influence (Sirles & Franke, 1989). In one study, a younger victim was believed more often than an older victim; the more severe the abuse (e.g., intercourse versus fondling), the less likely the mother was to believe the child; if the mother was home during an abusive incident, she was less likely to believe the child; if the perpetrator was a stepfather, only 56% of the mothers believed their children, but if he was a biological father, 86% believed their children; if the abuser was another relative, 92% believed their child's disclosure; if the child was also physically abused, the child was less likely to be believed; and if the perpetrator abused alcohol, the child was less likely to be believed (Sirles & Franke, 1989). Think about the logic behind these findings: Why might a mother be less likely to believe her child in the above instances? What would be the implications for her existence, family life, and well-being if she were to believe the child? On the other hand, think of the effects on the child: If a child is less likely to be believed when there is physical abuse and alcohol abuse on top of the sexual abuse, what are the implications of this "quadruple whammy" (not being believed, being sexually and physically abused, and living in a home with an alcoholic) on the child's mental health?

Short-Term Outcomes

The effects of CSA are pervasive; the mental health outcomes can affect almost every aspect of functioning. In the short term, sexual abuse has been found to affect a child's emotional, behavioral, cognitive, and interpersonal health. For instance, sexually abused children have been found to be more depressed, anxious, suicidal, and aggressive than nonabused children (e.g., Boney-McCoy & Finkelhor, 1995; Hotte & Rafman, 1992; Lanktree et al., 1991). They have lower self-esteem, suffer more often from post-traumatic stress disorder (PTSD), and engage in more sexualized behavior, such as mimicking intercourse, inserting objects into the vagina or anus, or exposing genitals (e.g., Boney-McCoy & Finkelhor, 1995; Friedrich et al., 2001; Wozencraft, Wagner, & Pellegrin, 1991). Cognitively, they achieve less well in school, perceive themselves as different from their peers, blame themselves for negative events, and distrust others (e.g., Mannarino et al., 1994). They are also less socially competent (Mannarino & Cohen, 1996).

Long-Term Outcomes

Usually, children's symptoms in reaction to sexual abuse improve over time; however, sometimes symptoms do not abate and may actually worsen into adolescence and adulthood (Kendall-Tackett et al., 1993). Individuals whose symptoms persist are more depressed and suicidal than nonabused adults (Briere & Runtz, 1987; Browne & Finkelhor, 1986). They seem to suffer from more anxiety disorders, including PTSD, than the general population (Saunders et al., 1999; Stein et al., 1988). They also tend to be more violent, angry, self-mutilating, and irritable than their nonabused counterparts (Briere & Gil, 1998; Briere & Runtz, 1987; Duncan & Williams, 1998). In addition, they may suffer from self-blame, low self-esteem, a negative attributional style, low self-efficacy, helplessness, and hopelessness (Jehu, 1988). They may externalize their emotional distress through such activities as self-mutilation (e.g., cutting, burning, hair pulling), heightened sexual activity, bingeing and purging, and alcohol or substance abuse (Briere & Gil, 1998; Briere et al., 1997; Piran et al., 1988). Finally, a history of CSA can greatly affect adult intimate relationships: Sexually abused females are more likely to divorce or never marry; they have fewer friends, are less satisfied with their relationships, and are more interpersonally sensitive than women without a history of abuse (Elliott, 1994; Gold, 1986; Russell, 1986).

Although all these possible consequences of CSA have been observed in adulthood, very few studies consider that many mediating factors, such as other aspects of family functioning, could also be contributing to the negative outcomes. In one study of the consequences of CSA, researchers compared the outcomes of twins who had been sexually abused as children with their co-twins who had not been abused (Nelson et al., 2002). They found that the nonabused co-twins of sexually abused twins functioned more poorly than those twin pairs in which neither twin had been abused, suggesting that many of the family background factors associated with CSA, namely parental alcohol abuse, physical and emotional abuse, and childhood neglect, must have contributed to negative outcomes. However, the nonabused co-twins fared better than their abused twins, suggesting that the CSA had an independent effect on the abused twins' poor outcomes. The outcomes that the sexually abused twins were at an increased risk for included major depression, suicide attempts, conduct disorder, alcohol dependence, nicotine dependence, social anxiety, rape after the age of 18, and divorce (Nelson et al., 2002).

Dissociation

One controversial outcome that has been attributed to CSA is dissociation. *Dissociation* has been defined as the mind's ability to remove itself from the reality of an abusive situation and the thoughts, feelings, and memories that go along with the abuse. Although several researchers have found a relationship between CSA and dissociation (e.g., Chu & Dill, 1990; Elliott & Briere,

1992; Zanarini et al., 2000), especially when it is combined with physical abuse (e.g., Mulder et al., 1998), others (e.g., Pope et al. 1998) have argued against the existence of dissociation and do not believe that a person can forget traumatic events. In one recent review of the relevant literature, Joseph (1999) argued that there is significant neurological support for the reality of dissociation as a phenomenon and that it exists as a possible consequence of CSA. In the review, Joseph presented evidence that memory deficits of emotionally traumatizing events are due to disturbed hippocampal activation and arousal. An emotionally traumatizing event can cause the release of corticosteroids, which suppress neural activity associated with learning and memory and can cause hippocampal atrophy. Because the hippocampus is involved in memory processes, it is possible that emotionally traumatizing events will not be remembered.

There is also anecdotal evidence that memories of CSA can be forgotten but later remembered. Consider the case in Box 5.3 of a woman who, in adulthood, remembered an incident of CSA. What is unique about this case is that the offender admitted to sexually abusing her; in most cases of this type, the offender denies the abuse, and, therefore, it is unknown whether the recovered memory of CSA is true or not, as there is also evidence that false memories can be induced in people (see e.g., Loftus & Ketcham, 1994).

Box 5.3   Can People Forget and Later Remember CSA?

David A. Hoffman, a former child psychologist . . . pleaded guilty in April [1994] to gross sexual exploitation . . . Hoffman was charged with the crime after a woman remembered being sexually abused during a two-year period, beginning when she was 8 and living in Columbus with her mother . . . The woman is now 26 and lives in Michigan. She had no recollection of the abuse until July 1992, said detective John Harris . . . "She worked in a probation office in Grand Rapids, Mich., typing reports," Harris said. "Her first memory of the abuse came when she was typing a report regarding a sexual abuse case. Then, whenever she had to type reports involving sexual abuse, she would become very distraught." . . . The woman sought therapy. She called Harris after her psychologist urged her to file a police report. In 1993 . . . [Hoffman] "admitted committing the molesting offenses," Harris said.

SOURCE: Medick, 1994, as cited in Freyd, 2002, p. 139.

Several questions arise in regard to this case. One, why would individuals forget such traumatic memories? Two, how do they forget? And three, how do they later remember? In answer to the first question, it has been theorized that children forget incidences of CSA because not remembering abuse by a trusted caregiver is necessary for survival (Freyd, 2002). Children need to preserve their relationships with their caregivers for their security and survival, and to admit that their caregivers are betraying them would lead to a disruption in that relationship and a probable disruption in their security. Indeed, studies show that children are more likely to forget abuse by a trusted caregiver than they are to forget abuse by a stranger (e.g., Williams, 1994). How does the child forget and later remember the CSA? One mechanism through which children forget is the neurological changes in response to traumatic events mentioned previously. Another means is through selective attention—that is, during a CSA experience children can focus their concentration on other events that are occurring simultaneously, such as music playing in the background. Concentrating on the music allows the child to avoid completely processing the traumatic event; however, because some of the event will necessarily be processed, cues—such as the same music playing again in the future—can prompt a child to later remember the CSA (Freyd, 2002). Consider again the case in Box 5.3. How did these factors play into (1) why the woman forgot her abuse; (2) how she forgot; and (3) how she remembered?

Not only is there anecdotal evidence of forgotten episodes of CSA, there is also empirical evidence. In one study, a group of 129 girls originally brought to an emergency room because of sexual abuse were interviewed 17 years later. Among the issues addressed were questions regarding childhood histories of sexual abuse (Williams, 1994, 1995). Of these women, 38% reported no recollections of CSA. An additional 10% more reported periods of forgetting and later remembering the abuse (Williams, 1994, 1995). Thus, close to one half of this group of women, in which there was documented evidence of CSA, had periods of time in which they forgot the CSA. Although women who were younger at the time of the abuse were more likely to forget it, 26% of those who were 11–12 years old, 31% who were 7–10 years old, and 62% who were 4–6 years old at the time of the CSA had no memories of the abuse (Williams, 1994).

Does CSA Really Have Negative Consequences?

Another controversy surrounding the consequences of CSA comes from the contentious article published in 1998 by Rind, Tromovitch, and Bauserman. They argued that (1) the majority of studies on the consequences of CSA are based on clinical samples, and therefore, the findings cannot be generalized to the population as a whole; (2) there is no evidence of a direct link between CSA and adult psychopathology, nor is there evidence that everyone who experiences sexual abuse as a child will suffer psychologically or socially; and (3) clinical studies suffer from a possible flaw of "effort after meaning." That is, when people seek clinical help for their current functioning, and search for reasons for their problems, they may pick out an event such as sexual abuse that happened earlier in their lives. To support their criticisms, Rind, Tromovitch, and Bauserman (1998) conducted a meta-analysis of 59 CSA studies done with college students. They found that although students who had a history of CSA functioned less well than those who did not report such a history, the effects were only slight. That is, non-CSA students were only slightly better adjusted than CSA students, and the sexually abused men showed fewer negative reactions to their experience than the sexually abused women.

The Rind et al. (1998) report created a huge public controversy in this country. Their results were lauded by pedophiles, subjected to intense criticism in the media, and condemned by Congress. Defense lawyers have used their results to suggest that the victims of CSA in court proceedings were not harmed by the CSA. Because of this backlash, several researchers (e.g., Dallam et al., 2001; Ondersma et al., 2001) have reassessed the conclusions made by Rind et al. by reanalyzing the data and critically evaluating the original studies used in the analysis. For instance, Ondersma et al. (2001) pointed out that the effect sizes that Rind et al. found for the relationship between CSA and 17 negative outcomes are similar to the effect size between smoking and lung cancer, but the smaller effect size between smoking and lung cancer does not lead people to assume that smoking has nothing to do with lung cancer, or to smoke more and more because there are no negative outcomes. This analogy can be taken one step further: Smoking is associated with more than just lung cancer (e.g., emphysema, heart disease, etc.), and not every person who smokes will get lung cancer, but the person is at risk for one of a whole host of outcomes. Similarly, CSA is associated with not just one particular negative outcome, and not every person who is victimized by CSA will suffer from one particular outcome, but the person is at risk for a whole host of outcomes.

Furthermore, Dallam et al. (2001) pointed out many inherent flaws in the Rind et al. analysis, including: (1) use of inappropriate statistics to detect effects sizes in various analyses, including the association of CSA with outcomes and of gender and CSA with outcomes; (2) misreporting of some key data, such as reporting the percentages of subjects in some studies who said they were negatively affected by CSA as being lower than they really were; (3) lack of inclusion or exclusion criteria for studies to be used in the meta-analysis—that is, they included studies that varied greatly in their definition of CSA, including one (Landis, 1956) whose definition was experiences with "sexual deviants" at any age (even over the age of 17), another (Sedney & Brooks, 1984) whose definition included any sexual experiences during childhood, and others (Greenwald, 1994; Landis, 1956; Sarbo, 1985) that included sexual experiences that occurred after age 17; (4) exclusion of other relevant outcomes shown to be highly related to CSA but not systematically studied in college students, such as PTSD, antisocial behavior, substance abuse, early and risky peer sexual experiences, and revictimization of sexual abuse in adulthood; and (5) the generalization of the results from studies of college students to the general population, even though college students are a highly functioning group of people, and fewer victims of CSA who have debilitating outcomes would be found among them, as CSA victims tend to suffer from academic difficulties and are less likely to finish high school.

In response to the fifth criticism, Rind and Tromovitch (1994) also published a meta-analysis on seven nationally representative studies of CSA and found the same results—i.e., more dysfunction among CSA victims but small effect sizes. However, many of the previous criticisms may apply to this meta-analysis, as well. The studies of Rind and associates do raise some important points: (1) differing definitions of CSA can cause problems when conducting research on its causes and consequences; and (2) when conducting research on the causes and consequences, we need to pay special attention to possible mediating variables. There may not be any direct route between CSA and adverse consequences, but CSA can certainly indirectly cause negative outcomes through these mediating factors.

Special Issue: Female Sexual Abuse of Children

Most of the above literature concentrated on one type of CSA: adult male to female child. This type of CSA was the first to come to public attention back in the 1970s and 1980s, when adult females victimized as children started coming forward. Then male victims of CSA also began to talk about their sexual victimization by men, and it seemed that young boys were at almost equal risk as young girls for CSA from adult males (Elliott, 1993). Currently, there is growing attention to another hidden perpetrator of CSA: that of the female child sexual abuser. Traditionally, it has been assumed that women cannot and would not sexually abuse children. After all, women do not have penises, so how can they sexually abuse someone? Furthermore, female-perpetrated CSA goes against our long-held, cherished views of women as nurturers of children; if females can and do sexually abuse children, it undermines our views of how females relate to children. Finally, female-perpetrated CSA stands in direct contrast to our traditional explanations for male-perpetrated CSA, those of male power and aggression: If females are also sexual abusers, then male power and aggression do not lie at the heart of CSA (Elliott, 1993).

It has been argued (e.g., Hetherton, 1999) that females, particularly mothers, perpetrate CSA much more than believed and that they may "hide" their CSA in the guise of maternal caretaking acts, such as bathing and dressing the child; during such acts mothers may fondle the child's genitals, for instance. Sometimes females' sexual abuse can become much more severe and overt, as is described in Box 5.4. Acts that have been reported include the insertion of objects into the anus and vagina; the rough handling of boys' penises in an attempt to get them erect; oral sex and masturbation of the child and forcing the child to reciprocate; forcing the child to watch adults having sex; and bestiality (Longdon, 1993).

Box 5.4   Female-Perpetrated Child Sexual Abuse

My father was absent most of the time due to his work. Mother slept with me nearly every night as far back as I can remember. The initial memories of abuse were that of being fondled, which probably began in infancy. By the time I was three years old, Mother was having me touch her as well. Later I was introduced to oral sex. This sort of behaviour occurred almost nightly until I was twelve years of age. This in itself was horrible enough; but by the time I entered school, Mother started torturing me in sexual ways.

The first time I remember being sexually tortured was when Mother took me into a wooded area and fondled me, had oral sex with me, and inserted her fingers into my vagina. I cried and screamed because of the severe pain. This only made Mother angry; so to shut me up and to threaten me, she picked up a large stick and shoved it inside my vagina. This incident taught me the lesson of silence and to turn off feelings of pain.

—Lynne Marie, now 40 years old.

SOURCE: Elliott, 1993, p. 131.

How common is female-perpetrated CSA? The current statistics indicate that over 90% of sexual abusers are male (Jennings, 1993). Two things need to be considered when interpreting these statistics: Before CSA itself became known as a social problem, it was assumed that one in 1 million children were the victims of CSA. Now that we are able to acknowledge that children can be sexually abused, we know that the numbers are closer to one in five children. Once we acknowledge that women can sexually abuse children, we may see a rise in reports of female-perpetrated CSA (Elliott, 1993). Moreover, if we accept that the current statistics are correct (that one in five children is sexually abused and that, of those, 10% are abused by females), then close to 6 *million* people in the United States today were, are being, or will be sexually abused by females.

Several researchers have tried to address the problem of female-perpetrated CSA by studying mother-son incest cases. In one of the first studies on this issue, the mother-son incest experienced by the eight men studied were typically cases where the mother attempted to satisfy her own emotional and physical needs for intimacy and security (Krug, 1989). She would seek out her son, in many cases when she was in conflict with the adult male in her life; in some cases, she made overt sexual overtures towards her son. Consider the case of mother-son incest in Box 5.5. What types of sexual acts occurred? Would you consider this sexual abuse? Why do you think the mother perpetrated the act? Do you think that the son was complicit and/or wanted the sexual activity? What were the effects on this man's life?

Box 5.5   A Case of Mother-Son Incest

A 29-year-old, lower middle class white male, named Bob*, who was in a methadone maintenance program, entered psychotherapy because of symptoms of depression which were not related to the methadone treatment. During treatment, Bob revealed incidents of sexual abuse by his mother, who was on the faculty of a prestigious university, had divorced his father when he was two years old, and never remarried. During his childhood, Bob served as both his mother's confidant and her advisor, and from the time of the divorce until Bob was in his mid-teens, Bob's mother continually slept with him. Starting when Bob was seven years old, his mother insisted that they have intimate sexual contact, which on some occasions included sexual intercourse. At 10 years of age, Bob started using recreational drugs, and by age 15, he began using heroin. At age 18, Bob married and left home, but the marriage lasted only three years, as both Bob and his wife were heavy drug users.

SOURCE: Krug, 1989.

NOTE: *This name has been fabricated to facilitate the telling of the story.

Krug (1989) found that the eight men in his study suffered many problems in adulthood that could be related to their mother-son incest experiences. Specifically, most were anxious, depressed, had extra-relationship sexual contacts, and abused substances. Kelly et al. (2002) reported similar symptoms in their clinical study of 17 cases of mother-son incest. They found that men who were abused by their mothers experienced more adjustment problems than men whose perpetrators were not their mothers, even though the abuse experiences were somewhat subtle (e.g., genital fondling) and possibly difficult to distinguish from normal caregiving.

The mother-son incest victims in the Kelly et al. (2002) study experienced more sexual problems, dissociation, aggression, and interpersonal problems than the father-son incest victims. Moreover, these problems were mediated by their initial perceptions of abuse: If they initially perceived the sexual abuse as nonabusive, they were more likely to report more PTSD symptoms and aggression. If they had any positive feelings about the abuse at the time it occurred, they were more likely to suffer from later aggression and self-destructiveness. Why is it that these positive feelings might be associated with more maladjustment? Might it be that experiencing an incest experience as somewhat pleasurable might lead to a later strong reaction of shame, guilt, and disgust? Is it possible that the victims may not understand how they could experience something so wrong as remotely pleasurable, and therefore might perceive themselves as being even more deviant for having these feelings? As an example of such a dynamic, consider this victim of mother-son incest:

> One male survivor described feeling that his most intense orgasm occurred at age 13 while having intercourse with his mother, who immediately ridiculed and verbally abused him after he ejaculated. He recalled running into the bathroom and scrubbing his genitals in a manic attempt to wash away the incest. He then cried and vomited when he realized he could not wash it away. In therapy, he stated that the abuse would have been less harmful if he had not experienced pleasure because "I am now associated with something that is disgusting that I liked. I have incest tendencies. I am a part of the incest. I am as screwed up as my mother. I am as tainted, I am as damaged, I am as dirty as she is . . . The incest has cheapened and dirtied my manhood." (Kelly et al., 2002, p. 437)

Consider this case further. What role do you think the mother's ridicule and verbal abuse played in his response to the experience? Why would the mother ridicule and verbally abuse him under the circumstances?

Several researchers and victims argue that female-perpetrated CSA may be more harmful than male-perpetrated CSA (e.g., Elliott, 1993), although there is as yet no empirical support for this proposition. One basis for this argument is that the experiences of survivors of female-perpetrated CSA are often silenced. Consider this victim's story:

My mother and grandmother sexually abused me from the time I was four. I hated going home and spent most of my time figuring out how to get out of the house . . . I begged my aunt to let me live with her. In my life I have taken most drugs, been alcoholic, had a nervous breakdown and have never had a happy relationship with a man or woman. Yet, I have managed to hide the real reason from everyone. I felt that I must be a complete freak of nature for this abuse to have happened to me. No one is sexually abused by a woman. I must be crazy. (Elliott, 1993, p. 1)

In addition, if survivors of female-perpetrated CSA do speak out, they may not be believed, even by their own therapists. In fact, one study showed that 78% of the victims of female-perpetrated CSA are not believed when they disclose their abuse experiences (Elliott, 1993). To receive help, many of these victims, both male and female, fabricate their abuse experiences and say that the abuser was male (Longdon, 1993). Therefore, the consequences of the CSA in and of itself are compounded by the fact that no one takes their abuse experiences seriously (Longdon, 1993). Consider how these victims' experiences influence our current estimated rates of female-perpetrated CSA. In what way would our estimates be affected? Finally, some survivors report that female-perpetrated CSA is worse than male-perpetrated CSA because it goes against the stereotype of what a mother is supposed to be. Survivors of female-perpetrated CSA experience the same possible consequences as discussed in the previous section of this chapter. Although there are no systematic studies, it has been documented that victims of female-perpetrated CSA have suffered from substance and alcohol abuse, suicide attempts, gender identity problems, difficulties in maintaining relationships, unresolved anger, shame, guilt, self-mutilation, eating disorders, depression, and agoraphobia (Elliott, 1993).

Why would females abuse children? Although systematic studies are rare, a few studies give some indication as to the dynamics of female-perpetrated CSA. First, female perpetrators are much more likely than male perpetrators to act in concert with another person, and many times the female is acting in order to please her "partner in crime." However, many females do act alone. Females tend to use violence less frequently than male perpetrators and instead resort to coercion. Females are more likely than males to know their victims, and they tend to have fewer victims than males. It appears that, like male perpetrators, females abuse girls more often than they abuse boys. However, this last finding may be merely a result of reporting bias—that is, because of certain social mores, males are less likely to report any type of abusive experience, and this tendency may be compounded when the abuser is female (Jennings, 1993).

There is an association between a past history of sexual abuse and females' current offending and between substance abuse and current offending. Moreover, female sexual abusers seem to be particularly dependent on men's attention for their self-esteem and their survival (Jennings, 1993). However, none of these traits fully describes every female sexual abuser, and some researchers have attempted to create typologies of abusers. One particularly interesting typology was created by Matthews (1993) from her experiences as a therapist to female sexual abusers. She postulates that there are three types of female sexual abusers: (1) *the teacher/lover offender,* who sees her victim as her partner and believes that the experience is a positive one for both of them; this type of abuser tends to be psychologically a child herself but responds very well to therapy; (2) *the predisposed offender,* who generally has a history of being sexually abused herself; these women act alone in their abuse and generally abuse family members; they have very deviant sexual fantasies and have problems with self-destructive and suicidal behaviors because they hate themselves and believe they were born evil; these women are often very hard to treat; and (3) *the male-coerced offender,* who is coerced into abusing children by her male partner(s); these women are often passive and powerless in their relationships, and fear violence and abandonment if they do not participate in the abuse; they do not feel loved or lovable and stay in these relationships in order to avoid being alone; many times and for whatever reason, they will initiate the sexual abuse of children themselves, though. Through her work with treating female sexual abusers, Matthews (1993) has found that the best interventions occur when long-term treatment is combined with short-term incarceration.

Prevention and Intervention

As you may have discerned from this discussion of CSA, many of the victims and perpetrators appear in great need of some sort of therapeutic intervention. Some workers in the field (e.g., Cohen & Mannarino, 2000) believe that the entire family of incest victims should undergo therapy. Most programs for intervention combine several, if not all, of the following elements: group therapy for the perpetrator; group therapy for the spouse of the perpetrator; group therapy for the child victim; dyadic therapy for the nonabusing parent and the victim; individual therapy for the victim; and eventual family therapy for the perpetrator, victim, nonabusing spouse, and any siblings. However, if the perpetrator is unwilling to admit to the abuse and/or if the nonabusing parent is unwilling to adequately protect the children from the perpetrator, families will not qualify for these programs. Unfortunately, this is the problem for the majority of sexual abuse cases (Cohen & Mannarino, 2000).

The most empirically validated individual treatment program for children who are victims of CSA is abuse-specific cognitive behavioral therapy, which involves psychoeducation (educating the child about sexual abuse and the therapy procedure); anxiety management (how to use relaxation and cognitive and emotional strategies to reduce anxiety-provoking thoughts about the abuse); exposure (talking, drawing, and writing about the abuse experience to reduce negative emotions and avoidance); and cognitive therapy (challenging cognitive distortions about the event and the consequent negative thoughts about the self and others) (Cohen, Berliner, & Mannarino, 2000).

Although treating the victims of CSA is necessary, a more effective approach to ending the suffering associated with it would be to prevent it from happening in the first place. All states, as mentioned previously, have laws against CSA; however, these laws are obviously not enough to deter potential perpetrators. The most widely used, and most controversial, CSA prevention programs in this country are programs that target children. These programs all tend to have the following goals: "educating children about what sexual abuse is; broadening their awareness of possible abusers to include people they know and like; teaching that each child has the right to control access to his or her body; describing a variety of 'touches' that a child can experience; stressing actions that a child can take in a potentially abusive situation, such as saying no or running away; teaching that some secrets should not be kept and that a child is never at fault for sexual abuse; and stressing that the child should tell a trusted adult if touched in an inappropriate manner until something is done to protect the child" (Reppucci & Haugaard, 1993, p. 312).

The controversy surrounding these programs stems from the fact that they were modeled on sexual assault programs for women, which sought to empower women by educating them about sexual assault and how to protect themselves in the event of an assault (Berrick & Gilbert, 1991). Therefore, these programs try to empower elementary school students by educating them about what sexual abuse is and ways in which they can ward off an assault (Repucci & Haugaard, 1993). The problem is this: Are all children developmentally mature enough to understand the distinctions of "good" versus "bad" touches and by whom and when can they be touched in private places? To give an example, can a four-year-old understand the following: "A good touch is when someone close to you gives you a hug. A bad touch is when someone touches you in a private place. Someone close to you can touch you in a private place and it may not be bad, such as when they are touching you there to clean you. However, someone close to you can touch you in a private place and it can be bad even if it is in the context of bathing."

If we find that some children, particularly older ones, are capable of understanding the intricacies of sexual abuse, as some (but not all) have been shown to be, should we then also be giving them the responsibility of protecting themselves? This seems to be the major question here. What are some of the implications of placing this responsibility on children? Is it possible that if they do subsequently become the victims of sexual abuse, the normal self-blame reaction that tends to follow an abusive experience may be compounded by their perception that they were supposed to have protected themselves? Can you think of ways to improve programs designed to help children protect themselves without giving them developmentally inappropriate responsibilities?

In response to such criticisms about typical prevention programs, Plummer (1993) countered that such programs have had some positive influence on the prevention of CSA. According to Plummer, millions of people, including adults, have become educated as to the extent and nature of the CSA problem, and children are now more willing to come forth and disclose sexual abuse that occurred to them. These certainly are positive results but do not resolve the problem of putting the responsibility of protection solely on the child. As Finkelhor and Strapko (1992) recommended, we should call these programs *disclosure programs* instead of *prevention programs*. Furthermore, as both sides (Plummer, 1993; Reppucci & Haugaard, 1993) emphasize, parents should be brought into these programs as the primary protectors of their children; unfortunately, as research has shown, parents do not attend such programs, nor do they learn very much about the prevalence, indicators, or appropriate responses to CSA when they do attend (Berrick, 1988).

Summary

Child sexual abuse by caregivers is a major problem in this country; however, its exact rate is difficult to ascertain, as the research is plagued with controversy as to which acts should be considered abusive. Our best estimates suggest that as many as one fourth of women and one fifth of men may have experienced sexual abuse as a child. Predictors of CSA come at most levels of the ecological model, but it is factors at the individual/developmental level that may be most important in figuring out why adults would find sexual gratification in children, particularly children who are related to them. CSA can have devastating consequences to its victims, both in the short and long term, and these victims (as well as their perpetrators and their families) are in need of psychological help. Most of the research and resources have focused on father-daughter incest; however, current research shows that mothers can sexually abuse their children, too, at a rate that is probably much higher than we expect. Future research should be directed towards not only males-as-perpetrators, but females-as-perpetrators as well. Consider again the case of Scott that opened this chapter. He was sexually abused by many members of his family, including both his mother and father. Do you think that each of his parents had different reasons for sexually abusing him? What were those reasons? What were the consequences for Scott?

6

Child Neglect and
Psychological Maltreatment

In October of 2003, Raymond and Vanessa Jackson were arrested for starving four adopted children in their home in New Jersey. The Jacksons had seven children altogether, but only the four boys, ages 9 to 19, were on the brink of starving to death. The case came to the attention of authorities when a neighbor found the eldest boy rummaging in their garbage at 2 a.m. looking for food. It was later revealed that the boys had been locked out of the kitchen and survived on a diet of uncooked pancake batter, peanut butter, and dry cereal. The boys said they sometimes gnawed on wallboard and insulation to stave off hunger. None of the boys weighed more than 50 pounds. Neighbors assumed the oldest boy, who was only four feet tall, was 10 years old, not 19. When the children were taken into state custody, the three youngest were put into foster care, and the oldest was hospitalized in a cardiac unit. What is particularly alarming is that the Jacksons had been visited 38 times over the previous two years by social workers for the state. How did these workers not notice the starving children? Did they see the children when they came to the house? Did they not ask for the children's medical records? How did these children go unnoticed by a state that had this family under surveillance? These questions were among those posed by reporters, legislators, and child advocates in the weeks following this horrific discovery. (McAlpin, 2003; Usborne, 2003)

This notorious case received widespread attention as an example of the failure of child protective services (CPS) to adequately protect children

133

from the horrendous behaviors that parents can sometimes perpetrate on their children. Only one year prior to this case, New Jersey's CPS came under fire because of their inadequate response to a family in which a seven-year-old boy, dead from starvation, was found in a plastic box in the basement, with his twin and four-year-old half brother alive, but emaciated, in an adjoining room. As a result of that discovery, New Jersey spent millions of dollars revamping their CPS system, only to have the Jackson story emerge one year later. New Jersey is not the only state to have failures come to light. The question that must be asked is why the system is unable to protect children from such terrible neglect. Is it possible that neglect is particularly difficult to identify, especially when it occurs behind closed doors? What warning signs might there be that a child is being neglected?

Like neglect, psychological maltreatment can be extremely difficult to detect. In Chapters 4 and 5, we discussed definitional problems that have plagued the literature on physical and sexual abuse; such definitional problems are even more pronounced for the two types of parental maltreatment discussed in this chapter. Consequently, arriving at prevalence statistics for neglect and psychological maltreatment has been difficult, and results of studies are sometimes widely divergent. Despite these difficulties, research on the predictors and consequences of these behaviors, as well as on prevention/intervention efforts, has proliferated.

Neglect

Scope of the Problem

Only in relatively recent decades have the profound negative consequences of neglect led to its public recognition as a severe form of child maltreatment. But what exactly is neglect? How should it be defined? Do we need to see immediate psychological damage to the child in order for his/her parents' behaviors to be considered neglectful? What if the psychological consequences do not appear until later in life, when a person is cognitively able to realize the neglectful practices of his/her parents? Should intentions be taken into consideration? Does it matter whether a child is neglected because stressors in the parents' lives made them unable to provide proper care for the child? Should the omission of adequate care be considered neglect only if parents deliberately—perhaps through actual hostility—ignore the care of their child?

Neglect is commonly viewed as a failure to meet minimal community standards of care; unlike physical and sexual abuse, neglect refers to acts of omission rather than of commission. Several different forms of neglect have been identified (e.g., Erickson & Egeland, 2002). *Physical neglect*, the most obvious and well-known form of neglect, occurs when parents fail to provide for the basic physical needs of their child, such as feeding, bathing, and providing shelter, or when they fail to protect the child from harm and danger. *Emotional neglect* occurs when parents fail to attend to their child's basic emotional needs. Examples are not picking up or attending to a crying infant and not comforting children when they are hurt; however, there is considerable controversy over exactly what a child's basic emotional needs are, and cultures vary in their views on emotional neglect. For instance, Korbin (1980) pointed out that the Western middle class practice of giving infants and young children their own rooms to sleep in at night would be considered emotionally neglectful by the standards of other cultures. Despite these cultural issues, it appears that in its more severe forms emotional neglect can lead to a medical condition called nonorganic failure to thrive (i.e., retardation in the rate of growth).

Medical neglect is failure to address the basic medical needs of the child, such as routine checkups, immunizations, recommended surgery, and prescribed medicine. As noted in Chapter 3, this type of neglect has been at the root of considerable controversy in regard to several religious communities in the U.S. *Mental health neglect* is the failure to provide mental health treatment to a child who obviously suffers from serious emotional and/or behavioral disorders; this type of neglect is not widely acknowledged as a form of maltreatment. Finally, *educational neglect* occurs when parents do not comply with state ordinances for the attendance of their children in school. Often, professionals also consider a parent's failure to cooperate or become involved in the child's education to be a form of educational neglect (Erickson & Egeland, 2002). Overall, at the core of conceptions of neglect is the assumption that parents are responsible for protecting their children from known hazards. As more hazards become known, the definition of neglect and its different forms will continue to evolve (Garbarino & Collins, 1999).

One of the basic problems facing scholars concerned with neglect is that it is frequently not investigated or analyzed as a construct independent of physical abuse—probably because legally they are both forms of maltreatment (Erickson & Egeland, 2002). Consequently, problems that may be due specifically to neglect often are ignored, perhaps because our culture is more preoccupied with violence and acts of commission than it is with things that cannot be so readily seen (acts of omission). Severely beaten children attract more attention because the wounds are evident, but the emotional scars that neglected children suffer from may be more insidious and traumatic (Garbarino & Collins, 1999).

It is ironic that physical abuse grabs so much of the public's attention because official statistics indicate that neglect is much more common than physical abuse. According to the U.S. Department of Health and Human Services (DHHS, 2001), neglect is by far the most common type of child maltreatment—59.2% of all maltreatment cases substantiated by CPS in 2001 were cases of neglect, and 7.1/1,000 children in the U.S. were victims of neglect. Moreover, even though CPS workers should know neglect is the most prevalent form of maltreatment, neglected children are 44% more likely than physically abused children to experience a recurrence of their maltreatment. CPS workers seem more concerned with protecting physically abused than neglected children, a concern that, according to research on the relative consequences of each type of maltreatment, seems unfounded.

Data from the Third National Incidence Study (NIS-3; Sedlak & Broadhurst, 1996; see Chapter 4) reinforce the findings from the DHHS—neglect is the most prevalent type of maltreatment in this country. However, as opposed to the fluctuations in the rate of neglect reported by the DHHS between 1997 and 2001, NIS-3 showed a significant increase in the incidence of neglect between 1986 and 1993. Perhaps this increase was due to the fact that neglect finally gained notice in the late 1980s and professionals were able to recognize it better. Whatever the cause, under the Harm Standard, the number of neglected children rose between 1986 and 1993 from 474,800 to 879,000; under the Endangerment Standard, the numbers rose from 917,200 to 1,961,300, an increase of 114%. Physical neglect was the most often reported type of neglect (338,900 in 1993 under the Harm Standard; 1,335,100 under the Endangerment Standard), but emotional neglect was also a significant problem (212,800 in 1999 under the Harm Standard; 585,100 under the Endangerment Standard) (Sedlak & Broadhurst, 1996).

Despite the great wealth and resources of the United States, neglect is a grave problem for children in this country. Moreover, even though it is commonly thought that children are at greater risk for fatal physical abuse, children are more likely to die from neglect than from any other form of maltreatment. According to the DHHS (2001), in 2001, 35.6% of maltreatment deaths were due to neglect, and an additional 21.9% of fatalities were due to neglect combined with physical abuse (DHHS, 2001). Moreover, the rates of fatalities due to neglect may be even higher than these statistics show because they are more difficult to investigate than fatalities due to physical abuse. There is also considerable controversy over how to define and assess fatalities due to neglect (Bonner, Crow, & Logue, 1999; Lung & Daro, 1996). Take the examples in Box 6.1. Which of the fatalities would you attribute to neglect?

Box 6.1   Possible Cases of Fatal Child Neglect

1. A 4-month old infant died from massive trauma after being ejected from the front seat of a car in a traffic collision. The child was not secured in an infant car seat (p. 156).

2. A 6-year-old boy fatally shot his 4-year-old brother with a handgun he found in the pocket of a coat hanging in the closet. The parents were in the next room watching television (p. 156).

3. A mother was giving her 2-year-old son, Bob, a bath. She left him in the tub and went to the kitchen for 10 minutes. She then sent her 8-year-old in to check on Bob, who was lying face-down in the tub. Bob was taken to the hospital, where he was unresponsive with a temperature of 90°F. After treatment, Bob began breathing again, and he was admitted. After 9 days in the pediatric intensive care unit, Bob had minimal brain functioning and was unresponsive to any stimuli. He died two days later. The cause of death was deemed to be pneumonia with anoxic brain injury after a near-drowning. The CPS worker told the mother that 10 minutes was too long to leave a 2-year-old unattended in a tub, and the mother replied that she did it all the time.

4. A 7-month-old infant named Roger was brought to a local ER by his parents because of respiratory problems. The police were called to the ER because of a fight between the parents, and the parents left before the infant saw a doctor. Roger was then brought to the ER the next day by his grandfather and had to be hospitalized because of severe respiratory distress and malnutrition. Roger was the size of a newborn. A CPS investigation found that both parents abused drugs and alcohol and that Roger's mother had used crack, marijuana, and alcohol while she was pregnant with Roger. There were also outstanding warrants for both parents' arrests, and both had been in prison at least once previously. Roger died five days after admission because of multiple organ dysfunction and pneumonia.

SOURCE: Bonner, Crow, & Logue, 1999.

As these examples indicate, the determination of "fatality by neglect" may be in the eye of the beholder. Many such cases may not be determined to be neglect-related; even if neglect is seen as the cause, the determination would not necessarily be noted on the death certificate (McClain et al., 1993). For all these reasons, neglect fatality rates should be assumed to be much higher than the reported figures indicate. Furthermore, although fatalities can result from a failure to provide supervision, protection, medical care, or nutrition (Bonner, Crow, & Logue, 1999), where exactly do we draw the line between a momentary lapse in supervision and chronic neglect? And, does this distinction necessarily make a difference when determining whether neglect was involved in a child fatality? Are any of the neglectful acts in Box 6.1 more egregious than the others? Can any be associated with merely a lapse in supervision, or do they all appear to reflect chronic neglect? Is there any sure way to tell the difference between momentary inattention and chronic neglect in any or all of these cases?

Special Issue: Prenatal Exposure to Alcohol and Drugs

Consider the last case example presented in Box 6.1, that of the mother who used drugs while she was pregnant. In recent years, prenatal drug exposure has been deemed a form of neglect because these mothers often neglect the well-being of their fetuses through lack of prenatal care and through their continual use of drugs during infancy. The most recent statistics from the National Institute on Drug Abuse (1994) show that as many as 221,000 infants are born each year exposed to illicit drugs prenatally, whereas the number of infants born prenatally exposed to alcohol far exceeds this estimate and cannot readily be determined. The effects on infants of prenatal exposure to substances may vary depending upon the substance used; many prenatally exposed infants are exposed to many different drugs (Frank et al., 1988). Following is a brief discussion of the prenatal effects of two of the most researched substances.

Cocaine. Much research has been devoted to the effects of prenatal exposure to cocaine, especially since the rise of cocaine and crack use in the 1980s. Preliminary research showed that cocaine-exposed infants had lower birth weights, were more likely to be born premature, and were more likely to experience intrauterine growth retardation and microencephaly. There was also evidence that these babies were irritable; hypersensitive; experienced movement disorders, stiffness, abnormal sleep and wake cycles, and fine motor deficits; and had disorganized states and an abnormal cry. In the long term, these children seemed to have problems with below-average intellectual functioning, attention deficits, decreased affective displays, hyperactivity, aggressiveness, anxiety, depression, and withdrawal (Harden, 1998).

However, there is controversy over the extent to which these findings directly reflect prenatal exposure to cocaine. After all, mothers who use cocaine during pregnancy often use other substances, have poor prenatal care, and are malnourished. After their children are born, they tend to raise them in chaotic and nonsupportive environments. Thus, the sequelae observed in cocaine-exposed infants may reflect poor pre- and postnatal environments (Richardson, Day, & McGauhey, 1993). A recent review of the literature showed that, after controlling for confounders, there were no consistent, negative effects of prenatal exposure to cocaine on physical growth, developmental test scores, or language development (Frank, Augustyn, Knight, Pell, & Zuckerman, 2001). However, a recent, rigorous study of cocaine-exposed infants showed that they seem to have significant cognitive deficits and developmental delays at age two, even after controlling for possible confounds (Singer et al., 2002).

Alcohol. Of all drugs, it is the legal drug, alcohol, that has been shown to have the direst consequences for prenatally exposed infants. Moreover, it is the only substance of abuse that has been shown to have irreversible negative effects, such as mental retardation, neurological deficits, facial malformations, and growth retardation (Kelley, 2002). The most serious effects of prenatal alcohol exposure fall under the label fetal alcohol syndrome (FAS). Approximately 12,000 infants each year are born with FAS (Kelley, 2002). This syndrome is characterized by pre- and postnatal growth retardation, microencephaly, abnormal facial features, mental retardation, and other possible behavioral problems, such as hyperactivity and poor motor coordination, which are indicative of central nervous system dysfunction (Day & Richardson, 1994). Less serious effects from alcohol exposure fall under one of two categories: alcohol-related birth defects (ARBD) and alcohol-related neurodevelopmental disorder (ARND). The extent to which infants and children suffer from each of these is unknown because the criteria for each are not always clear (May & Gossage, 2001). Children with ARBD have alcohol-related physical abnormalities of the skeleton and some organ systems, without the facial abnormalities. Children with ARND exhibit retardation or learning difficulties and other behavioral problems (May & Gossage, 2001).

Studies of the effects of prenatal exposure to alcohol have consistently shown that the worst outcomes are for children whose mothers were binge drinkers. In one recent study, infants whose mothers consumed at least five drinks per occasion at least once per week experienced deficits in birth weight, psychomotor development, play skills, and processing speed (Jacobson, Jacobson, Sokol, & Ager, 1998). These deficits were not observed in infants whose mothers also drank during pregnancy, but not in a bingelike fashion. In a longitudinal study of the short- and long-term effects of binge drinking during the month prior to pregnancy recognition, it was found that at age 7½, this type of binge drinking was the best predictor of the children's deficits in attention, memory, cognitive processing, and problem-solving skills. These children were also more likely to have learning problems, be below average academically, and be hyperactive and impulsive. At age 11, they had problems with distractibility, restlessness, and lack of persistence. Finally, prenatal exposure to binge drinking was the best predictor of these children's problems with response inhibition and attention span at age 14. In addition, the more drinks the mother drank during her binge in that first month of pregnancy, the worse the outcome for her child in terms of math and reading skills (Streissguth et al., 1990, 1994a, 1994b).

Child Welfare Concerns for Prenatally Exposed Infants. Many states in this country have laws that label prenatal exposure to drugs as child abuse (Chasnoff & Lowder, 1999). From a child welfare perspective, it is noteworthy that the environment into which these infants are born is fraught with risks for their development. The behavioral problems of drug-exposed infants often lead to them being "hard to parent"; that is, these infants exhibit irritability, frequent crying, poor feeding patterns, poor consolability, frequent startles, and frantic sucking (Schutter & Brinker, 1992). This situation would be difficult to handle for even the best of parents, as these children tend to elicit parenting reactions that are punitive, hostile, or ignoring (Cuskey & Wathey, 1982). However, when the mother is a substance abuser, the situation becomes even worse. Substance abusing mothers tend to be rigid and overcontrolling of their children. They display little emotional involvement or responsiveness, and they receive little pleasure from interacting with their children (Burns, Chethik, Burns, & Clark, 1991). They also tend to have low self-esteem and few positive social supports or parenting models (Cuskey & Wathey, 1982). It should come as no surprise, then, that these children are significantly more likely to be abused and neglected than demographically matched controls (Jaudes, Ekwo, & Voorhis, 1995; Kelley, 1992). In one study, as many as 60% of drug-exposed children were subject to subsequent substantiated reports of child maltreatment (Kelley, 1992).

Given the risks for maltreatment to these children, it is not surprising that a significant portion are removed from their homes. In one study, as many as 42% of cocaine-exposed children had been placed in foster care (Kelley, 1992). Moreover, these children do not fare well in the foster care system: Children prenatally exposed to drugs are more likely to require multiple foster care placements than other children, and they tend to remain in foster care for longer periods of time (Curtis & McCullough, 1993).

The inclusion of prenatal drug exposure as a form of neglect has been considered a contributor to the relatively high rates of reported neglect in this country (Green, 2000). What do you think? Should prenatal drug exposure be considered a form of neglect? If so, in what way is it a form of neglect? If it is not neglect, then what type of maltreatment should we consider it to be? Should we consider it a type of maltreatment at all? Should mothers who abuse drugs during pregnancy have their children taken away at birth? If not, do you think there are other steps that can/should be taken to ensure the well-being of the infant?

Predictors and Correlates

Macrosystem

The Third National Incidence Study and other studies on child neglect (e.g., Hamburg, 1992) consistently show that poverty is the largest contributor to neglect in this country. Specifically, NIS-3 found that compared to families who made over $30,000 per year, families with an annual income of less than $15,000 were 44 times more likely to be identified as neglectful, with children in these families identified as 56 times more likely to experience educational neglect, 40–48 times more likely to experience physical neglect, and 27–29 times more likely to experience emotional neglect (Sedlak & Broadhurst, 1996). In addition, NIS-3 showed that of all types of maltreatment, neglect was the most clearly associated with poverty. Why might there be such a strong relationship between income and neglect? Could it be that children who are in poor families are also more likely to be reported for neglect because there is a bias that poor families are inherently more neglectful of their children? Can you think of any other reasons for such a strong association? Recall that poverty is associated with unemployment, limited education, social isolation, large families with many children, and teenage pregnancy. What role might these factors play in neglect?

The associations between neglect and poverty are so strong that some researchers (e.g., Hamburg, 1992) have argued that we should coin another term: "societal" or "collective" neglect. Researchers also argue that eliminating poverty is essential to eliminating child neglect (e.g., Crittenden, 1999). As Hamburg (1992) noted, the United States' inability and/or unwillingness to provide adequate health care, child care, education, and policies to help all families and children in this country contributes to the epidemic of child neglect. Indeed, even in the research domain, this country suffers from "neglect of neglect" (Dubowitz, 1999). However, it is also important to note that eliminating poverty may not completely eliminate child neglect, as wealthy families are also known to neglect their children, particularly emotionally (Crittenden, 1999). Thus, having an abundance of resources is not sufficient to ameliorate the inabilities of some parents to appropriately address their children's needs.

Exosystem

Although poverty imposes enormous stresses on families, not all poor families neglect their children. As a matter of fact, the majority of poor families do not neglect their children; thus, poverty is neither a necessary nor sufficient cause of neglect. The majority of families throughout human history have been poor, and by historical and cross-cultural standards, most "poor" U.S. families would not be considered impoverished. Therefore, a lack of material goods cannot be reason enough for parents to neglect their child (Crittenden, 1999). As Crittenden (1999) suggested, to understand the nature of neglect more thoroughly, we must compare those poor families who do not neglect their children to poor families who are neglectful.

In Crittenden's (1999) view, there must be a third variable that causes both poverty and neglect in poor families that neglect their children. She postulated that the problems of being unemployed, unmarried, and socially isolated may be the key variables that separate poor neglectful families from poor nonneglectful families. According to her formulation, most parents throughout human history had to work for their resources, had to stay committed to each other for the family's survival, and had to live and raise their children in communities where other families were doing the same. If many of today's poor, neglecting families are unable to do what humans throughout history have done, Crittenden argued that they must suffer from an inability to form and maintain enduring, successful, and productive human relationships.

There is evidence to support Crittenden's conceptualization of the neglectful poor. First, neglectful families tend to relocate a lot, which contributes to their sense of social isolation (Polansky, 1979). Compared to nonneglectful mothers, neglectful mothers often have no social networks (Dubowitz, 1999) or feel surrounded by unsupportive people (Polansky et al., 1985). Neglectful mothers may also have poorer relationships with their own mothers than nonneglectful mothers (Coohey, 1995). Even when neglectful mothers maintain ties with their mothers and rely on them for help with child care and money, they neither perceive them as a source of emotional support nor give much support in return. Thus, these relationships are characterized by a low level of exchange because the neglectful mothers perceive their mothers in a negative light (Coohey, 1995).

In addition to perceiving little support from their own mothers, neglectful mothers seem to have poor relationships with their extended families (Giovannoni & Billingsley, 1970), and their interpersonal relationships may be characterized by pervasive feelings of futility, emotional numbness, clinginess, and loneliness (Polansky et al., 1981). Also, neglectful parents are more likely to be single parents, have a poor quality marriage (e.g., Brown et al.,

1998), not live with their partners, and have shorter relationships with their partners (Coohey, 1995). In addition to these problems, or perhaps due to them, several researchers have found that neglectful families tend to be the poorest of the poor (e.g., Bath & Haapala, 1993), and often suffer from unemployment (Dubowitz, 1999).

Microsystem

Characteristics of the Child. What kinds of characteristics might children have that could contribute to poor relations with their parents and thereby lead to neglect? It appears that boys (Sedlak & Broadhurst, 1996), irritable or fussy children (e.g. Thomas & Chess, 1977), and disabled or premature children (e.g., Belsky, 1980) are more likely to suffer from neglect than other children—though not all investigators believe these links have been adequately established (Biringen & Robinson, 1991). Brachfield, Goldberg, and Sloman (1980) assert that ultimately it is the parents' responsibility to respond to their children in a sensitive manner, regardless of any difficulties the children may have. What is your view of this issue? Can you think of child characteristics that may make child care difficult? Should parents who do not cope well with those characteristics, and thereby neglect their children, be reported for child maltreatment? If so, what would be the best intervention methods for such cases?

Characteristics of the Family. NIS-3 identified characteristics of the family that may be risk factors for child neglect. Children from single-parent families had a 165% greater risk of being physically neglected, a 64% greater risk of being emotionally neglected, and a 220% greater risk of being educationally neglected than children from two-parent homes. Educational and physical neglect were more likely to be identified in families with four or more children, followed by only-child families, with two- to three-child families least likely to be categorized as suffering from educational or physical neglect (Sedlak & Broadhurst, 1996). Why do you think some of these associations exist? Why might children from single-parent families and large families be more likely to be reported for neglect than their counterparts? Why might an only child be more likely to suffer from educational or physical neglect than a child from a family with two or three children?

Individual/Developmental

Neglectful mothers tend to be poorly educated, young, angry, dissatisfied, and hostile, and to have an external locus of control and low self-esteem

(Brown et al., 1998). They are more likely to suffer from some sort of psychopathology, including depression (Lahey et al., 1984), poor impulse control especially under stress (Altemeier et al., 1982), and even sociopathy (Brown et al., 1998). Neglectful fathers are more likely to show little or no warmth and to suffer from psychopathology or sociopathy (Brown et al., 1998). Neglectful parents are also more likely to abuse substances (Pelton, 1994).

In addition to problems affecting their ability to respond to their children, many neglectful parents have a history of abuse or neglect in their own child-hood. Pianta et al. (1989) suggested that neglectful mothers are trying to resolve their own issues of trust, dependency, and autonomy—issues that are related to their history of abuse and neglect. They also argued that rather than trying to meet their children's needs, these mothers are seeking to satisfy their own needs through their relationship with their children, and unrealistic expectations of their children probably led to an unhappy parent-child relationship. Of all types of maltreating mothers, neglectful mothers seem to have the most negative interactions with their children (Burgess & Conger, 1978). They also tend to have poor coping mechanisms for dealing with problems with their children. For example, instead of confronting their child (as abusive parents do, although their confrontational strategies are also not constructive), neglectful parents often do not demand anything from their children and actually avoid them during stressful events (LaRose & Wolfe, 1987). Perhaps because no one is psychologically available to these mothers, they cannot be psychologically available to their children (Garbarino & Collins, 1999).

Consequences

There is some evidence that neglect, in comparison to other types of mal-treatment, can have the most pervasive and negative outcomes. Neglected children are the most at risk for fatalities, and emotional neglect can lead to nonorganic failure to thrive (Drotar et al., 1990), which can result in death or pervasive physical and psychological problems. Several investigators have come to the conclusion that neglected children, in comparison to other mal-treated children, also suffer the worst effects socially, emotionally, and aca-demically. For example, one longitudinal study showed that at age 24 months, neglected children, in comparison to physically abused and control children, had the least enthusiasm in problem-solving tasks and were the most angry, frustrated, and noncompliant children (Egeland, Sroufe, & Erickson, 1983). At 42 months, they had the worst impulse control, the least flexibility and creativity, the lowest self-esteem, the most incompetence on school tasks, and the most difficulty coping. They were also the most withdrawn, dependent, and angry of all the groups of children.

Other problems have been identified in neglected children. For example, they tend to have poor expressive and receptive language skills, a low overall IQ, cognitive dysfunctions, and academic difficulties (Cahill, Kaminer, & Johnson, 1999; Culp et al., 1991; Gowan, 1993). In social interactions, they tend to be passive, withdrawn, and unaffectionate; they tend to initiate inter-actions less frequently than other children (Crittenden, 1992) and have fewer positive play interactions with other children (Lewis & Schaffer, 1981)—possibly because they tend to display mostly internalizing, withdrawn, and aggressive behaviors (Erickson et al., 1989; Manly et al., 2001). Their aggres-sive tendencies also emerge with their neglectful mothers, with whom they tend to be physically and verbally aggressive (Bousha & Twentyman, 1984).

The low IQ and poor academic performance characteristic of these children (Eckenrode et al., 1993) can extend into adulthood. At age 28, even after controlling for several possible confounds, adults who had a childhood history of neglect still had a lower IQ and poorer reading abilities (Perez & Widom, 1994). The problems of social incompetence, depression, withdrawal, and other behavior problems also seem to continue throughout neglected children's childhoods (Aber et al., 1989; Herrenkohl et al., 1991).

When neglected children become adolescents, they have a higher likeli-hood than controls of being delinquent and engaging in violent crime (Widom, 1989; Zingraff et al., 1994). Neglected children are also at risk for running away from home and becoming prostitutes (Kaufman & Widom, 1999; Widom & Kuhns, 1996), and for developing personality disorders and psychological problems such as depression, anxiety, self-mutilation, and suicidal ideation (Johnson et al., 1999; Johnson et al., 2000; Lipschitz et al., 1999). As adults, individuals with histories of neglect may neglect their own children; several researchers have found that some neglectful parents report a history of neglect in their own childhoods and many report a history of feeling unwanted or unloved as children (Gaudin et al., 1996; Polansky et al., 1981).

Consequences of Neglect in Combination with Other Forms of Maltreatment

Although these studies show that neglected children are at risk for many developmental problems, the results are subject to debate on a number of methodological grounds. For one, many investigators fail to distinguish between the various types of neglect. Second, because there is no consensus in the literature on the definition of neglect, most investigators study only neglected children whose neglect was substantiated by CPS. The problem here is that because CPS is an overburdened institution, only the most severe cases get substantiated (Gaudin, 1999), and it may be only these very severe cases that show the most negative outcomes. Third, as with other types of maltreatment, the outcomes of neglect depend upon a variety of mediating factors. Several factors that may impact the outcome, particularly poverty, were discussed in the previous section on predictors and correlates of neglect. The more severe and more frequent these factors, the more likely the child will have a poor outcome. Other factors, such as the age of the child and the chronicity and severity of the neglect, have also been found to influence outcomes. If the neglect begins when the child is an infant, the child is more likely to become developmentally disabled. In one study, children who were neglected as infants, in comparison with those whose neglect began in early childhood, performed less well cognitively; were less confident, assertive, and creative; and had poorer self-esteem and emotional health (Erickson et al., 1989).

The combination of neglect with other forms of maltreatment also seems to contribute to negative outcomes. As mentioned in Chapter 4, most forms of child maltreatment do not occur in isolation, and the consequences of experiencing multiple forms may increase the risk for poor outcomes. For instance, in comparison to either type of maltreatment alone, neglect in combination with sexual abuse has been associated with worse reading and math skills, more suspensions from school, and increased delinquency (Eckenrode et al., 1993; Gaudin et al., 1996). Neglect in combination with physical abuse appears to be related to poorer English skills (Eckenrode et al., 1993). On the other hand, one study found that children who experienced only neglect, as compared with children who experienced both neglect and physical abuse, had worse expressive and receptive language skills—that is, the combination of maltreatment types actually led to a *better* outcome in the language domain (Culp et al, 1991). Why might this be the case? Do you think it is because children whose parents physically abuse them are at least getting some attention, even if it is negative, and that this aggressive attention is better than having no attention at all? Are there any alternative explanations you can think of?

Consequences of Neglected Children in Comparison to Other Abused Children

The Culp et al. (1991) study just mentioned had to deal with one of the problems discussed earlier in this chapter—that the legal system tends to lump all abused and neglected children together under the term "abused" or "maltreated." However, as can be seen from the discussion thus far, neglect and physical abuse are conceptually and experientially different acts with potentially different consequences. Some studies, such as those done as part of the Minnesota Mother-Child Project (e.g., Erickson et al., 1989), attempt to address these possible differences by comparing neglected children not only to non-maltreated controls but also to physically abused children. After all, neglect and abuse are two different phenomena—one involves acts of commission, the other involves acts of omission.

Several investigators who compared neglected and physically abused children have found clear differences in outcomes for these two types of maltreated children. For example, compared to physically abused children, neglected children reported more fear when judging how others would react to the moral transgression of an unfair distribution of resources; they also thought that the perpetrators of this transgression would feel less sadness (Smetana et al., 1999). There have been some findings that neglected children internalize moral values more than physically abused children; however, when interacting with their mothers, neglected children were more likely to express their anger, whereas physically abused children were more likely to inhibit it (Crittenden & DiLalla, 1988; Koenig, Cicchetti, & Rogosch, 2000). If these are reliable differences, how would you account for them?

In related research, physically abused and neglected children were equally likely to display anxious and disorganized attachment to their caregivers, but neglected children seemed to suffer more socially from this attachment style (Barnett, Ganiban, & Cicchetti, 1999); moreover, there is evidence that neglected children have poorer views of themselves, their mothers, and the social world than physically abused or other children (McCrone et al., 1994; Toth et al., 1997). Theoretically, these more insecure attachment styles may be contributing to the poor interpersonal skills found in studies of neglected children. That is, in comparison to physically abused and nonabused children, neglected children are more passive, withdrawn, avoidant, and isolated in their peer and other relationships (e.g., Erickson et al., 1989). In addition, although they are less aggressive, noncompliant, and uncooperative than physically abused children (Crittenden, 1992), they are more aggressive, noncompliant, and uncooperative than non-maltreated children (Erickson et al., 1989). Consequently, they tend to be unpopular with their peers (Erickson et al., 1989).

Neglected children also have more problems in their emotional development when compared to physically abused or nonabused children. They have more problems with regulating their own emotions; are less able to discriminate the emotions of others; are more hopeless in stressful situations; have poorer coping strategies; display less agency, ego control, self-esteem, positive affect, and sense of humor; and they display more dependency and negative affect (Egeland et al., 1983; Egeland & Sroufe, 1981; Erickson et al., 1989; Pollack et al., 2000).

Prevention and Intervention

There is no overarching social policy for the issue of child neglect. However, because neglect can have different causes and consequences than physical abuse, there is a need to develop prevention and intervention programs that are specific to neglect and informed by neglect research. This issue is becoming more pressing because, as the issue of neglect becomes better publicized, more cases are being reported, such as the one that introduced this chapter. Furthermore, recent social policy changes in this country could lead to a further increase in neglect cases. For example, welfare reform legislation enacted in 1996 could lead to more parents being burdened with the dual responsibilities of working at low wages and taking care of children without any affordable child care assistance available or without any government aid. This scenario could conceivably lead to more cases of neglect (Gelles, 1999).

Furthermore, current CPS policies regarding neglect are quite vague. For example, there are no clear-cut guidelines for what is considered less-than-optimal parenting and what is considered neglect (Gelles, 1999). Consequently, usually only the most extreme forms of neglect are addressed in intervention programs. When intervention programs are mandated for neglectful families, programs tailored for child neglect seem to be effective with fewer than 50% of the families who are mandated to use them, and the most effective ones are those that are long-term and very comprehensive (Gaudin, 1993; Holden & Nabors, 1999). It is difficult to develop effective programs when the individual/developmental problems that make it difficult for parents to connect with their children are the same problems that make it difficult for parents to connect with service providers (Erickson & Egeland, 2002).

Because of the specific problems that neglectful families face, several researchers have forwarded suggestions to improve the effectiveness of intervention programs. For example, DePanifilis (1999) has argued that such programs should be family-based, not individual-based; they should help to create an alliance between the family and agencies or people in their community who are sources of support; they should empower families to manage several co-occurring stressors in their lives effectively in order to solve their own problems without relying on the social service system; and they should work with the families' strengths, instead of trying to fix their weaknesses (DePanifilis, 1999). Because of evidence that the most effective intervention programs are tailored to the individual family's needs (Wekerle & Wolfe, 1993), DePanifilis (1999) suggested that successful neglect intervention programs should encompass some or all of the following, depending on the families' needs: (1) provision of concrete resources; (2) provision or mobilization of social support networks; (3) developmental remediation, such as therapeutic day care for the neglected child, peer groups at school that are geared towards developmentally appropriate tasks, public health visiting to the family with the goal of providing assistance with attachment needs of each family member, mentors for the parents to provide nurture, recreation, and role modeling, and/or individual assistance with parenting skills; (4) cognitive-behavioral therapy to provide instructions about basic child care, social skills training to handle child care tasks, stress management such as relaxation techniques or other coping mechanisms, and cognitive restructuring to address dysfunctional and self-defeating thoughts; (5) individual-focused therapy to help individual family members deal with other related problems, such as alcohol or drug abuse, depression, and developmental delays; and (6) family systems therapy to aid with family functioning, communication skills, roles, home management, and responsibilities. To what levels of the ecological framework are these goals directed? What will be needed at the macro- and exosystem levels to establish such programs?

Psychological Maltreatment

Scope of the Problem

According to the DHHS (2001), psychological maltreatment is the second least prevalent form of child maltreatment (medical neglect is the least prevalent), accounting for 6.8% of all maltreated children. Although the DHHS identified only 0.9/1,000, or 61,778 children as victimized by psychological maltreatment in 2001, the NIS-3 (Sedlak & Broadhurst, 1996) indicated that 532,200 children were psychologically victimized in 1993. Why the differing numbers? Why does one study based upon CPS reports find incredibly low estimates for psychological maltreatment, while another study based on cases identified by mandated reporters finds moderately high estimates? Historically, the concept of psychological maltreatment has had the most definitional and conceptual problems of all the child maltreatment types. Therefore, it has been very difficult for legal and medical services to provide accurate estimates or services for families suffering from this elusive phenomenon. To give an example, imagine what would happen if a child called the police because his parents called him a dirty name. Has he been psychologically maltreated? What is the likely legal response? What would your response be?

Society has traditionally subscribed to the belief that physical and sexual maltreatment are more harmful than psychological maltreatment (Brassard et al., 2000; Kashani & Allan, 1998). However, studies have shown that people who experienced both physical and emotional abuse at the hands of their caregivers consistently say that they were bothered more by the emotional trauma than the physical trauma (e.g., Jurich, 1990). Take, for example, the case in Box 6.2. Do you think the emotional scars that Cinderella suffered will heal as quickly as the physical scars suffered by the child in Box 4.1 of Chapter 4? What messages were conveyed to Cinderella? How easily can the effects of these messages be overcome?

Box 6.2   Cinderella: A Case of Psychological Maltreatment

The teenage half brother of Cinderella, an emotionally abused child, lived at home with her until she was four years old and recounted a series of maltreatments against Cinderella by her stepmother. Cinderella was never allowed to play outside, while the other kids in the household were. Cinderella was left in the car alone when her stepmother and the other kids went visiting or shopping. Her stepmother called her "stupid" and "ugly" on many occasions. She also yelled at Cinderella a lot and gave her a large share of the household chores. The stepmother spanked Cinderella many times, sometimes even with a spatula or a wooden spoon, and while the other children received little corporal punishment from the mother, Cinderella often had scratches or bruises all over her body.

One major source of conflict between Cinderella and her stepmother was over Cinderella's eating habits. If the stepmother did not feel that Cinderella was eating properly, she would forcefully stuff food into Cinderella's mouth or slap her. Many times Cinderella would also be sent to her room for this behavior and, on her way, the stepmother would kick her hard enough to hurt her. Cinderella was kept in her room so often that the bedroom smelled of urine. The other children were rarely sent to their rooms. The half brother recounted that Cinderella cried often, was always very sad, and never smiled. However, after Cinderella was removed from the home and the half brother visited her in her new home, he stated that for the first time Cinderella was happy, smiling, and affectionate—her entire appearance and demeanor had changed.

SOURCE: Binggeli, Hart, & Brassard, 2001.

Several researchers (e.g., Garbarino et al., 1986) have argued that psychological maltreatment is at the core of all other forms of child maltreatment, including physical and sexual abuse and neglect. According to Binggeli, Hart, and Brassard (2001): "a) psychological maltreatment is probably embedded in nearly all other acts of abuse and neglect as the psychological meaning of those acts, b) it appears to be the strongest predictor of the impact of child maltreatment, and c) psychological maltreatment may have the longest-lasting and strongest negative effects on survivors of child abuse and neglect" (p. 15). In addition, psychological maltreatment can occur on its own. Thus, as opposed to being one of the *least* prevalent forms of child maltreatment, as reported by DHHS, it is more likely *the most* prevalent.

In 1995, the American Professional Society on the Abuse of Children (APSAC) attempted to resolve some of the definitional crises plaguing the construct of psychological maltreatment by offering a basic definition and conceptualization. Two of the most widely accepted statements include:

Statement 1: " 'Psychological maltreatment' means a repeated pattern of caregiver behavior or extreme incident(s) that convey to children that they are worthless, flawed, unloved, unwanted, endangered, or only of value in meeting another's needs" (APSAC, 1995, p. 2).

Statement 2: "Psychological maltreatment includes (a) spurning, (b) terrorizing, (c) isolating, (d) exploiting/corrupting, (e) denying emotional responsiveness, and (f) mental health, medical, and educational neglect" (APSAC, 1995, p. 4).

Take another look at the case in Box 6.2. Does this case fit this conceptualization and definition of psychological maltreatment? Based on the six factors in Statement 2, what types of psychological maltreatment did Cinderella suffer from? Did the perpetrators convey the messages described in Statement 1? How?

Based upon the APSAC conceptualization, a better estimate of the incidence of psychological maltreatment may be gained by looking at studies that include measures for the factors in Statement 2. Several researchers have attempted to achieve a better estimate of the incidence of psychological maltreatment by either using adults' retrospective reports of behaviors reflecting the above definition of psychological maltreatment or asking parents the types of behaviors they use with their children. Based on adults' retrospective reports of psychological maltreatment, somewhere between 25%–37% of people experienced it in childhood (Buntain-Ricklefs et al., 1994; Gross & Keller, 1992; Moeller, Bachman, & Moeller, 1993). These percentages are likely to be an underestimate because the participants may not accurately recall childhood psychological maltreatment episodes.

Perhaps a more accurate estimate comes from Vissing et al.'s (1991) nationally representative survey of 3,458 parents. These parents were asked how many times in the previous year they had been verbally or symbolically aggressive with their children—for example, name-calling and swearing at or insulting their children, stomping out of the room, threatening to hit or throw something. More than 60% of these parents said they had used at least one type of verbal/symbolic aggression on their children the previous year; the average number of verbally/symbolically aggressive acts was 12.6. Vissing et al. (1991) estimated that if ten acts per year were used as a cutoff for psychological maltreatment, then 26.7% of children in this country are victims of psychological maltreatment per year. However, even this study is likely to have underestimated the incidence of psychological maltreatment, because parents are likely to underreport their own use of verbal/symbolic aggression and because only a limited number of verbally/symbolically aggressive behaviors were assessed.

Predictors, Correlates, and Consequences

Little research has been conducted on the specific predictors of psychological maltreatment, probably because most cases of psychological maltreatment never become known to the public. According to the NIS-3, children in the lowest income bracket (less than $15,000/year) were 13–18 times more likely to experience emotional abuse than those whose families made more than $30,000 per year, and mothers and fathers were equally likely to perpetrate emotional abuse (Sedlak & Broadhurst, 1996). To our knowledge, the only other specific correlate that has been well researched is the co-occurrence of psychological maltreatment with other forms of maltreatment. In both a sample of CPS cases of physical abuse and physical neglect and a control sample, approximately 90% of the physically abused and 90% of the physically neglected children also had co-occurring psychological maltreatment (Claussen & Crittenden, 1991).

The consequences of psychological maltreatment have been much more widely researched than its predictors and correlates. Although not specifically addressed in studies, many mediators probably impact the severity of the adjustment problems of psychologically maltreated children. Factors such as the severity of the abuse, its duration and frequency, age, child's gender, and socioeconomic status probably influence how much the child will be affected by the maltreatment. Furthermore, the presence of a supportive adult in a psychologically maltreated child's life can greatly mediate any negative consequences (e.g., Rutter, 1985).

It is likely that the negative impact of psychological maltreatment can be found in almost every aspect of a child's life. Furthermore, because children who experience it are often not reported to authorities, they may experience the maltreatment throughout their childhood and still feel its effects well into adulthood. The effects of psychological maltreatment are often better and more powerful predictors of negative outcomes than any other form of maltreatment (e.g., Brown, 1984; Crittenden, Claussen, & Sugarman, 1994; McCord, 1983; Ney, Fung, & Wickett, 1994; Vissing et al., 1991).

In early childhood, psychological maltreatment has led children to feel unloved and inadequate. They may suffer from low self-esteem, anxious attachment to their caregivers, negative emotions, hyperactivity, and distractibility. They may lack impulse control and have difficultly learning and solving problems. Some lack enthusiasm, persistence, and creativity in their schoolwork. Furthermore, many display angry and noncompliant behavior, which frequently leads to physical aggression (Egeland & Erickson, 1987; Egeland, Sroufe, & Erickson, 1983; Erickson, Egeland, & Pianta, 1989; Herrenkohl, Egolf, & Herrenkohl, 1997). This aggression has been shown to continue into the third grade (Lefkowitz et al., 1977) and on into adolescence, where it is reflected in a higher degree of juvenile delinquency and assaultive behaviors (Brown, 1984; Herrenkohl, Egolf, & Herrenkohl, 1997; Loeber & Strouthamer-Loeber, 1986; Vissing et al., 1991). How many of these effects can you see in the case of Cinderella in Box 6.2? In her case, do you think it is possible to separate the effects of the physical maltreatment from those of the psychological maltreatment?

Along with a higher rate of juvenile delinquency and aggressiveness, adolescents with a history of psychological maltreatment also seem to have problems with their interpersonal relationships (Vissing et al., 1991). These problems may result from negative perspectives on their possibilities for enjoyment in life, purpose in life, prospects for a future life, chances of having a happy marriage, and expectations for being good parents (Ney, Fung, & Wickett, 1994). These adolescents also have higher rates of conduct disorders, attention problems, anxiety and withdrawal, psychotic behaviors, and motor excesses (Crittenden, Claussen, & Sugarman, 1994).

As college students, individuals with a history of psychological maltreatment have higher rates of anxiety, depression, interpersonal sensitivity, dissociation, low self-esteem (Briere & Runtz, 1988, 1990), and bulimia (Rorty, Yager, & Rossotto, 1994). As adults, those with a childhood history of psychological maltreatment are more likely to physically abuse their children and spouses (DeLozier, 1982; Dutton, 1995). Outside their families, adult males with childhood histories of psychological maltreatment have been shown to have higher rates of criminal behavior, emotional instability, and substance abuse (McCord, 1983). To what extent do these outcomes seem to be the same as the outcomes of the other forms of maltreatment? How would you account for any differences in outcomes associated with psychological maltreatment as compared with the outcomes of the other forms of maltreatment? In what ways can the similarities in outcomes be accounted for?

Special Issue: Children Witnessing Interparental Aggression

Many researchers conceptualize the witnessing of interparental aggression as a form of emotional abuse. One interesting finding from the National Family Violence Survey (NFVS) was that the more violence husbands committed against their wives, the more violence wives committed against their children. Wives who were victims of beatings had the highest rates of child abuse perpetration, but even victims of minor husband violence had twice the rates of severe assaults on their children as nonabused women (Straus, 1990b). This evidence that abuse of wives often involves the children, at least indirectly, has prompted many researchers to study the effects of wife abuse on the children in the home. Consider the case of Maria presented in Box 6.3. How do you think this witnessing of the abuse of her mother might affect her in the short run? In the long run?

Box 6.3	Children Witnessing Interparental Aggression

Once married, conflict between Mr. and Mrs. P. increased dramatically as a result of Mr. P.'s infidelities and verbal abuse, and worsened with the birth of their first child, Maria. The first violent incident occurred 3 months after Maria was born and involved Mr. P. pushing, slapping, and hitting his wife. The second incident occurred 4 months later in which Mr. P. repeatedly punched his wife's face with his fists. Following this, Mrs. P. was battered severely every few months with considerable family tension between incidents. Each battering incident left Mrs. P. with bruises on her face or torso and necessitated calling in sick to the hospital where she worked as a nurse. Their second and third children were born at intervals of two years, and although each child was planned and ostensibly desired by both parents, Mr. P.'s beatings were especially severe shortly after each birth. Maria [now age 7] witnessed the most extensive violence between her mother and father, but there were times when Mr. P. would force all his daughters to watch while he beat their mother.

SOURCE: Rosenberg, Giberson, Rossman, & Acker, 2000, p. 261.

In a review of the relatively recent studies on children exposed to interparental violence, Wolak and Finkelhor (1998) found that most of the research involved children of battered women in shelters. Most, if not all, of these children had frequently witnessed very severe violence by their fathers against their mothers; moreover, these children were not just passive viewers of the violence—they often interceded in the abuse or were victims of abuse themselves. Wolak and Finkelhor's review revealed that such children tend to have problems in five areas of functioning: (1) *behavioral*—they tend to be more aggressive and delinquent than children not exposed to violence, to be cruel to animals and truant to school, and to suffer from attention deficit hyperactivity disorder more often than non-exposed children; (2) *emotional*—they tend to be more anxious, angry, depressed, and withdrawn than non-exposed children, to suffer from lower self-esteem, and to have symptoms of post-traumatic stress disorder (PTSD); (3) *social*—they tend to have poor social skills, be rejected by their peers, and have an inability to empathize; (4) *cognitive*—they may suffer from language and other developmental delays and have academic problems; and (5) *physical*—they may suffer from difficulty sleeping and eating, have poor motor skills, and display psychosomatic symptoms (e.g., bed-wetting).

Children may suffer both direct and indirect effects from witnessing interparental violence. Direct effects include dealing with the immediate physical danger and emotional problems resulting from witnessing a traumatic incident as well as learning violent behaviors modeled by parents. Indirect effects may occur when the negative impact of abuse on the mother's emotional and physical health result in poor parenting skills (e.g., inconsistency, distraction, irritability, unresponsiveness or neglect, and/or abuse), which in turn can affect the psychological functioning of her children. Whatever the pathways, the negative effects of witnessing parental violence seem to follow children into adolescence, when they are at greater risk for running away and using drugs and alcohol. As adults, former child witnesses of interparental violence are more likely than nonwitnesses to exhibit trauma symptoms, anxiety, depression, low self-esteem, aggression, poor social skills, stress, and alcohol and drug problems. Furthermore, they are more likely to perpetrate spousal and child abuse in their own families and are more likely to be victimized by spousal aggression (Wolak & Finkelhor, 1998).

Like witnessing physical aggression between parents, the witnessing of interparental psychological aggression may also have negative impacts; the more children witness psychological aggression between their parents, the more psychological aggression they use and/or receive in their later dating relationships (Hines & Malley-Morrison, 2003). However, more research needs to be done to validate these results and test other types of negative outcomes. In addition, research is lacking on the effects of children's witnessing common couple violence and violence against their fathers.

Prevention and Intervention

To our knowledge, there is no literature addressing prevention and intervention efforts that are specific to psychological maltreatment. There may be several reasons for this. First, some researchers have found that CPS agencies are reluctant to pursue cases of psychological maltreatment in the absence of other forms of abuse (especially physical or sexual abuse). Second, for agencies in most states to substantiate cases of psychological maltreatment, they must have evidence of both the parental behavior and mental injury to the child (Brassard et al., 2000). Furthermore, there is a lack of federal funding opportunities for investigating psychological maltreatment and, consequently, research into its prevention and intervention is lacking (Binggeli, Hart, & Brassard, 2001).

It is possible that because no prevention or intervention effort has explicitly targeted psychological maltreatment, any efforts to address it in the context of other maltreatment types have failed. In fact, psychological maltreatment appears to be the most resistant to intervention efforts (Daro, 1988). In one study, psychological maltreatment had the highest recidivism rate: 75% of parents continued to abuse their children in this manner, and 10% of clients actually became more abusive of their children. In addition, clients' compliance with treatment did not relate to their success rates (Ney, Fung, & Wickett, 1994). What are your thoughts about the best ways to reduce or end psychological maltreatment of children? Can anything be done to persuade parents that responding to behaviors in their children that they do not like by calling the children names, putting them down, or swearing at them may not be the most effective means of fostering healthy development?

Summary

This chapter covered the two most elusive types of child maltreatment: neglect and psychological maltreatment. The study of both has been plagued with definitional problems, but they may be the two most prevalent types of child maltreatment. Researchers argue that psychological maltreatment is at the core of all types of child maltreatment, and official statistics show that neglect is the most common type of maltreatment. In addition, it is likely that all studies with data on the prevalence of these two types of maltreatment are underestimating their true extents. Possibly because neglect and psychological maltreatment are difficult to spot, children such as the Jackson boys and Cinderella are slow to receive intervention efforts. However, even if their families did receive intervention, such interventions are usually not tailored to these two types of maltreatment, and parents who maltreat their children in these ways are notoriously resistant to treatment. What do we do to help these children, then? As this chapter has shown, both neglect and psychological maltreatment may have the worst consequences of all types of maltreatment; therefore, it is important to understand why parents neglect or psychologically abuse their children so that we can properly intervene. Research on the predictors of neglect shows that the causes of these behaviors come at every level of the ecological model and include such variables as poverty, social isolation, unemployment, certain child characteristics, parental psychopathology, and substance abuse. Perhaps if we target these predictors we can ameliorate the problem of child neglect in this country. The situation, however, for psychological maltreatment is much more complex. Probably because of the difficulties in defining this behavior, little research has been done on its predictors. Therefore, because of the limited knowledge we have of this elusive phenomenon, we are unable to properly treat either the perpetrators or the victims.

7

Wife Abuse

I met Mr. Wonderful on May 1st and he moved in on May 31st. Two weeks
later he tore the earlining to my right ear. After repeatedly hitting me, forcing a
broom handle up inside me and spitting in my face I finally had him arrested.
Oh yes, he lit my hair on fire and put lit cigarettes on my face and arms,
too . . . [A]fter three marriages and divorces I didn't want another failure
so when he got out of jail after six months and came to see me I let him in the
house. He hit me again but the District Attorney dropped the charges I made
against him this time, even though he had broken probation and harassed me.
(Account of Nan G., from Norwood, 1988, p. 98)

C onsider the above case of wife abuse. What factors do you think contributed to the violence Nan experienced? How common do you think
such violence is for women in this country? Why do you think the district
attorney dropped the charges in the latest incident? What could be done to
help Nan get out of this relationship? The current chapter deals with such
issues, as we consider the incidence of wife battering and less physically violent relationships, including those characterized by emotional and sexual
abuse. We also consider the vast research done in the previous few decades
on the predictors and consequences of wife abuse and review several types
of prevention and intervention efforts aimed at healing the female victims
and changing the behaviors of the male perpetrators.

During the 1960s, the women's movement enlightened the general public
in the United States on the problems of sexism and male domination. This
public education program was followed in the 1970s by an anti-rape cam-
paign dramatizing the extent to which women could be the direct victims of

male violence. These women's movement initiatives led to the establishment of networks where women could share their victimization experiences (Stacey, Hazlewood, & Shupe, 1994). Shelters for battered women began to open in the mid-1970s, in the face of opposition from much of the public and the government. Nonlethal intimate violence was thought of as characteristic of the poor, mentally ill, and socially deviant, and therefore not worthy of funds. Moreover, intimate violence was often attributed to the women's movement, with the implication that the female victims got what they deserved (Straus, 1980). Fortunately, researchers in this area did not always agree with government officials. In 1974, starting with Richard Gelles's groundbreaking study on the incidence and possible causes of spousal abuse, the systematic study of spousal abuse began.

Physical Maltreatment

Scope of the Problem

The physical abuse of wives has been the most extensively studied form of adult maltreatment. Perhaps because of widespread attention to the problem in both the scientific community and the media, the major prevalence studies have shown steady declines in physical wife abuse since the 1970s, even though different methods of assessing the prevalence of physical abuse of wives sometimes lead to vastly different estimates. For example, according to the United States Department of Justice (DOJ), in 2001, there were 588,490 reported cases of women who were the victims of assault by an intimate partner (Rennison, 2003), which was a decline from the 1993 estimate of 1.1 million women. These numbers represent a 50% decline, from ten women per 1,000 to five women per 1,000.

These DOJ statistics reflect only the incidents of wife abuse deemed criminal behavior by the respondents; another survey obtaining physical abuse rates from a nationally representative sample of couples answering questions on various tactics used to resolve conflicts shows similar declines. In the 1975 National Family Violence Survey (NFVS), 12.1% of wives reported that they had been the victims of some sort of violence (e.g., slapping, pushing, punching, and/or beating up) from their husbands within the previous year. Nearly 4%, or 2.1 million wives nationwide, reported that they had been the victims of severe violence, which the researchers labeled "wife beating" (Straus & Gelles, 1986). In the 1985 resurvey, there was a 27% decrease in the rate of severe violence by husbands, which projected to 432,000 fewer cases of wife beating and a national incidence rate of 1.6 million wives (Straus & Gelles, 1986).

The National Violence Against Women Survey (NVAW), a telephone survey administered to 8,000 men and 8,000 women between November 1995 and May 1996, did not provide data across time on the incidence of wife abuse but is still an important source of information on the problem. According to this survey, 20.4% of women were victims of physical assaults from a current or former intimate partner over the course of their lifetimes, and 1.4% had been victimized within the prior 12 months. The average frequency of attacks was 7.1 incidents, and the abuse tended to last an average of 3.8 years. Moreover, 38% of the women who had been assaulted reported life threats and/or fear of bodily injury, and 45% reported fear that their attacker would seriously harm or kill them or someone close to them (Tjaden & Thoennes, 2000). The prevalence rates reported in all these studies show that despite the apparent decline in rates of physical wife abuse over the past few decades, physical wife abuse continues to be a major social problem in the U.S.

Gaining an understanding of the problem of physical wife abuse is complicated by the fact that there seem to be two main types of wife abuse: (1) *common couple violence,* which is typically reported in studies of the general population and is marked by more or less "minor" reciprocal violence between husbands and wives; and (2) more extreme *terroristic violence,* which is mostly found in studies of shelter populations of battered women and criminal justice surveys; women experiencing terroristic violence are usually subject to systematic, serious, and frequent beatings, and any violence by the women would usually be in self-defense. These two types of violence against women are most likely non-overlapping (Johnson, 1995).

It is difficult to come by exact estimates of this second type of violence, but, in comparison to common couple violence, it is quite rare. As mentioned in the Preface, relationships characterized by low to moderate levels of reciprocal aggression are much more common than those characterized by extreme violence on the part of one of the partners. However, terroristic violence has been the fascination of the public and researchers alike, and, thus, more research has been conducted on this type of wife abuse. Consider the case of terroristic wife abuse presented at the beginning of this chapter. Does it sound like ones you have heard about in the media? What dynamics do you perceive? Given the information presented, what are possible causes of such types of abuse? What are possible consequences?

Predictors and Correlates

Studies on the predictors and correlates of wife abuse have come at every level of the ecological model but concentrate primarily on the characteristics of the wife, the husband, and the family. The studies focusing on predictors have all been correlational—thus, we do not know for certain whether the risk factor preceded the wife abuse, the wife abuse preceded the risk factor, or a third variable caused both the risk factor and the wife abuse.

Macrosystem

Poverty. Low income, particularly poverty, seems to be a major risk factor for wife abuse in this country. According to statistics from the DOJ, wife abuse is highest in the poorest income category and becomes lower as income increases (Rennison & Welchans, 2000). In fact, low income is consistently viewed as one of the major and most important risk factors for the abuse of wives (e.g., Hotaling & Sugarman, 1986; Straus, 1990c). In addition, although *poverty* is a predictor of wife abuse, several studies have shown that being blue collar, or working class, is also a strong predictor (e.g., Hotaling & Sugarman, 1986; Kaufman Kantor & Straus, 1990). Finally, it is not just income that predicts wife abuse but also having few economic opportunities (Follingstad, Wright, Lloyd, & Sebastian, 1991), having a low occupational status (Straus, 1990c), and having low educational attainment (Hotaling & Sugarman, 1986). Why might these associations exist? Why might wife abuse be more prevalent among the poor? Do you think the associations could be due solely to class-related reporting biases? Is it possible that wife abuse is equally likely no matter what one's social class is? Are stressors associated with poverty likely to contribute to wife abuse even when possible reporting biases are considered? Could wife abuse be essentially absent in wealthier classes? Might there be some problems that wealthier victims of wife abuse would have that poorer victims would not?

Attitudes. Several researchers have studied attitudes that may be related to wife abuse, particularly attitudes relating to traditionalism and the approval of violence against wives. It appears that both husbands and wives in marriages characterized by wife abuse are more traditional and conservative in their belief systems (Follingstad, Wright, Lloyd, & Sebastian, 1991; Rosenbaum & O'Leary, 1981). The belief that husbands should be dominant in marriage, whether held by the wife or the husband, can lead to wife abuse if the marriage is under stress (Straus, 1990c). Moreover, traditional and conservative beliefs lead both men and women to be less assertive with each other, which is also related to the abuse of wives (Follingstad, Wright, et al., 1991; Hotaling & Sugarman, 1986; Rosenbaum & O'Leary, 1981). Husbands who abuse their wives are also more likely to approve of violence in interpersonal relationships (Kaufman Kantor & Straus, 1990). They view violence toward wives as a legitimate means of interaction

(Straus & Gelles, 1990b), so much so that the belief that hitting a wife is appropriate behavior has been deemed as one of six major risk factors for wife abuse (Straus, 1990c).

Power and Stress. Although it is commonly believed that husbands have all the power in marriages characterized by wife abuse, some researchers have shown that this may not always be the case. In fact, it seems that status incompatibilities, wherein *either* the husband *or* the wife holds all the power, may predispose the couple for wife abuse as a means to either legitimize or regain the husband's power in the relationship (Straus et al., 1980; Yllo & Straus, 1990). These husbands may have a great need for power (Dutton & Strachan, 1987) and feel little control over the events in their lives (Prince & Arias, 1994); thus, if the wife is somehow the more powerful partner, the husband may use violence to take that power away from her.

All the risk factors discussed so far have one thing in common: stress. Poverty, attitudes regarding marriage and violence against wives, and status incompatibilities would not necessarily directly lead to violence if stress was not underlying all of them. In fact, stress, particularly when it is combined with feelings that the marriage is unimportant (Straus, 1990c), is one of the strongest predictors of a husband hitting his wife (O'Leary, 1988).

Exosystem

Two variables, neighborhood poverty and social isolation, have been the focus of research assessing exosystem level predictors of wife abuse. Neighborhood poverty has been shown to be a significant predictor of wife abuse (Cunradi, Caetano, Clark, & Schafer, 2000), and the more isolated a family is and the less social support it has, the greater the likelihood of wife abuse (Follingstad, Brennan, et al., 1991). In fact, one of the six factors that put men in the NFVS at risk for hitting their wives was social isolation—men who had sources of support tended not to be abusive, even if they were under a great deal of stress (Straus, 1990c).

Microsystem

Characteristics of the Victim. Research on characteristics of wives that may make them vulnerable to abuse is somewhat controversial because it may appear to be a form of "blaming the victim." However, it is important to make the distinction between causation and blame. Whether a victim's behavior played a causal role in a victimization incident is a scientific question. Whether the victim's behavior is blameworthy is a moral question for the legal system and for public opinion (Felson, 2002). It is important for social scientists to investigate causality in incidents, while the determination of blame is a matter for the courts.

The research on victim-level characteristics associated with wife abuse concentrates primarily on demographic characteristics putting women at risk. Being young seems to be a particularly strong risk factor in both community samples (e.g., Fagan & Browne, 1994; O'Leary et al., 1989; Suitor, Pillemer, & Straus, 1990; Vest et al., 2002) and samples of battered women (e.g., Rennison & Welchans, 2000). According to the DOJ, intimate violence peaks in the 20- to 24-year-old age range, and then steadily declines (Rennison & Welchans, 2000). Furthermore, the link between youth and wife abuse is so strong that the commonly held belief that pregnancy is a risk factor for wife abuse does not hold once the age of the wife is accounted for (Gelles, 1990).

Another, more controversial, victim-level predictor for wife abuse is the intergenerational transmission of physical abuse victimization. Several studies have shown that women exposed to aggression in their families of origin, either through witnessing or experiencing it directly, were more likely to becomes victims of abuse by their husbands (e.g., Cappell & Heiner, 1990; Cascardi, O'Leary, Lawrence, & Schlee, 1995). Furthermore, there may be genetic influences on this intergenerational transmission of physical abuse victimization (Hines & Saudino, 2004a). Although a genetic role in being victimized may seem illogical, there is consistent evidence that people are not merely passive receivers of their environments. That is, people actively choose their environments based upon their genetically influenced traits (Scarr & McCartney, 1983).

Other victim-level risk factors include physical and mental health problems such as depression (Cascardi et al., 1995; Vest, Catlin, Chen, & Brownson, 2002). The wife's alcohol use may also be a risk factor for being abused. In clinical samples of alcoholic women, there is often a history of violent victimization in relationships (Kaufman Kantor & Asdigian, 1996). Ultimately, the abuse seems to be a function of the *male* partner's drinking behavior, but that does not explain why the alcoholic women are involved with abusive men.

Characteristics of the Relationship. Couples who are unable to communicate or resolve conflicts in rational, nonjudgmental ways are at risk for wife abuse (Douglas, 1991). Moreover, the use of psychologically abusive tactics predisposes couples to wife abuse; that is, couples tend to attack each other's vulnerabilities prior to an episode of physical wife abuse (Gelles &

Straus, 1988). It is precisely these types of interactions that may lead to marital discord, which has been shown to correlate with psychological abuse of wives, which in turn predicts later physical abuse of wives (O'Leary, Malone, & Tyree, 1994). Such interactions can also result in marital dissatisfaction, another strong predictor of physical wife abuse (O'Leary, 1988).

Physical abuse of female partners occurs not only when couples live in the same household but also prior to marriage and when couples are separated or divorced. Among dating couples, as many as 30% of college students and young adults are involved in a physically aggressive relationship (e.g., Hines & Saudino, 2003; Morse, 1995), and physical aggression prior to marriage is a strong predictor of aggression during marriage (O'Leary et al., 1989). Compared to dating and married couples, cohabiting couples appear to have the highest rates of physical aggression (Stets & Straus, 1990a), and being separated or divorced from one's abuser is also a major risk factor for being physically abused (e.g., Rennison & Welchans, 2000; Vest et al., 2002). The most dangerous time for a battered wife seems to be after she leaves her batterer (Walker, 2000).

Individual/Developmental Level

Alcohol and Substance Use. An important individual/developmental level predictor of wife abuse is the alcohol and substance use of the abusers. A high percentage of male batterers evidence alcohol problems (e.g., Gondolf, 1988; Hamberger & Hastings, 1991; Hotaling & Sugarman, 1986; Saunders, 1992), and a high percentage of male alcoholics have problems with partner violence (e.g., Gondolf & Foster, 1991; Murphy, O'Farrell, Fals-Stewart, & Feehan, 2001; Stith, Crossman, & Bischof, 1991). When male alcoholics remit from using alcohol, their rates of partner violence mirror those of population-based samples (O'Farrell, Fals-Stewart, Murphy, & Murphy, 2003). Alcohol use is characteristic not only of men who use terroristic violence but also of men who engage in common couple violence. The 1985 NFVS showed that alcohol was involved in half of the violent couple interactions, and the use of alcohol by a man put him at two to three times the risk for hitting his wife. Furthermore, it seems that the combination of several influences—being a blue collar worker, drinking, and approving of violence—put a man at greatest risk of hitting his wife (Kaufman Kantor & Straus, 1990).

Arguments abound about whether alcohol and substance use *cause* wife abuse (e.g., Flanzer, 1993; Gelles, 1993). Some argue that alcohol abuse causes wife abuse because the physiological effects of alcohol (e.g., lowering inhibitions) lead a husband to aggress against his wife while under the influence (e.g., Gelles, 1993). However, approximately 80% of the heaviest drinkers do not hit their wives, and several studies have shown that alcohol is involved in only approximately one half of battering incidents; thus, alcohol use is neither a sufficient nor necessary cause of wife abuse. Even among batterers with drinking problems, their use of alcohol during battering incidents is variable; that is, they may use a lot of alcohol during one incident, some in another, and none in still another (Walker, 2000). On the other hand, recent research has shown that the probability of a man assaulting his wife during a drinking incident is as much as 11 times greater than when he has not been drinking (Fals-Stewart, 2003). Some researchers have argued that perhaps a third variable, such as impulsiveness or some sort of abuse-prone personality, is responsible for both the alcohol abuse and wife abuse (e.g., Collins & Messerschmidt, 1993). One study of possible third variable effects showed that male alcoholics who were partner violent showed significantly higher levels of antisocial personality traits than those who were not partner violent; after controlling for antisocial personality traits, partner violence was no longer associated with alcohol use, suggesting that alcohol problem severity was largely redundant with antisocial personality traits (Murphy, O'Farrell, et al., 2001).

Intergenerational Transmission. Another major individual/developmental risk factor for wife abuse is exposure to violence in the family of origin. Among abusive husbands in therapy (e.g., O'Leary, Malone, & Tyree, 1994), nationally representative samples of men (e.g., Kalmuss, 1984; Straus, 1990c), and batterers (e.g., Holtzworth-Munroe & Stewart, 1994), either being abused as a child and/or witnessing interparental violence put men at risk for hitting their wives. In fact, the male's exposure to family-of-origin violence is often considered the strongest predictor of wife abuse (Hotaling & Sugarman, 1986).

The most popular theory to explain this intergenerational transmission is social learning theory (e.g., Eron, 1997); this theory posits that, because aggression against intimates runs in families, children learn how to behave aggressively through watching their parents and being reinforced for their own aggression. However, recent research suggests that this intergenerational transmission may be at least partly genetically influenced (Hines & Saudino, 2004a). That is, children seem to inherit genes from their parents that predispose them for aggressive behaviors; however, their eventual use of these behaviors may not depend upon whether their parents behave aggressively (i.e., being exposed to an aggressive familial environment). Instead, children's genetic predisposition to behave aggressively may influence them to seek out aggressive peer groups or to associate with other children in the family who may also be genetically predisposed to behave aggressively. Therefore, their eventual use of aggressive behaviors may have little to do with their parents' actual use of aggressive behaviors but may have everything to do with inheriting a genetic predisposition from their parents and being exposed to aggressive models in their peer groups (Hines & Saudino, 2004a).

An important caveat to the intergenerational transmission of violence is that most men exposed to violence in their families of origin do not abuse their wives (Widom, 1989). Some protective factors against intergenerational transmission have been identified, including: older age at first exposure to violence, a less difficult temperament, higher intelligence, and "proper" cognitive appraisals of the violence. Examples of the cognitive appraisals that can have a protective influence include not idealizing or protecting the images of violent parents, not choosing sides in parental fights, and having better coping mechanisms to deal with parental violence (Caesar, 1988).

Other Factors. Other individual/developmental risk factors include mental disorders and the personality characteristics of abusive men. For instance, among batterers in court-mandated treatment, over 25% had severe mental disorders, such as major depression, delusional disorder, and thought disorder; 39% had anxiety symptoms; and about 50% had personality problems, such as narcissistic, passive-aggressive, depressive, and/or antisocial personality problems (Gondolf, 1999). Many batterers also show generalized antisocial tendencies, as over one half of batterers may have been arrested for other offenses in addition to their intimate violence.

Even among men who use common couple violence, a generalized antisocial tendency may account for their use of aggression against their wives. For example, men who hit their wives in the NFVS were more aggressive both outside the home and against their own children (Hotaling, Straus, & Lincoln, 1990). In addition, among husbands in a marital therapy program, both aggressive (i.e., willingness to hurt others, enjoyment of conflicts and arguments, a desire to get even) and defendant (i.e., a readiness to defend oneself, suspicion of others, tendency to be offended easily) personality traits prior to marriage predicted psychological aggression at 18 months after marriage, which in turn predicted physical aggression at 30 months of marriage (O'Leary, Malone, & Tyree, 1994). Because the association between these aggressive personality traits and wife abuse are so strong, some have argued that aggressive personality is a major risk factor for the physical abuse of wives (e.g., O'Leary, 1988).

Consequences

Victims of wife abuse may suffer from two types of injuries: physical and psychological. For psychological consequences, because the research is correlational, we cannot make any definitive conclusions that the observed consequences are true effects of the abuse. However, for physical injuries we are better able to make definitive conclusions as to causes and effects.

Physical Injuries

In the 1985 NFVS, 7.3% of the women who reported being severely assaulted by their spouses needed medical attention (Stets & Straus, 1990b). Cascardi, Langhinrichsen, and Vivian (1992) found that 15% of the women who reported experiencing minor spousal abuse and 11% of the women who reported experiencing severe spousal abuse indicated that they had suffered broken bones, broken teeth, and/or injury to a sensory organ. Among women who reported abuse to the Department of Justice, 50% were injured by their husbands (Rennison & Welchans, 2000). In addition to being injured, sometimes seriously, many women are also at risk for being killed by their partners. Furthermore, the flip side is also a possible consequence of wife abuse: That is, sometimes the abused wife will kill her husband in self-defense. These issues are discussed further in Box 7.1 on spousal homicide.

Box 7.1   Spousal Homicide

Although rare in comparison to the rates of spousal abuse, the homicide of intimate partners accounts for approximately 9.7% of all murders in this country (FBI Uniform Crime Reports, 2001). The homicide of intimate partners seems to comprise three different types: (1) an abusive husband kills his wife; (2) an abused wife kills her husband in self-defense; and (3) an abusive wife kills her husband. Although abused husbands have been known to kill their wives in self-defense, this type of homicide is quite rare and has never been systematically investigated. Furthermore, there is little research on the third type of homicide (i.e., an abusive wife kills her husband) outside of just reporting how often it happens.

The murder of both husbands and wives has been dropping steadily in this country since 1976, when 1,357 men and 1,600 women were killed by their intimate partners (Rennison, 2003). The most recent statistics released by the FBI show that in 2001, 295 men and 1,034 women were killed by an intimate partner. This dramatic decline in spousal homicide, particularly the killing of husbands, is thought to reflect an increase in the laws and services offered to battered women. Indeed, states that have the most laws and resources for battered women (e.g., shelters, crisis lines) have the lowest rates of partner homicide (Stout, 1989).

Most, if not all, of the cases of male-perpetrated intimate homicide involve men who either are afraid that the wife will leave him or upset that she has, or are jealous that the wife may be involved with another man (Cazenave & Zahn, 1992). The first motive, that of feared separation, accounts for the fact that men are more likely to kill estranged spouses than spouses with whom they currently reside (Cazenave & Zahn, 1992). Moreover, fear of separation seems to be more of a problem in homicide-suicide cases than jealousy is. Although homicide-suicide occurs in approximately 15% (Block & Chrisakos, 1995) to 27% (Morton, Runyan, Moracco, & Butts, 1998) of all partner homicides by men, in all cases the men killed their spouses and themselves because the wife had left the relationship (Rasche, 1988, as cited in Saunders & Browne, 2000).

What distinguishes abusive men who kill their partners from abusive men who do not? In a review of the literature, Aldarondo and Straus (1994) found that the following individual/developmental level characteristics of the abuser predicted life-threatening violence: dependency, violent behavior outside the home, exposure to violence in the family of origin, possession of weapons, and killing or abusing pets. Characteristics of the abusive situation also predicted life-threatening violence, including high frequency of violence, physical injuries, rape, threats to beat or kill, and controlling and psychological maltreatment.

When women kill their spouses, two situations are equally possible: either an abusive woman kills her spouse or a battered woman kills her abuser in self-defense (Cazenave & Zahn, 1992). The first situation has rarely been systematically investigated; we do not know what distinguishes abusive women who kill their spouses versus abusive women who do not. Furthermore, we do not know their motives behind killing their partners (e.g., separation, jealousy, as with abusive men who kill), nor do we know any characteristics of the relationship or the abuser that distinguish them. The little research that has been done on this issue will be discussed in the following chapter on

(Continued)

(Continued)

husband abuse. We do, however, know some information about battered women who kill their abusers in self-defense. In a review of research on battered women who murder their abusers, Browne (1987) found seven markers that distinguished battered women who kill from those who do not: frequent intoxication of the abuser; drug abuse by the abuser; high frequency of wife abuse; more severe injuries of the wife; being raped or sexually assaulted by the abuser; being subjected to threats of murder by the abuser; and suicide threats by the wife. Thus, the nature of the abuse is different for battered women who kill versus those who do not: Their abuse is more severe, sadistic, and frequent, and it involves sexual assault. These women remark that at the beginning of the relationship their husbands were the most romantic, attentive lovers they ever had and that the abuse did not begin until after they made some sort of major commitment to one another. At this point, the husband's attention became an obsession, and he controlled and restricted her behavior. Although the battered women did not report the first battering incident, subsequent attempts to get help from authorities often failed or did not stop the violence. As the violence escalated in these marriages, there occurred a steady decrease in contrition on the part of the husband—expressions of remorse became less common, and the severity and frequency of the beatings increased. Often, the wife was prompted to kill her batterer because his violence reached a new level: Either he threatened to kill their children or he gave signs that he was ready to follow through on previous threats to kill her (Browne, 1987).

Psychological Injuries

Most of the relevant research on the psychological consequences of wife abuse has been done with clinical samples of battered women. Overall, it appears that battered women may experience anxiety, depression, anger and rage, addictive behaviors, nightmares, dissociation, shame, low self-esteem, somatic problems, and sexual problems (Giles-Sims, 1998). These reactions seem to be a function of the duration and severity of the abuse they experience, with women whose experiences are longer and more severe suffering more psychological problems (Follingstad, Brennan, et al., 1991). Even when the effects of all relevant social and demographic variables (e.g., marital conflict, age, and income) are considered, it seems that wife abuse still has negative outcomes (Gelles & Harrop, 1989). Furthermore, many battered women indicate that their physical and mental health seemed to decline just prior to the physical abuse, continued to decline during the abuse, and only started to improve after they had left the abusive relationship (Follingstad, Brennan, et al., 1991). It seems as if the initial declines in health were a function of the psychological maltreatment experienced just prior to the first incident of physical abuse. Moreover, even though the abused women's health improved after leaving the relationship, it did not return to pre-relationship levels.

Battered women seem to suffer from a range of psychological problems as a result of their abuse. As many as 10%–23% of battered women may attempt suicide, and as many as 50% may contemplate it (Pagelow, 1984; Stark & Flitcraft, 1988). In several studies of battered women, both subclinical and clinical levels of psychological problems were found (Follingstad, Brennan, et al., 1991; Follingstad, Wright, et al., 1991). For instance, the majority of battered women in one sample reported feeling angry, emotionally hurt, fearful, and anxious as a result of the abuse (Follingstad, Wright, et al., 1991). Over one fourth had symptoms of depression and anxiety, and over one half presented with symptoms of psychosomatic reactions to stress, including persistent headaches, back and limb problems, and stomach problems.

Subjective reports of distress by battered women and the extent of the abuse they suffered seem to predict their levels of depression, anxiety, and general psychopathology (Kemp, Rawlings, & Green, 1991). In a comparison study of battered women from a shelter, battered women from a community sample, and nonbattered women, both groups of battered women were significantly more likely than the nonbattered group to suffer from major depression, psychosexual dysfunctions, phobias, obsessive-compulsive disorders, anxiety, dysthymia, alcohol and drug abuse, panic disorders, and antisocial personality disorders (Gleason, 1993). Most of the battered women were diagnosed with major depression (63%–81%), psychosexual dysfunctions (87%–88%), and phobias (63%–83%).

Although the majority of battered women in Gleason's (1993) study did not suffer from alcohol and drug abuse, a substantial minority did abuse alcohol (23%–44%) and drugs (10%–25%), rates that were significantly greater than the rates for nonbattered women. Although alcohol and drug abuse appear to be risk factors for being abused, many battered women do not start drinking until after their abuse begins. Stark et al. (1981) found that 16% of the battered women in their study were alcoholic, and 74% began drinking after the onset of their battering experiences. Furthermore, battered women, in comparison to nonbattered women, were at nine times the risk for abusing drugs following the onset of the battering.

Battered Woman Syndrome

According to Lenore Walker (1993), battered women may experience intrusive memories in dreams, flashbacks, or dissociative experiences. They may also experience psychogenic amnesia, in which they cannot remember the incidents of abuse at all, and they may experience sleep problems, eating problems, hypervigilance to danger cues, exaggerated startle response, irritability or anger responses, and/or psychosomatic symptoms such as gastrointestinal problems, headaches, and chronic illnesses. Walker labeled this specific constellation of symptoms *battered woman syndrome,* a syndrome that is analogous to post-traumatic stress disorder (PTSD) in that it encompasses the re-experiencing of trauma, intrusive recollections, generalized anxiety, low self-esteem, and social withdrawal that victims of trauma may experience.

The existence of a battered woman syndrome has been the subject of much controversy. One of the primary criticisms is that not all battered women experience the symptoms Walker describes, and the symptoms are not unique to battered women; thus, labeling it a syndrome is problematic (Ferraro, 2003). Critics have also argued that this notion of a syndrome does not encompass the range of issues related to the nature and dynamics of battering, the effects of violence, battered women's responses to violence, and the social and psychological context in which intimate violence occurs (Dutton, 1996). However, there has been support for the notion that battered women suffer PTSD-like symptoms: As many as 60%–84% of battered women may have PTSD symptomatology (Kemp, Rawlings, & Green, 1991; Saunders, 1994), and the level of distress and extent of the abuse are related to the level of symptomatology (Kemp et al., 1991). Even among community samples of women, the level of PTSD symptomatology is higher in abused (not necessarily battered) women than in maritally discordant and happily married controls (Cascardi, O'Leary, Lawrence, & Schlee, 1995). Based upon this discussion, do you feel that a diagnostic category of battered woman syndrome should exist? If so, why? If not, what would you propose to label the set of symptoms that battered women sometimes experience?

Victims of Common Couple Violence

Although the bulk of the research has been on battered women, there is evidence that women in community samples also suffer psychologically from partner aggression. For instance, in a sample of newlyweds from Buffalo, physical abuse prior to marriage predicted frequency of wives' drinking episodes after marriage (Testa & Leonard, 2001). Other data indicate that women who suffer from violence by their husbands exhibit higher levels of depressive symptomatology than nonabused women (Cascardi, Langhinrichsen, & Vivian, 1992; Cascardi & O'Leary, 1992). In the 1985 NFVS, increases in the frequency of both minor and severe violence against wives led to increases in psychological distress, as measured by the number of days in bed from illness, psychosomatic symptoms, stress levels, and number of depressive symptoms (Stets & Straus, 1990b).

Psychological Maltreatment

Scope of the Problem

As compared to physical maltreatment, there is a dearth of research on men's psychological maltreatment of wives. This is surprising, given that: (1) psychological maltreatment predicts physical maltreatment in relationships (e.g., Murphy & O'Leary, 1989); (2) psychological maltreatment tends to co-occur with physical maltreatment (e.g., Stets, 1990); (3) many times it occurs in the absence of physical maltreatment (e.g., Stets, 1990); and (4) most battered women, when interviewed about psychological maltreatment, state that the psychological maltreatment was much worse than the physical abuse they received (e.g., Follingstad, Rutledge, Berg, Hause, & Polek, 1990). Consider the case in Box 7.2. What types of psychologically abusive acts did Nancy's husband use? What were the consequences?

Box 7.2   Emotional Abuse of Nancy

As someone who suffered abuse throughout a 15-year marriage and beyond, I would like to share what I believe to be some sure signs of emotional abuse. Emotional abuse is more insidious than other abuses and just as damaging. Through this type of persecution, my partner attacked my very soul—using words and mannerisms that caused much pain and suffering. Over time, he systematically eroded my self-confidence and self-worth and created hurt so deep I could no longer bear his presence in my life.

My partner never took responsibility for his own actions. He blamed me incessantly, even for his own abusive behaviour. When confronted, he always had some excuse to justify himself. At his hands, I was

(Continued)

(Continued)

subjected to insults, put-downs, shouting, threats and sarcasm. I was criticized, called names, humiliated, intimidated and given ultimatums. Sometimes, he disguised his snide or cutting comments as humour. I found that even his subtlest comment could hurt me as much as his stronger, louder and more obviously denigrating statements. He typically ended his verbal assaults by accusing me of provoking his abuse or telling me that I deserved it. He shunned my explanations and what I might say in my own defense . . .

[M]y partner habitually chose to walk or stand in front of, rather than beside, me when we were out together. The messages I got were that he couldn't care less about me, was somehow better than me . . . that I could never be his equal anyway. Often he verbalized these sentiments too . . . I learned the hard way that living under the cloud of emotional abuse does affect one's health and well being. Because I believe that relationship partners can and should discuss ways to ensure that their words and actions do not inflict discomfort on one another, I made many attempts to alert him to how his words and actions made me feel. Sadly, he rejected them all, telling me repeatedly that whatever I had to say was not worth his time or attention.

SOURCE: Globus-Goldberg, 2001, paragraphs 1–3, 6. www.womanabuspreven tion.com/html/nancy_s_story.html. Reprinted with permission.

There seem to be three reasons for the lack of research on psychological maltreatment: (1) physical abuse and its consequences are much more apparent than psychological maltreatment and its consequences; (2) there is an implicit (and probably mistaken) assumption that the consequences of physical abuse are more severe than those of psychological maltreatment; and (3) research on psychological maltreatment has been plagued by definitional problems; indeed, it has been even harder for investigators of psychological maltreatment to come to a consensus on its definition than it has been for investigators of physical abuse (Arias, 1999).

The issue of definitional problems can be illustrated through describing some of the different studies addressing psychological maltreatment. The first measure to explicitly ask about psychological maltreatment was the original Conflict Tactics Scales (CTS) administered in the 1975 NFVS. The CTS operationally defined psychological maltreatment with six acts that either verbally (e.g., insulted or swore at other person) or symbolically (e.g., threw, smashed, hit, or kicked something) hurt the other person (Straus, 1990d). Walker (1979), in her study of battered women, determined that in addition to verbal and symbolic hurt, psychological humiliation and verbal harassment were also components of psychological maltreatment. In later work with battered women, Follingstad et al. (1990) identified six distinct types of psychological maltreatment: (1) verbal attacks (i.e., ridicule, verbal harassment, name-calling); (2) isolation (social and/or financial); (3) jealousy and possessiveness (even with family, friends, and pets); (4) verbal threats of harm, abuse, and/or torture; (5) threats to divorce, abandon, and/or have an affair; and (6) damage and/or destruction of personal property. In contrast to these six types of psychological maltreatment, Murphy and Cascardi (1999) identified five subcategories of psychological maltreatment in their clinical investigations of battered women: (1) isolating and restricting the partner's activities and social contacts; (2) attacking the partner's self-esteem through humiliating and degrading comments; (3) withdrawing in hostile ways; (4) destroying property; and (5) threatening harm or violence.

Most of the work on psychological maltreatment indicates that if a woman is physically abused she also tends to be emotionally abused; however, the reverse is not necessarily true. For instance, the 1985 NFVS showed that in only 0.2% of the cases was a woman physically abused but not emotionally abused, and 65% of the male respondents reported emotionally abusing but not physically abusing their wives (Stets, 1990). Thus, physically abused women are likely to be emotionally abused, but most emotionally abused women are not also physically abused. In addition, the rates of psychological maltreatment of women appear to be quite high. The 1985 NFVS showed that only 25% of the couples reported no emotionally abusive acts in the previous year (Stets, 1990). Even higher rates of psychological maltreatment of wives were found among couples attending marital therapy: 94% of the men insulted or swore at their wives; 94% did or said something to spite their wives; 90% stomped out of the room, house, or yard; and 97% sulked or refused to talk about an issue (Barling, O'Leary, Jouriles, Vivian, & MacEwen, 1987).

To measure psychological maltreatment, all the studies just cited used the CTS, with its relatively restricted set of psychologically abusive acts. The few studies that used the most comprehensive psychological maltreatment assessment tool, Tolman's (1989) *Psychological Maltreatment of Women Inventory (PMWI)*, suffer from a different problem: They tend to have samples restricted to battered women. Among 200 battered women in a shelter, over 75% said their partners had perpetrated the vast majority of the 58 behaviors listed in the scale within the past six months, including swearing at her (95%), blaming her for his violence (90%), ordering her around (89%), calling her names (86%), insulting her in front of others (85%), monitoring her time (85%), and not allowing her to socialize with friends (79%) (Tolman, 1989).

Predictors and Correlates

Even though the majority of research on the psychological maltreatment of wives has been conducted with battered women, the limited research on predictors has been done with community samples and concentrates on only two levels of the ecological model: microsystem and individual/developmental. On the microsystem level, psychological maltreatment has been associated with increased marital conflict and dissatisfaction and fewer children in the home (Stets, 1990; Straus & Sweet, 1992). At the individual/developmental level, the psychological maltreatment of wives is associated with several characteristics of the husband perpetrator, including higher frequency of alcohol intoxication, the use of verbal aggression outside of the home, and younger age (Stets, 1990; Straus & Sweet, 1992). Furthermore, there is preliminary evidence for the intergenerational transmission of psychological maltreatment (Hines & Malley-Morrison, 2003), and this transmission may have genetic influences (Hines & Saudino, 2004a).

Consequences

The bulk of the research on psychological maltreatment of wives has been on its consequences, and the majority of this research has been with battered women. Of six types of psychological maltreatment assessed in 234 battered women, jealousy/possessiveness and isolation were the most frequent types; however, verbal attacks and verbal threats of harm, abuse, and torture were the most painful types (Follingstad et al., 1990). The women who felt that the psychological maltreatment was worse than the physical abuse experienced particularly severe consequences, including higher passivity; greater social isolation; greater fear, shame, and depression; lower self-esteem; and greater acceptance of responsibility.

Greater psychological maltreatment is also associated with more serious and chronic illnesses, more visits to a doctor, more use of mental health services, more use of psychotropic medicines, lower relationship satisfaction, fewer feelings of power and control in the relationship, and more attempts to leave the relationship (Marshall, 1996). In one community sample of women, psychological maltreatment by husbands was associated with both depression and problem drinking in the wives (Arias, Street, & Brody, 1996, as cited in Arias, 1999).

Most research on psychological maltreatment of battered women has assessed the relative impact of physical and psychological maltreatment on psychosocial functioning. For instance, among battered women, psychological maltreatment—not physical abuse—predicts a woman's intent to leave her partner (Arias & Pope, 1999), marital dissolution (Jacobson, Gottman, Gortner, Berns, & Shortt, 1996), low self-esteem (Aguilar & Nightingale, 1994), psychological maladjustment and distress, alcoholism (Kahn, Welch, & Zillmer, 1993), and PTSD (Arias & Pope, 1999; Kahn et al., 1993). Also, psychological maltreatment (i.e., jealous control, ignoring, ridiculing traits, criticizing behavior) is a stronger predictor than physical abuse of fear among battered women; each abuse type contributes equally to low self-esteem, and, even though physical abuse contributes more to depression, psychological maltreatment is still a significant predictor (Sackett & Saunders, 2001).

Overall, these studies show that among battered women, the consequences of psychological maltreatment may be more severe than the consequences of physical abuse. Moreover, the psychological maltreatment of battered women seems to impede them from seeking help for their relationship problems because it instills fear, increases dependency on the abuser, and damages self-esteem, which in turn lessens the strength that is necessary to break up a relationship (Murphy & Cascardi, 1999). Even when the battered woman and her husband do seek help, greater psychological maltreatment by either partner heightens the chances of their dropping out of treatment (Brown, O'Leary, & Feldbau, 1997).

Sexual Abuse

Scope of the Problem

Sexual abuse of wives has been the least studied type of wife abuse— perhaps because until recently most people assumed that it was impossible, by definition, for a husband to sexually assault his wife. To this day, there are many people whose conceptions of sexual abuse exclude the possibility of sexual abuse in marriage. Another reason for lack of attention to wife sexual abuse is problems in defining it. Consider the cases presented in Box 7.3. Which of these would you consider "sexual abuse"? Why? Are there any cases that you do not consider to be sexual abuse? Why? Consider also that most researchers who study the sexual abuse of wives focus on "wife rape." Which cases in Box 7.3 should be considered "wife rape"? Why?

Box 7.3   Possible Cases of Sexual Aggression Against a Wife

1. *Mrs. Fisher:* "It happens very often. If I refuse he will go to other women. Then it would be my fault and a sin. Whether I like it or not I have to give in" (Russell, 1990, p. 83).

2. *Mrs. Carter:* "Sometimes when I didn't want to have intercourse and he did, he pushed it. He used verbal tactics. He laid guilt trips on me. Like when we would go for a long time without any sex he would demand it, and I would feel like I had to have sex though I didn't want it. I'd just be passive and let him have his way. I felt I had no way out" (Russell, 1990, p. 77).

3. *Mrs. Morgan:* "He wanted me to have sex with him when I didn't want to. I had no desire. I didn't want him to touch me, so he forced me. He said I was his wife and I had to do it . . . He said he was going to beat me. He called me names. He said I was his wife and he had a right to sleep with me when he wanted to" (Russell, 1990, p. 44).

4. *Mrs. Palmer:* "One time I had the flu and I didn't feel like having sex, but he forced me to anyway. He used his arms and body to pin me down so I couldn't move. With all of the violence that had occurred before—him beating me all the time—I was afraid of him when he told me I better not move . . . After the incident when I had the flu he only had to talk to me, to use verbal force, because I was afraid of him" (Russell, 1990, p. 94).

5. *Mrs. Atkins:* "We had come back from a party and he had drunk too much . . . When we got home I went to bed immediately. He came in, got in bed, and wanted to make love. I wasn't in any mood to do that. He forced himself, by using his strength to pin my arms down. He started kissing me, touching me, aggressively grabbing me, all the time holding my arms down . . . I fought him for a long time. I tried to free my arms, but I couldn't. I yelled abuse at him, told him to leave me alone and to stop it. Finally, because it was obvious I wasn't going to get my way, I ended up just lying there" (Russell, 1990, p. 165).

According to the DOJ, 41,740 women were raped or sexually assaulted by an intimate partner in 2001, a number that projects to 2 per 10,000 women (Rennison, 2003). However, there are a number of problems with these numbers. First, "intimate partners" can be boyfriends, dates, ex-boyfriends, or ex-husbands, not just spouses. Also definitions of "rape" or "sexual assault" were not clearly defined: They were left to the interpretation of the respondent.

Lack of explicitness is a problem because many women are reluctant to label sexual aggression by an intimate partner as "rape." Women are more likely to talk about stranger rape to their friends, relatives, and professionals than to talk about rape by an intimate partner (Ullman & Siegel, 1993), and battered women are more willing to talk about their battering experiences than about sexual abuse experienced from their husbands (Browne, 1987). Consider the woman whose husband demanded sex three times a day, became physically abusive if she did not agree, forced her to have sex in front of their child four times, and once threatened to rip out her vagina with pliers. When asked if she had ever been raped, she replied, "No" (Finkelhor & Yllo, 1985).

Our best estimates of the extent of wife rape come from two large representative surveys of women from San Francisco (Russell, 1990) and Boston (Finkelhor & Yllo, 1985). The researchers in these studies asked if the women had ever experienced "unwanted sex" from their husbands and, if so, how the husband was able to get them to have sex. If the woman was forced in some way (even just the threat of force) or was unable to consent (e.g., was drugged or asleep at the time), she was considered raped by the researchers. Of the 930 San Francisco women, 14% of the married women reported that their husbands had forced them to have sex (Russell, 1990). In Boston, 10% of the 326 married women had been forced to have sex with their husbands (Finkelhor & Yllo, 1985). Based on these two representative surveys, one out of every seven to ten women is raped by her husband.

These figures are considered underestimates of the actual rate for several reasons (Russell, 1990). First, the researchers did not ask about forced oral or anal sex (or digital penetration); although some of the participants spontaneously volunteered that information, many probably did not. This omission is an important one because victims of marital rape are more likely than victims of acquaintance rape to have experienced forced oral or anal intercourse (Peacock, 1995). Second, several respondents refused to disclose any incidents. Third, the surveys included only those women who lived in households; thus, they excluded women who were homeless or residing in some sort of institution (e.g., hospital, mental health facility, jail) and who may be at an increased risk for marital rape. Finally, some women felt that it was their wifely duty to submit to sex; therefore, because these women did not feel that they could say no, they were in a sense "unrapeable" (Russell, 1990). Russell therefore proposed that a better way of estimating the prevalence of marital rape would be to calculate the percentage of women who were forced into sex but also felt that they had the right to refuse.

This last point is important to consider. Reread the cases in Box 7.3. According to Finkelhor and Yllo (1983), four types of coercion are used during marital rape: (1) social coercion, in which the wife feels it is her duty to submit; (2) interpersonal coercion, in which the husband uses threats to leave, cheat on, or humiliate his wife to get her to submit; (3) threats of physical force; and (4) actual physical force. Which cases in Box 7.3 show which type of coercion? Does it matter to our definitions of rape which type of coercion a husband uses to get his wife to have sex with him? Is Mrs. Fisher "unrapeable" because she accepts societal mores obligating wives to have sex with their husbands as part of the marriage vows? What happens if a woman does not physically resist her husband because she knows that either she will lose the battle or that if she resists she will be hurt worse, as is the case with Mrs. Morgan? Does her lack of resistance mean that the act is not "rape"? Consider the account of Mrs. Atkins. If she passively gave in to her husband's demands in the future because of her experiences in this particular instance, can her future submissions be considered "rape"?

Although the prevalence rates of wife rape extrapolated from these studies in San Francisco and Boston may be underestimates, both studies found that wife rape was almost three times as likely as stranger rape (Finkelhor & Yllo, 1985; Russell, 1990). Furthermore, the more intimate the relationship between the perpetrator and the victim, the more likely the rape was to succeed and the more frequently the victim was raped (Russell, 1990). Indeed, 33%–50% of wives who are raped are raped more than 20 times during their marriages (Finkelhor & Yllo, 1985; Russell, 1990). In the majority of cases (58%), the husband uses only the minimal force necessary to carry out the rape, but in many instances more severe force is used. For instance, in 17% of the San Francisco cases, a weapon was used; in 16%, the husband hit, kicked, or slapped his wife, and in 19% of the cases, the husband beat or slugged his wife (Russell, 1990).

As you review the cases in Box 7.3, consider these questions: Can wife rape occur in the absence of battering? Since the rapist many times needs to use force to carry out his rape, does this necessarily make the rapist a batterer, too? Among the San Francisco women, wife rape did occur in the absence of battering—about 14% of the wife rape incidents were perpetrated by husbands who did not batter their wives (Russell, 1990). Thus, the forms of wife abuse were classified as follows: 23% were either wife rape only or primarily wife rape; 22% were cases in which wife rape and battering were equal problems; and 54% were cases in which the primary problem was battering or in which battering was the only form of abuse (Russell, 1990).

Studies of battered women also reveal high rates of wife rape: As many as 59% of battered women may also be raped (Campbell & Alford, 1989). Because battered women are at higher risk for wife rape than nonbattered women, most of the research on the predictors and consequences of wife rape, as well as on prevention efforts, has been done with battered women who were also raped. Mostly, workers in the field have assumed that wife rape is just another type of abuse that battered women have to endure; therefore, the women who are raped but not battered are virtually ignored in both research and social service interventions for this problem (Russell, 1990). When you read the next sections, keep in mind that most of this research, other than the two representative surveys in Boston and San Francisco, was conducted on battered women, usually battered women in shelters. What are the implications of this fact for the generalizability of the results?

Predictors and Correlates

Little work has been done on the predictors of wife rape, and the little that has been done has looked at the characteristics of the woman and her perceptions of the characteristics of the husband as they relate to wife rape. Consequently, much work still needs to be done in this area, and the results presented need to be replicated before being fully accepted.

Macrosystem Influences

Four macrosystem level influences on wife rape have received the most attention in the literature: poverty, ethnicity, religion, and attitudes towards wife rape. Studies on economic influences show that women who do not complete high school, who are unemployed, and who make under $10,000 per year are at greater risk for wife rape than their counterparts (Finkelhor & Yllo, 1985). Women who are or were raised Protestant (Finkelhor & Yllo, 1985; Russell, 1990) or have no religious affiliation (Finkelhor & Yllo, 1985) are also at greatest risk for wife rape. Catholic and Jewish women seem to have the lowest risk (Finkelhor & Yllo, 1985; Russell, 1990). In addition to religious affiliation, ethnicity also seems to predict wife rape: Asian women seem to be at the lowest risk, Blacks at the highest risk, and Whites and Latinas do not differ from the other races in their risk for wife rape (Russell, 1990). Why do you think Protestants, those with no religious affiliation, and Blacks are at increased risk for wife rape, whereas Catholics, Jews, and Asians experience the lowest rates? Could the associations be solely due to reporting biases and/or cultural ideas as to whether wife rape could even occur? Are there aspects of the cultures that may make wife rape more or less likely among certain religions and ethnicities?

In assessing the motives behind wife rape, Bergen (1996) found that many of these men use rape to punish their wives and to assert power and control over them. Furthermore, many of the husbands believe it is their conjugal right to have sex with their wives whenever they want to. If that right is threatened, such as when the wife cannot have sex for medical reasons (e.g., she just had a baby or is sick, as with Mrs. Palmer in Box 7.3), a husband's sense of entitlement might be threatened and he may ultimately force her to have sex with him.

Microsystem

Characteristics of the Victim. Age seems to be an important victim-level predictor of wife rape. In comparison to nonraped women, women who are under 30 or over 50 are more likely to be raped by their husbands than women between 30 and 50 (Finkelhor & Yllo, 1985). Prior sexual victimization also appears to be a risk factor: In comparison to women who did not experience wife rape, women raped by their husbands were more likely to report previous sexual abuse by an authority figure or blood relative, prior to the age of 14 (Russell, 1990). Also, they were more likely to have experienced rape or attempted rape by someone other than their husbands (Painter & Farrington, 1998; Russell, 1990) and more likely to have had unwanted sexual experiences with a woman and/or a non-blood relative (Russell, 1990). Consider why some of these associations may be occurring. Why are women who are under 30 and over 50 at greater risk for wife rape? Why would having a history of sexual assault make a woman more vulnerable to wife rape?

Characteristics of the Family and Relationship. In comparison to men who "just" batter, battering men who also rape their wives have more children in the home and are more dominant in the relationship (Frieze, 1983). Although some researchers have found no association between wife rape and other forms of abuse (Meyer et al., 1998), several other researchers have found that husbands who both batter and rape their wives are more severe and more frequent in their battering than husbands who "just" batter their wives (Frieze, 1983; Shields & Hanneke, 1983).

Individual/Developmental Level

On an individual/developmental level, raped wives often report that their husband-rapists may have a problem with drinking. In the San Francisco cases, 20%–25% of the husbands were drinking at the time of the rape (Russell, 1990), as was the case with Mrs. Atkins in Box 7.3. Furthermore, in comparison to men who only batter, batterers who also rape their wives have more drinking problems and are more violent both in and outside the home (Frieze, 1983). Finally, there is evidence that there may be an intergenerational transmission of sexual abuse: Rapist husbands tend to come from families in which there was physical abuse of a spouse and/or sexual dysfunction (Bowker, 1983).

Consequences

Consider what you have read so far about wife rape, including the cases in Box 7.3. Now think about the possible consequences of wife rape, especially in comparison to other victimized women such as those who are "just" battered and those who experience rape from a stranger. Who do you think experiences the worst psychosocial consequences—the victim of wife rape, of stranger rape, or of battering? Consider Mrs. Atkins's case in Box 7.3. What if the perpetrator in that case was a man she just met at the party? Might the rape result in different psychosocial outcomes? The popular belief seems to be that wife rape victims will not suffer as much as stranger rape victims. Do you agree? Consider this quote from Karen, a wife rape survivor:

> It was very clear to me. He raped me. He ripped off my pajamas, he beat me up. I mean, some scumbag down the street would do that to me. So to me it wasn't any different because I was married to him, it was rape—real clear what it was. It emotionally hurt worse [than stranger rape]. I mean you can compartmentalize it as stranger rape—you were at the wrong place at the wrong time. You can manage to get over it differently. But here you're at home with your husband, and you don't expect that. I was under constant terror [from then on] even if he didn't do it (Bergen, 1996, p. 43).

Although one study (Kilpatrick et al., 1988) has shown that victims of wife, date, and stranger rape do not differ in their rates of physical injuries, perceived life threats, threats of serious injury, or rates of mental disorders, several researchers have found that victims of wife rape suffer *worse* psychologically than victims of stranger rape. For example, in Russell's (1990) study, 77% of the wife rape victims reported being very or extremely upset about the rape, and 49% stated that it had a great effect on their lives. These reactions were much more severe than those reported by victims of stranger rape. Moreover, women raped and beaten by husbands had more severe anxiety, paranoia, and psychoticism, and were less likely to enjoy sex, than women raped by strangers. Wife rape may be worse than stranger rape for several reasons: The wife's ability to trust is disturbed, which leads her to feel more isolated and powerless, and even worse, she has to live with her rapist. Because she must live with her rapist, she is prone to being victimized again and again, and, as shown previously, husband rapists are more likely to complete the rape and more likely to rape repeatedly than other kinds of rapists (Russell, 1990).

Wife rape victims experience a wide range of outcomes, including physical ones (vaginal stretching, miscarriages/stillbirths, bladder infections, infertility, soreness, bruising, genital injuries, muscle tension, headaches, fatigue, and nausea and vomiting) and psychological ones (anxiety, shock, intense fear, depression, and suicidal tendencies) (Adams, 1993; Campbell & Alford, 1989; Russell, 1990). Wife rape victims have been shown to suffer from severe, sometimes prolonged, depression and suicidal ideations; they may experience flashbacks and nightmares of the incidents; they may have an inability to trust other men and a fear of intimacy; and they may have long-term sexual dysfunctions (Bergen, 1996).

In assessing possible consequences of wife rape, not only have researchers compared raped wives to victims of stranger rape but they have also compared women who are both battered and raped by their husbands to those who are battered only. Women who are both battered and raped are more likely than "just" battered women to use violence on their own children (Frieze, 1983), score significantly higher on scales measuring anxiety, paranoia, and psychoticism, and are less likely to enjoy sex (Shields & Hanneke, 1983). Furthermore, women who are both battered and raped by their husbands, in comparison to battered women, have more severe injuries, are at greater risk for being murdered or of murdering their husbands, are at greater risk for being beaten during pregnancy, and have significantly more health problems (e.g., sexually transmitted diseases, urinary tract infections, decreased sexual pleasure and desire, hemorrhoids, and other gynecological problems). In addition, they have lower self-esteem and a worse body image (Campbell & Alford, 1989).

Prevention and Intervention

Although victims of wife rape and wife battering may use the criminal justice system to try to prevent further abuse at the hands of their abuser (see Chapter 12), there are other options available to them and their abusers. In the 1970s, after the problem of wife battering came to public attention, several grassroots movements began founding shelters for battered women and their children. Then, in the 1980s, intervention programs for men who batter began to appear. Finally, some mental health professionals also advocate couples' counseling for violent couples.

Battered Women's Shelters

Since the original shelters appeared in the 1970s, these resources have proliferated. There are currently over 2,000 shelters in the United States for battered women and their children, many of which are part of the National Coalition Against Domestic Violence (NCADV, 2003). The NCADV was formally organized in 1978 and is the only national organization of shelter and service programs for women. Many of these shelters are similar to the Center for Domestic Violence Prevention (CDVP, 2003), which offers the following services: a 24-hour support line; a community support group for abused women; art and play therapy for children exposed to intimate violence; short-term crisis intervention counseling; clinical counseling by trained staff under the supervision of a licensed therapist; and legal assistance in obtaining restraining orders, preparing for court, and attending hearings. Shelter services may continue for 6–8 weeks and include individual counseling and support groups; assistance with job and permanent housing searches; child care; tutoring and counseling programs for children; emergency outreach in collaboration with the local police; and a teen outreach and counseling program for victims and potential victims of dating violence. CDVP also offers transitional housing for up to one year for completers of the shelter program.

Although not all shelters have as comprehensive a program as the CDVP, they usually offer varying combinations of the above services. The main questions about these shelter programs concern the extent to which these programs help their clients. Do abused women leave their batterers as a result of utilizing such services? Or, do they return to their batterers or end up with another man who abuses them? Even worse, does the violence increase when the victims seek assistance? Finally, do the women (and children) who utilize them find them helpful?

In answer to the question of whether women leave their batterers as a result of these programs, the evidence is mixed. Some studies find that 25%–50% of shelter women eventually return to their violent homes (Gondolf, 1988; Strube, 1988); those who do not return tend to have a higher economic status and husbands who do not seek help for their battering problems (Gondolf & Fisher, 1988). Contact with post-shelter advocates may decrease the likelihood of the woman being further victimized (Sullivan et al., 1992), but many women eventually return to their batterers even after using these services. Critics argue that an emphasis on the woman leaving the relationship is problematic because leaving can increase the likelihood that she will be battered and/or killed and because she may not be psychologically or financially ready to leave her abuser (e.g., Hamby, 1998). Women who use only shelters (i.e., not in combination with other legal or social support systems) may actually be at an *increased* risk for violence from their husbands, probably in the form of retaliation. However, women who combine the use of shelters with other services fare better, at least for the first six weeks after seeking help (Berk, Newton, & Berk, 1986).

Research on whether women perceive the services they receive from shelters as helpful is quite limited. Some women have reported that they find them somewhat more helpful than other service agencies (Donato & Bowker, 1984), but there have also been reports that many battered women feel alienated by the services offered at shelters. These women tend to be the many women who experience not only battering but also wife rape (Bergen, 1996). Several surveys of shelters reveal that wife rape is often not viewed as a domestic violence issue because it is not life-threatening; rather, it may be perceived as a sexual assault issue that should be dealt with by rape crisis centers. On the other hand, surveys of rape crisis center personnel reveal that they view wife rape as a domestic violence issue. Because rape crisis centers do not have the appropriate facilities to handle women who have been both raped and battered, they tend to refer cases of wife rape to shelters (Bergen, 1996; Russell, 1990; Thompson-Haas, 1987). Therefore, wife rape victims who call shelters for assistance are often referred to rape crisis centers, where they are referred back to shelters. What are the possible ramifications of a system that cannot seem to handle the problem of wife rape? Consider the victim of wife rape and battery who finally finds the courage and time away from her abuser to contact one of these agencies for help, and then is referred to the other. What are the chances that she will make the next phone call? What if she only had the time to make one phone call before her husband returned? What if it took much mental energy just to make that one phone call and she is referred elsewhere? Can she muster up enough mental energy to call another place? If she does make that next phone call, what will happen if she is referred out again?

Critics argue that both types of centers need to have policies in place to aid victims of wife rape; otherwise, the women who call them could be put in dangerous situations. Shelters, for instance, need to recognize that a substantial portion of their clientele are also victims of wife rape and that the needs of women who are both sexually assaulted and battered may be different than those who are "just" battered. Many women who are both battered and raped consider rape to be the more significant problem but feel the problem is ignored in a shelter setting (Bergen, 1996). If programs for sexual assault are offered, these women are more likely to feel that their experiences have been validated. Consider what happened when researcher Thompson-Haas offered a weekly support group at a shelter for battered women who had also been sexually assaulted—nearly every woman attended (Russell, 1990). Because of the increasing attention paid to the problem, it is likely that both battered women's shelters and rape crisis hotlines are starting to address the issue of wife rape (Wellesley Center for Women, 1998).

Programs for Batterers

The first intervention program for batterers started in 1977. Since then, hundreds of programs have proliferated across the United States. Their popularity has become so widespread that judges often order batterers to enter them as part of their sentences. The majority of clientele are thus court-mandated and tend to be the most severely violent batterers (Austin & Dankwort, 1998; Hamby, 1998). In addition, a large percentage of states have developed standards for batterer programs. These standards require periodic evaluations; participation in a coordinated community response to intimate violence; reporting threats of violence to the authorities; maintaining staff with certain levels of training and experience; provision of group intervention and avoidance of both individual and couples' interventions; and inclusion of issues of power and control in program content. Such state mandates have come under fire from many mental health professionals who believe the mandates interfere with their rights to treat their patients as they see fit and argue that ordering certain practices (e.g., no individual or couples' therapy, an emphasis on power and control issues) without any empirical evidence showing that these practices work (Hamby, 1998) is problematic and indefensible (Austin & Dankwort, 1998).

The goals of batterer programs tend to be ensuring the victim's safety, changing the perpetrator's attitudes towards violence, getting the perpetrator to assume responsibility for his violence, and teaching him nonviolent methods of resolving conflicts. The majority of these programs tend to follow one of two main orientations: (1) a focus on power and control; or (2) a focus on anger management. Programs focusing on power and control (approximately 20% of the programs) grew out of the shelter movement and are organized around the feminist notion that male battering is a social problem stemming from the patriarchal organization of society. The male batterer is seen as intentionally committing violence to maintain, within the relationship, a patriarchal structure in which he is superior and she is inferior. To address this macrosystem value orientation, the emphasis is on getting men to admit what they have done and to take full responsibility without resorting to denial, minimization, or victim blaming (Hamby, 1998). The state mandates for batterers' programs seem to subscribe to this orientation.

Programs emphasizing anger management (approximately 25% of programs) are organized around the notion that men who batter have problems with anger: They anger easily, dwell on their anger, escalate in their anger during confrontations, and spend much of their time in an angry state of mind. These programs emphasize that men should take time-outs to cool themselves down when a conflict occurs and use positive self-talk during arguments (e.g., "This isn't worth going to jail over."). They tend also to emphasize treating concurrent substance abuse problems. Furthermore, because the focus is on the use of appropriate skills during arguments, they often teach assertive communication skills, emotional expressiveness, stress reduction, and relaxation techniques. Thus, in contrast to the *power and control* method, which focuses on macrosystem risk factors, this *anger management* method focuses on individual/developmental risk factors (Hamby, 1998).

How successful are batterers' programs? There is considerable argument as to what should be the criterion of success. Must all intimate violence be eliminated? Or is reduction the goal? If we consider the elimination of violence the goal, how long should the batterer be nonviolent before we consider him a treatment success? One month? One year? A lifetime? What if physical violence is eliminated but is replaced with severe psychological abuse? Should we also take into consideration increases in positive behaviors, such as caring behaviors towards the wife and children? Consider your answers to these questions in relation to the design of a treatment evaluation study. For instance, if we consider a lifetime of nonviolence without a concomitant increase in psychological abuse but with a concomitant increase in caring behaviors as our goal, how might a study be designed to measure treatment success?

Assessments of program success are likely to vary depending on the type of program. For instance, the feminist philosophy underlying power and control programs has been adamantly opposed to "pathologizing" either the victims or the batterers. Therefore, interventions with a focus on treating either the victim or the batterer in a psychological way have been discouraged because the main issue is assumed to be one of power and control. Mental health professionals, who are proponents of the anger management programs (and other programs, such as individual and couples' therapy), argue that the psychological outcomes of violence need to be addressed both in the victims whose spouses battered them and in the batterers who were probably victimized at some other point in their lives (Hamby, 1998). Consider the power and control method of treating batterers. How might treatment success be defined for that program? Now consider the anger management method. Would treatment success be defined any differently based upon their differing philosophy? If so, in what ways?

Barriers to program evaluations include the many problems inherent in the programs themselves. For instance, dropout rates are incredibly high: In a comprehensive study of dropout rates in batterer programs, of 200 initial phone calls requesting information about the program, only 27% of the men showed up for an initial assessment; only 14% ever went to a program session, and only 1% actually completed treatment in the 32-week program

(Gondolf & Foster, 1991). "Success rates" are usually determined by following the 1% who actually completed the treatment; thus, any statements of success based on this small fraction of batterers are flawed at best. It seems contradictory, though, that dropout rates are so high when treatment is usually court mandated. Why aren't these men staying in treatment when it is part of their sentence? The answer is that most jurisdictions do not follow up the progress of the men they order into treatment, nor are penalties imposed for not attending treatment (Gondolf, 1990; Harrell, 1991).

Success rates are also difficult to measure because of a lack of comparison groups. Although the use of comparison groups (e.g., men who are randomly assigned to no treatment) is unethical, some studies do compare men who complete treatment with either those who never show up or those who drop out. What would be problematic about studying success in this way? Treatment completers, some studies show, tend to be better educated, less severely violent, less likely to be unemployed, and less likely to have substance abuse problems or a criminal history than those who drop out (Rooney & Hanson, 2001). How would this fact affect the results of these kinds of evaluation studies?

One last problem with evaluating success rates is that most programs do not have any discharge criteria other than attendance. Batterers can come to these programs, not participate, not make any efforts to change, and still be discharged. This method of treating batterers is in stark contrast to treatment methods for alcoholics and people with other mental health problems, who are discharged from treatment only after clinical judgments have been made as to their progress in the programs. Therefore, if a batterer is discharged from a program just because he attended it, he may be just as violent as when he entered because he did not fully participate or gain anything from it (Gondolf, 1995). Does that mean that the program itself was at fault, or is the batterer himself to blame? Can we fault the program when the batterer refuses to participate?

Overall, reviews of research on batterers programs show there is no evidence that men who complete treatment are any different in their future rates of violence than men who do not. Furthermore, there is no evidence that one type of treatment is better than another. It also has been shown in community samples of less-violent men that their use of violence tends to decline naturally over time—which may occur in severely violent men as well (see Hamby, 1998). What are the implications of these findings? Should we drop all batterers programs as we know them? What, if anything, can be done to make sure that severe batterers do not continue to victimize women? Consider again that many states mandate that batterers enter these treatment programs and also mandate that the programs have certain treatment methods (i.e., power and control methods). What are your views now on these mandates?

Couples' Therapy

One last—and controversial—means of intervening in violent relationships is couples' therapy, which usually takes a cognitive-behavioral approach and addresses such issues as appropriate communication skills. These interventions tend to focus on building protective factors against violence rather than getting rid of risk factors, as other treatment options do (Hamby, 1998). They also tend to adopt a no-blame or shared-blame approach to the violence, which is why they have come under fire from workers who believe that the batterer should take full blame for his actions. Critics argue that couples' therapy ignores the safety of the victim and implicitly blames her for what happened. Furthermore, by treating the couple together, they may make the woman less likely to speak about certain issues for fear of the possibly violent repercussions (Bograd, 1984).

Although these are certainly valid concerns, many mental health professionals still use couples' therapies in certain cases of violent marriages. For instance, Holtzworth-Munroe et al. (1995) developed a preventive form of couples' therapy program. Specifically, they recruited engaged couples who had already experienced a first incident of physical aggression. The first occurrence of physical aggression tends to be unlike the battering usually seen in shelter residents but may be the beginning of a pattern of escalation: The perpetrator of the aggression learns that using aggression is rewarded (e.g., he/she gets what he/she wants) and is therefore more likely to use it in the future. When, in the future, the same levels of aggression no longer get the desired response, escalation of the aggression ensues. Holtzworth-Munroe et al. argued that by recruiting and treating nonbattering, physically aggressive couples, therapists can teach more appropriate conflict resolution skills and prevent more severe violence from ever occurring.

Holtzworth-Munroe et al.'s (1995) treatment program includes exercises and lectures on understanding the negative consequences of physical aggression, reducing the spouses' tolerance for violence in the relationship, and helping couples examine their risk for using violence. Discussions on conflict resolution, anger management, and jealousy are included, and work and home stressors and their possible influences on violence are addressed. A lecture on alcohol use as a risk factor for violence is also part of the program. The program consists of seven weekly sessions of 2.5 hours each, and couples are assigned homework between the sessions. Couples meet as part of a group of three to five couples, and they are also given individual counseling.

Initial evaluations show that dropout rates are low and couple satisfaction is quite high.

Thus, although couples' therapy may not be appropriate for some severely violent couples, it may be appropriate for couples for whom the violence is just beginning or for those couples who are characterized by a low level of violence. These couples, as discussed previously, are much more prevalent than couples in which battering occurs. One benefit of couples' therapy is that it realizes that the form of aggression we discuss in the next chapter, husband abuse, may actually occur. Shelters and batterers programs have traditionally overlooked the male victim and female perpetrator of partner violence, an omission that can have grave consequences for male victims of partner violence.

Summary

Over the past 30 years, the increasing attention to the problem of wife battering has led to significant decreases in its incidence. However, wife battering is still a major problem in this country, and much work needs to be done to prevent it from occurring and to intervene when it does. In addition, more research and efforts need to address the other problem behaviors discussed in this chapter; namely common couple violence, sexual abuse, and psychological maltreatment. More wives are the victims of common couple violence than of battering, yet the majority of the research is focused on battered women. Although battered women are certainly in need of more services than female victims of common couple violence, eliminating common couple violence would also result in a vast improvement in the psychosocial health of this nation. In addition, many women are the victims of sexual assault by husbands who are not necessarily battering them as well. Much work needs to be done to address the issue of wife rape, particularly when it occurs without battering. Finally, researchers still need to address more fully the problem of psychological maltreatment, a problem that is much more pervasive than the physical or sexual maltreatment of wives. According to our best studies on this issue, at least a substantial minority of wives, if not the majority, experience psychologically aggressive behaviors from their partners, and the predictors, correlates, and consequences need to be more fully understood.

8

Husband Abuse

Within the first six months of marriage, Allen's wife starting beating him, calling him names, and attacking him. After four years of constant emotional and physical abuse, and subsequent apologies, Allen decided to file for divorce. During the two years of his divorce proceedings, Allen presented evidence of his wife's violence to the judge, who did not act upon the accounts or believe him. The events that occurred during the divorce proceedings and afterwards are testimony to the plight of many men who are abused by their wives. The judge prohibited Allen from entering the state in which the divorce and custody hearings were being held—thus, he could not attend his own hearings. His ex-wife won custody of their children and kept them from him. He was deemed responsible for medical expenses she incurred through a fraudulent workman's compensation claim and other medical claims about which she never informed the insurance company. He was also financially responsible for the mortgage payments and utility bills for a house he was not allowed to enter. On February 18, 1991, after years of bitter legal battles with his ex-wife, Allen killed himself, leaving behind his children, a fiancée, and many friends. (Silence Whispers, 2003)

Consider the above case of husband abuse. Based upon the evidence given, would you determine that Allen had been abused by his wife? If so, in what ways? Did anyone else abuse Allen? Were his experiences with the legal system abusive? Why do you think the judges in his divorce and custody battles ruled the way they did? Why do you think Allen ultimately killed himself?

One of the most controversial issues in the literature on family violence has been the abuse of husbands. The greatest controversy has centered on the finding of Gelles (1974) and many other researchers (e.g., Morse, 1995; O'Leary et al., 1989; Straus, Gelles, & Steinmetz, 1980) that wives are not the only victims of spousal abuse. Findings on abuse of husbands have been challenged by some feminists, who argue that violence by wives is in self-defense, and, therefore, there is no such thing as an "abused husband" (e.g., Pleck, Pleck, Grossman, & Bart, 1977–78). Consequently, much of the literature on husband abuse through the past three decades has been primarily an argument over whether or not there is any such thing. Recently, several researchers concluded that there is ample evidence that husband abuse occurs (to what extent is arguable) and have proposed moving beyond the debate and studying husband abuse in its own right—its dynamics, causes, and consequences, and the prevention efforts needed to ameliorate it (e.g., Hines & Malley-Morrison, 2001). Because the research on husband abuse is still in its infancy, our review of the literature in this chapter is not confined to research on husbands. Rather, in several places, we draw on research on dating violence against men to make inferences as to what abused husbands may be experiencing.

Physical Abuse

Scope of the Problem

The primary concern of many researchers in the field of family violence is that if we deem husband abuse a significant problem, we will divert attention and resources away from abused wives, who are the real victims of spousal abuse (Mignon, 1998). However, acknowledging abuse by wives does not necessarily translate into fewer resources for female victims of marital violence, and ignoring the existence of husband abuse does not make it any less real. In addition, if our goal is to eliminate all violence in family relationships, we must consider and acknowledge maltreatment perpetrated by females. Consider the two cases presented in Box 8.1 of husbands talking about the physical assaults they experienced from their wives. Do you consider these cases abusive? Why or why not? Are the circumstances and dynamics similar in any way to the ones experienced by abused women? For example, do the participants present a simple reversal of the roles played in your picture of the typical case of spousal abuse? Are there ways in which these cases seem different from those of abused wives? Are there problems that appear to be unique to husbands? Are there problems that abused husbands may never experience but abused wives would? What might such problems be? Finally, consider these cases again, but reverse the roles of the abuser and victim. Do you now view the cases as more, less, or equally abusive? Why?

Box 8.1   Can Husbands be Physically Abused?

Richard C, an upper-income man working in financial services, says he was attacked 50–60 times by his wife in their 14 years together, most of the time while she was drinking: "A lot of times, I would be working on some papers and there would be a coffee cup there, and she would intentionally spill the coffee; she went from that to throwing the coffee, and then throwing the cup and the coffee. She would throw hot scalding coffee in my face. It was a gradual thing that built over a three-year period, until it got to the point where she would physically strike me . . . She would physically attack me, tear the glasses off, kick me in the testicles five, six, seven times . . . You couldn't control her. A couple of times, I would wrestle her to the ground, pin her arms around her, and wrap my legs around her, and tell her to calm down, calm down. She'd say, 'O.K. I'm calm now, I'm under control now.' And you let her go, and she'd be right back at you, doing it again" (Cook, 1997, pp. 39–41).

Jake T., a six-foot-tall construction worker, married three years to a woman who was five feet, three inches: "I think a lot of her problems had to do with the drug use. I mean, I could never tell when she might come unglued. It would happen all of a sudden, usually in the bedroom. For example, one night I was sitting on the side of the bed, taking off my shoes, and she just came at me, kicking and swinging, no warning, nothin.' Just bang, she starts in. That's the way it was with her. She would never say why. One time, she did throw a knife at me; it missed. But most of the time, she would hit with her fists and kick. I'd just either hold her arms, or put up my arms, and then leave, till she had a chance to settle down" (Cook, 1997, p. 49).

Rates of violence by wives have come from many of the sources from which we obtain rates of violence by husbands. First, crime statistics from the U.S. Department of Justice show that in 2001, over 103,000 males were the victims of assault by an intimate partner (Rennison, 2003). In addition, in contrast to the dramatically declining rates of reported partner violence against females, the rates for males did not decline quite so precipitously between 1993 and 2001. However, crime surveys are likely to provide an underestimate of intimate violence victimization because many people, both men and women, are unwilling to label the physical violence they receive at the hands of an intimate partner a "crime." This reluctance may be even more pronounced in men than in women because the man is supposed to be the physically dominant and aggressive partner; consequently, admitting to being victimized by a woman and labeling it a "crime" may be viewed as emasculating (Steinmetz, 1977). Indeed, studies show that men are not only reluctant to report assaults by women but also unlikely to report assault by other men, even when severe injuries result (Henman, 1996).

A second source of data on violence by wives comes from the recent National Violence Against Women Survey (NVAW), which showed that 7% of male respondents reported being physically assaulted by a current or former wife or cohabiting partner over the course of their lifetime, and 0.8% of the men reported being physically assaulted in the previous year (Tjaden & Thoennes, 2000). This survey may also underestimate the amount of violence against intimate partners for several reasons. For example, the respondents were asked first if they were assaulted by anyone and then subsequently asked who was the perpetrator; however, when thinking about assaults, most people (men and women) neglect to think about their family members' violence, which could also lead to underestimates of the true incidence of intimate partner violence.

A final source of data on violence against men comes from studies using the Conflict Tactics Scales (CTS). In contrast to the NVAW survey, the instructions for the CTS prompt the participants to think about their relationships first, and conflicts that may be occurring within those relationships, and then report the number of times specific violent acts were used. The National Family Violence Survey (NFVS) of 1975, which used the CTS, showed that 11.6% of the husbands reported having experienced some sort of violence from their wives within the previous year (Straus, 1980), and 4.6% of the husbands reported having been the victims of severe violence by their wives, which projected into 2.6 million husbands nationwide. Wives indicated that they had committed a median of three violent acts per year and their most frequent types of aggression were throwing things at or kicking their husbands. Straus (1980) argued that these findings on husband beating show that violence cannot be understood in terms of a single factor, such as sexism, aggression, lack of self-control, or mental illnesses, as previously asserted. The 1985 replication of the NFVS showed that of 6,002 couples, 12.4% of the husbands reported some level of physical assault by their wives in the previous year. In addition, 4.8% reported being the victims of severe violence, which projected again into 2.6 million male victims—a marked contrast to the apparent declines in the rates of husband-to-wife physical violence indicated in these surveys (Straus & Gelles, 1988).

These apparently equal, and sometimes higher, rates of female violence in intimate relationships as shown in the NFVS and other studies using the CTS (e.g., Hines & Saudino, 2003; Morse, 1995; O'Leary et al., 1994) have been the subject of much debate. One major criticism is that the CTS do not measure motives, and most of these women, if not all, may be acting out of self-defense (e.g., Pleck, Pleck, Grossman, & Bart, 1977–78). Data from several studies on violence by *battered* women support this proposition (e.g., Saunders, 1986; Walker, 2000). However, the bulk of the research on motivations for violence in intimate relationships shows that self-defense is not the motivation for women's violence in either common couple violence or female-only violence (Hines & Malley-Morrison, 2001). For instance, the major reasons reported by college women for using physical force against their partners were not attempts at self-defense but rather efforts to show anger, to retaliate for emotional hurt, to express feelings that they had difficulty communicating verbally, and to gain control over the other person (Follingstad, Wright, Lloyd, & Sebastian, 1991). Jealousy (Makepeace, 1981), anger and confusion (Cate et al., 1982; Henton et al., 1983), and dominance and control (Felson & Messner, 2000; Rouse, 1990) have also been reported as primary motives for female violence. Finally, one group of researchers showed that women report hitting their male partners because they know that their partners either will not or cannot hit them back (Fiebert & Gonzalez, 1997).

In addition to the research on motives, other data from the NFVS also fail to provide support for the interpretation that women's violence is mostly in self-defense. Specifically, Straus and Gelles (1988) asked respondents who hit whom first and found that in approximately half of the cases, the wives hit first. In addition, other studies have shown that females report using significantly more violence against their male partners than their male partners use against them (e.g., Hines & Saudino, 2003). Although critics argue that many times abused women will initiate their own violence in order to control the timing and place of violence by men, it appears that not all violence by wives can be considered simply a form of self-defense or retaliation.

These studies show that female-perpetrated violence does indeed exist in relationships and cannot always be dismissed as merely self-defense. Although females are more likely than males to use violence in self-defense, many women acknowledge that they have other motives for violence against their partners. In addition, while self-defense may be a motivation for some women in mutually violent relationships, several studies (e.g., Hines & Saudino, 2003; Morse, 1995; O'Leary et al., 1989) show that in at least one fourth of violent couples the violence is committed only by the females—in these relationships, the violence is obviously not a matter of hitting out of self-defense.

Only one study to our knowledge has systematically addressed the issue of female-perpetrated severe violence against husbands, violence that could arguably be labeled "battering." These were relationships in which the husbands were the primary victims of many controlling and physically aggressive behaviors from their wives, and felt the need to call a helpline that specifically addresses the needs of male victims of partner violence. All of these men experienced physically aggressive behaviors, and a significant percentage of them were victims of punching, kicking, and choking. Oftentimes, these men reported that their wives' aggressive behaviors were targeted at their groins (Hines, Brown, & Dunning, 2004).

Predictors and Correlates

Although not extensively researched, the predictors and correlates of husband abuse tend to be very similar in nature to the predictors and correlates of wife abuse and fall at several levels of the ecological model.

Macrosystem and Exosystem Levels

The most widely studied macrosystem-level predictor of husband abuse concerns socioeconomic status. For example, similar to wife abuse, low income (Cazenave & Straus, 1990; Rennison & Welchans, 2000) and unemployment of the husband (Newby, Ursano, McCarroll, Martin, Norwood, & Fullerton, 2003; Straus, Gelles, & Steinmetz, 1980) are associated with husband abuse. In the one study of an exosystem-level predictor of wife-to-husband physical abuse, Cunradi, Caetano, Clark, and Schafer (2000) found that neighborhood poverty was a significant predictor of the physical abuse of husbands.

Individual/Developmental

Most of the researched predictors and correlates of husband abuse can be found at the individual/developmental level. For instance, the younger a female is, the more abuse she perpetrates against her husband (Suitor, Pillemer & Straus, 1990; Rennison & Welchans, 2000). Moreover, women who use aggression in their relationships may suffer from anxiety, depression, or post-traumatic stress disorder (PTSD); may tend to avoid coping with problems in their relationships; and have problems controlling their anger (Swan & Snow, 2003). In addition, women who use severe physical aggression and controlling behaviors against their husbands may have problems with alcohol and drugs, mental illnesses, suicidal ideations, and histories of childhood trauma (Hines et al., 2004).

One of the most robust predictors of females' use of intimate partner aggression is a history of aggressive behavior perpetration. For example, longitudinal studies show that early delinquency and a history of conduct problems predict females' intimate partner violence perpetration, even after controlling for the male partner's use of aggression (Giordano et al., 1999; Moffitt et al., 2001). In addition, there is evidence that females who use aggression in their intimate relationships are aggressive in other relationships (e.g., peer and other family relationships) as well (Malone, Tyree, & O'Leary, 1989) and that females with an angry self-concept and who were viewed as troublemakers in their adolescence perpetrate physical intimate partner violence (Giordano et al., 1999).

Personality has also been shown to be a good predictor of female intimate partner aggression. Specifically, the personality traits of defendence (i.e., a readiness to defend oneself, suspicion of others, tendency to be offended easily), aggression (i.e., willingness to hurt others, enjoyment of conflicts and arguments, a desire to get even), and impulsivity (i.e., tendency to act on the spur of the moment, volatility in emotional expression) assessed prior to marriage have been shown to have an indirect association with females' physical aggression towards their husbands at 30 months of marriage (O'Leary et al., 1994). Other researchers (e.g., Hines & Saudino, 2004b; Robins, Caspi, & Moffitt, 2002; Sommer, Barnes, & Murray, 1992) have found that high neuroticism and low agreeableness contribute to females' use of both minor and severe physical aggression in relationships; in other words, females who tend to be anxious, emotionally unstable, hostile, self-centered, spiteful, and jealous are at higher risk for using physical aggression against their male intimates.

Finally, the most consistent predictor for women's use of aggression against their husbands is their exposure to violence in the family of origin. The experience of both corporal punishment and physical abuse during childhood have been shown to predict women's use of physical aggression against men (e.g., Kalmuss, 1984; Swan & Snow, 2003), as has witnessing interparental violence in childhood or adolescence (Kalmuss, 1984). Furthermore, there may be genetic influences on this intergenerational transmission of intimate partner violence (Hines & Saudino, 2004a).

Consequences

Physical Injuries

The majority of studies that have assessed the victimization of men in marriages compared abused men to abused women. Mostly, researchers attempt to ascertain whether abused women experience more physical injuries than abused men. Overall, the studies clearly show that women are at a higher risk for physical injury than abused men are (e.g., Stets & Straus,

1990b). This finding is logical, considering the relative size of the average man versus the average woman. Men can inflict more harm with their fists than women can, and they are more able to restrain an abusive partner than women are (Straus, Gelles, & Steinmetz, 1980).

It should be emphasized, though, that these studies also show that abused men are at risk for physical injury. According to NVAW, for example, female-perpetrated violence accounted for 40% of all injuries due to intimate partner violence in the previous year, 27% of all injuries requiring medical attention, 38% of all victims who lost time from work, and 31% of all victims who feared bodily harm (Tjaden & Thoennes, 2000; Straus, 2004). Thus, although men may cause more bodily harm than women in violent intimate partner relationships, female-perpetrated violence is not insignificant, as it accounts for a large proportion of physical injuries and deaths in the United States (Straus, 2004).

Emergency room doctors have reported treating many types of injuries to male victims, including ax injuries, burns, smashings with fireplace pokers and bricks, and gunshot wounds (McNeely, Cook, & Torres, 2001). Burns, in particular, have been shown to be a greater concern for male victims of spousal abuse than for female victims; men comprise a high percentage of family violence victims in burn units because their wives or female partners threw boiling liquids on them (Duminy & Hudson, 1993; Krob, Johnson, & Jordan, 1986).

Reports of men being physically injured at the hands of their female partners are also evident in the literature on community samples of couples. For example, Cascardi, Langhinrichsen, and Vivian (1992) found that 2% of men who reported experiencing minor or severe spousal abuse reported suffering broken bones, broken teeth, and/or an injury to a sensory organ. Similarly, data from the 1985 NFVS showed that 1% of the men who reported being severely assaulted needed medical attention (Stets & Straus, 1990b). Morse (1995) and Makepeace (1986) found even higher rates of injury among males: between 10% and 20% of the men reported some type of injury.

Box 8.2   When Women Kill

To our knowledge, the only systematic research that addressed women who kill their intimate partners was part of a larger study on women who kill in general. In this study, Mann (1996) researched the details of a representative sample of all known female-perpetrated homicides in six cities (Atlanta, Baltimore, Chicago, Houston, New York, and Los Angeles) between 1979 and 1983. Almost 50% of these homicides were cases in which the females killed an intimate partner. In one third of those cases, the victim was a husband; in another one third, the victim was a common-law husband, and in the remaining one third of the cases, the victim was a lover.

Several characteristics of the male victims of intimate partner homicide were identified. The majority of the victims were African American (83.4%); Whites constituted 8.3% of victims. The average age at death was 37.9 years, and approximately three fourths of the victims had prior arrest records, over half of which were for violent offenses. The majority of the victims had been drinking prior to their deaths, and a substantial percentage of the men precipitated the female offender—that is, they incited her in some way, either through insults, arguing, or physical means.

Women who killed their partners also had some distinguishing characteristics. On average, female homicide perpetrators were 33.6 years old and were predominantly African American (84.1%). They were below the national average in educational attainment, were unemployed or working in semiskilled or laborer positions, and had at least one child. Approximately half of the females had a prior arrest record, and one third had been arrested previously for a violent offense. A substantial minority of the women had used alcohol or drugs prior to the homicide, and just over half claimed they killed in self-defense. Furthermore, over half of the female perpetrators premeditated the homicide, and over 95% used either a knife or gun to kill their intimate partners.

In comparison to women who killed other family or nonfamily members, women who killed their intimate partners received, on average, a lower bond, a lighter sentence (if any), and a shorter probationary period. As Mann (1996) stated, the assumption in the legal system was probably that the male victims got what they deserved, and, therefore, the female perpetrators received lighter punishment. However, as Mann (1996) also pointed out, the data show that there was seldom any evidence of frequent male-perpetrated violence toward these women, and when violent domestic encounters did occur, the females often were victorious. Furthermore, the majority of the women were not married to their male partners and, therefore could have left the relationship at any time. In addition to these reasons and because over half of the homicides were premeditated and over half of the women had criminal histories, this study provides preliminary evidence that most women do not kill because they were battered.

Several studies also show that women's violence against men can be lethal. In the case that introduced this chapter, Allen killed himself because of the abuse he sustained at the hands of his wife and the subsequent legal battles. Women have also been known to kill their husbands and, as discussed in Box 8.2, many times these murders are not in self-defense. Take, for example, the case of the "Black Widow Murderer." In 1990, Blanche Taylor Moore was on trial for murdering her longtime lover with arsenic. Blanche had come under suspicion a few months prior after honeymooning with her second husband, an ordained minister named Dwight Moore. Dwight had to be taken to the hospital when he suddenly became sick, and the doctor discovered arsenic in his system. Because this first dose did not kill him, Blanche gave him a few more poisoned milk shakes; however, Dwight managed to survive. When this story broke out, people who knew Blanche called the police because they remembered that Blanche's first husband had died from arsenic poisoning and that a longtime boyfriend had "died of a heart attack." When police exhumed the body of her boyfriend, they discovered a toxic dose of arsenic in his system. She was convicted of that murder in 1990 (her second husband testified against her) and is now on death row in North Carolina (Farrell, 1993).

In addition to discovering arsenic in her boyfriend's system, exhumations also revealed toxic doses of arsenic in the bodies of both of her parents, and many people had called the police to report that they had reasons to believe that Blanche poisoned some of their relatives as well. As it turned out, Blanche Taylor Moore was suspected of killing several people over the course of a quarter century, all in the same small community (Farrell, 1993). Why do you think Blanche's murders went undetected for so long? Would Blanche have come under suspicion earlier if she were a man? What could possibly have influenced Blanche to commit these murders? During the trial, it was revealed that Blanche was the victim of sexual abuse from her father. Do you think this sexual abuse experience could be the cause for her killings? Do all child victims of sexual abuse kill several people over the course of a lifetime? Are her murders justifiable because of the abuse she experienced?

Psychological Injuries

As was discussed with respect to wife abuse, maltreatment of a partner can have psychological as well as physical effects, and several studies show that the physical abuse of husbands can have psychological effects. For example, in the wake of physical abuse, approximately 75% of the abused men in one study reported experiencing anger; nearly 40% reported being emotionally hurt; nearly 35% reported experiencing sadness or depression;

nearly 30% reported seeking revenge; nearly 23% reported feeling the need to protect themselves; approximately 15% reported feeling shame or fear; and approximately 10% felt unloved or helpless (Follingstad, Wright, et al., 1991). Furthermore, in one longitudinal study, 9.5% of younger abused males and 13.5% of older abused males reported experiencing fear in their violent relationships (Morse, 1995), and over half of the male callers to a domestic violence helpline for men expressed fears that their wives would severely abuse them if they found out about the call to the helpline (Hines et al., 2004).

Stets and Straus (1990b) researched the extent and severity of depression, stress, and psychosomatic symptoms manifested in both male and female abuse victims. For both abused men and abused women, the higher the level of violence experienced, the more severe the depression, stress, and psychosomatic symptoms. In addition, abused men were significantly more likely than nonabused men to experience psychosomatic symptoms, stress, and depression—a result similar to that of Cascardi, Langhinrichsen, and Vivian (1992), who reported that abused husbands had significantly higher levels of depression than nonabused husbands.

In a study assessing psychological outcomes among college men experiencing physical abuse in their present or most recent relationships, psychological distress and depression were significantly greater in men who reported being recipients of physical abuse than in men reporting no abuse (Simonelli & Ingram, 1998). In addition to experiencing stress and depression, the more physical aggression men experience in their relationships, the more post-traumatic stress symptoms and alcohol use they experience (Hines, 2001).

Overall, preliminary research shows that men who experience abuse at the hands of their female partners are also exhibiting signs of psychological suffering. However, there are several flaws inherent in this research to date. Can you think of what some of these flaws might be? What types of research designs were used in the research described? Were there any longitudinal studies? Can we definitively conclude that the abuse these men experienced *caused* their psychological suffering? Could their psychological suffering have contributed in any way to the abuse they experienced? Alternatively, what third variables could be operating to influence both the distress and the aggression they sustained? In short, we just do not know the temporal nature of the associations found in the available research, and longitudinal studies are needed to figure out what the ultimate causes of the distress are. Another flaw of the research to date is that most of the analyses did not separate men who only experienced abuse from those who both experienced and perpetrated abuse. The men who are the sole victims of violence in their intimate relationships should be assessed separately from men involved in mutually abusive relationships because the psychological ramifications could be quite different for the two groups.

Psychological Maltreatment

Scope of the Problem

An even less-researched area than the physical abuse of husbands is the psychological maltreatment of husbands. Moreover, in contrast to research on the psychological maltreatment of wives, the psychological maltreatment of husbands tends not to be studied within the context of battering relationships; that is, when researchers study the prevalence, predictors, and consequences of the psychological maltreatment of husbands, they tend to do so in community samples, not in battered men. Furthermore, most of these community samples utilize convenience samples, such as college men; therefore, psychological maltreatment of men tends to be studied in dating samples, not in married relationships. However, these studies do give insight into the possible dynamics of the psychological maltreatment of husbands.

Reports on the prevalence of psychological maltreatment of men in dating relationships support the conclusion that men can be abused in their relationships (Kasian & Painter, 1992; Molidor, 1995; Simonelli & Ingram, 1998; Hines & Malley-Morrison, 2001; Hines & Saudino, 2003). These studies estimate that at least half, and as many as 90%, of men are the recipients of some type of emotionally abusive act in their relationships. However, the same definitional issues that plague the study of psychological maltreatment against wives also plague the study of psychological maltreatment against husbands. Indeed, some researchers (e.g., Walker, 1990) have argued that women cannot abuse men, even emotionally, and, therefore, creating a definition of psychological maltreatment against husbands would be a moot point. What do you think? Should we even be concerned with coming to a consensus about a definition for the psychological maltreatment of husbands? Are women who are emotionally abusing their husbands merely verbally defending themselves against physical assaults? Or can women initiate the abuse without any triggers from their husbands? Consider the case of Jerry in Box 8.3. Although we are receiving only his side of the story, do you think his wife was acting out verbally in self-defense?

Box 8.3   Emotional Abuse Against Husbands: The Case of Jerry

My wife said that I was ugly, skinny, and my jaws were sunken in. She said that I should grow a mustache to hide my ugly upper lip, which she called "the world's ugliest." She wanted me to grow a beard to hide my sunken jaws. She wouldn't walk next to me but would walk ahead or behind because she was ashamed to be seen with me in public. She threatened to kill me during the night or castrate me while I was asleep. . . . She taught my son to call me "dummy." She called me "dummy" or "wimp." When I wanted to hug her in the afternoon, she accused me of wanting sex. She always put what I did in the worst light, like she'd always try to find something bad about anything I did. When I bought flowers for her out of my "mad" money, she criticized me for not saving the money . . . She found negative things about whatever I did. If I confronted her, it would escalate and her criticism would get worse and worse. If I said that a criticism about me wasn't true, then she'd say that "everyone thinks that about you." That leads to confusion. . . . She didn't like it when I had friends, like Steve and Laura who helped me with my work. She accused me of having affairs with Steve, Laura, and their teenage daughter.

SOURCE: Smith & Loring, 1994, p. 2.

Reports from male callers to a domestic violence helpline for men also show that wives sometimes use controlling behaviors against their husbands. Specifically, at least half of the male callers to this helpline reported that their wives used at least one of the following controlling behaviors against them: threats and coercion (e.g., threatening to kill themselves or their husbands, to call the police and have the husband falsely arrested, or to leave the husband); emotional abuse (e.g., making the victim feel bad about himself, calling him names, making him think he is crazy, humiliating him); intimidation (e.g., making him feel afraid by smashing things, destroying his property, abusing pets, or displaying weapons); blaming the men for their own abuse or minimizing the abuse; manipulating the system (e.g., using the court system to do such things as gain sole custody of the children or falsely obtain a restraining order against the victim); isolating the victim (e.g., keeping him away from his family and friends and using jealousy to justify these actions); controlling all of the money and not allowing the victim to see or use the checkbook or credit cards; and using the children to control him (e.g., threatening to remove the children from the home) (Hines et al., 2004).

Predictors and Correlates

Microsystem

The only studies to our knowledge that investigated microsystem predictors of psychological maltreatment against men assessed characteristics of the male victim. Specifically, the more a man witnessed his mother emotionally abusing his father when he was a child, the more psychological maltreatment he received at the hands of his girlfriend in college (Hines & Malley-Morrison, 2003). There may be genetic influences for this intergenerational transmission of psychological aggression victimization (Hines & Saudino, 2004a).

Individual/Developmental

Several studies have assessed possible individual/developmental characteristics of the females who emotionally abuse their male partners. These studies show that women who are high on the personality traits of Neuroticism, Extraversion, and Conscientiousness, and low on Agreeableness perpetrate more psychological aggression in their dating relationships (Hines & Saudino, 2004b), and that women who inflict psychological maltreatment tend to exhibit a preoccupied attachment style (O'Hearn & Davis, 1997). Furthermore, women who use Restrictive/Engulfment psychological maltreatment tend to display anxious and insecure attachment styles; those who use Hostile/ Withdrawal psychological maltreatment tend to be cold, vindictive, and domineering and may have suffered from separation protest (e.g., are anxious about separating from their partners); those who use Denigration tend to suffer from separation protest and compulsive care-seeking and to be vindictive and domineering; and those who use Dominance/Intimidation also suffer from separation protest (Murphy & Hoover, 2001). Finally, preliminary evidence shows that females' use of psychological aggression tends to run in families (Martin, 1990) and that this intergenerational transmission may be genetically influenced (Hines & Saudino, 2004a).

Consequences

The impact of physical abuse on men has been systematically studied much more than the impact of psychological maltreatment. The bulk of this research has been done with women, and published case studies, such as the one presented in Box 8.3, are rare indeed. What types of psychological maltreatment did this man, Jerry, experience? What possible effects do you think these behaviors had on him? Why do you think he stayed with this woman? According to Smith and Loring (1994), Jerry experienced some severe negative consequences from this psychological maltreatment: He felt frightened for his life, blamed himself for everything, and lost 31 pounds. Why did he stay with this woman? He states, "There were times that she bought me gifts and said she loved me; I occasionally felt a little kindness, and I thought maybe she would change. It was enough to keep me clinging to her" (p. 2). This statement contains an important insight, as the researchers believe that this man suffered from traumatic bonding, in which the abuser alternates abusive behavior with kindness, creating a bond that involves intermittent positive reinforcement, a type of bond that is difficult to break.

Although the Smith and Loring (1994) case study is an important contribution to the literature, it does not tell us much about the effects of psychological maltreatment in the general population of men. Only three studies have provided some indication of the possible psychological effects of psychological maltreatment against men in general. Vivian and Langhinrichsen-Rohling (1994) found that among couples receiving marital therapy, there were no significant gender differences in reported depressive symptomatology among the couples experiencing mild to moderate physical and psychological maltreatment. Simonelli and Ingram (1998) showed that the 90% of men in their sample who reported experiencing psychological maltreatment also reported depression and psychological distress. Finally, Hines (2001) found that the more psychological maltreatment men experienced in their relationships, the higher their symptom counts for PTSD, alcoholism, depression, and stress.

These four studies show that men suffer psychologically from the psychological maltreatment they receive at the hands of their intimate partners. However, this research is only a first step in determining the effects of psychological maltreatment against men. What other outcomes do you think we need to study in emotionally abused men? What other methods do we need to utilize to gain a better understanding of the effects of psychological maltreatment against men? What possible confounds need to be controlled in future studies? Does it seem like an avenue that is even worth pursuing? Why or why not?

Special Issue: Sexual Abuse of Husbands?

Although to our knowledge there are no studies on sexual victimization of men by their spouses, there is anecdotal evidence that husbands can be sexually assaulted by wives and more systematic evidence from studies on college students that men are victimized by sexual aggression. Preliminary studies of college females shows that as many as one third report perpetrating sexual coercion against males (e.g., Anderson, 1998; Hines & Saudino, 2003), whereas approximately one fifth of men report being sexually coerced by a woman (e.g., Hines & Saudino, 2003; Struckman-Johnson, 1988). Although percentages differ based on the exact operational definition of "sexual coercion," several of the men and women in these studies reported incidents of women forcing men into sexual intercourse against their will. Most of the coercive tactics used by the females in these encounters were verbal; however, a few females and males indicated that females sometimes use force to achieve their sexual goals (e.g., Anderson, 1998; Struckman-Johnson & Struckman-Johnson, 1998).

Preliminary studies on the predictors of sexual aggression by females show that on the individual/developmental level, they tend to have the personality traits of low Agreeableness, high Conscientiousness, and high Extraversion (Hines & Saudino, 2004b). They also tend to be high self-monitors who have lots of dating experience, have sex with their partners very early in relationships, and have relationships that are characterized by violence and game playing (Craig Shea, 1998). Furthermore, they seem to be more aggressive and power-oriented than women who are not sexually coercive, and they are less traditional in their views about women and relationships. Specifically, they feel that women have the right to express their sexual desires, and they see relationships as a means of gaining power, not as a means of expressing tenderness and love (Craig Shea, 1998). In addition, females' use of sexual aggression in relationships is significantly predicted by their use of psychological aggression (Hines & Saudino, 2003).

Little work has been done on the psychosocial consequences of sexual aggression against men, but it seems as if the effects are less than those experienced by female victims of sexual coercion. In one study on the impact of sexually coercive experiences, about one third of the men reported that they were not upset at all about what happened to them. However, there were several important variables that contributed to whether the man became upset, and there were certainly some men who were very upset about the experience. Variables that are possibly relevant to the study of the sexual abuse of husbands are: being drunk at the time of the incident; not being in a romantic relationship with the woman; and the woman using threats or demonstrating the capacity to harm the man. These variables were all related to a man's negative reaction to the sexual aggression in at least one previous study (Struckman-Johnson & Struckman-Johnson, 1998). Although all of these studies were on college men, they do give evidence that men can be victimized by sexual aggression from their intimate partners, with negative effects that are worthy of investigation.

Prevention and Intervention

> I was once in a county courtroom in 1995 that was hearing cases for restraining orders. A man and a woman went before a judge. The plaintiff and the defendant were at their proper tables and the judge addressed the man and told him he had to stay away from the woman. When the judge asked the man if he understood these conditions, the man replied, "No, Your Honor, I don't. I am seeking a restraining order against *her* for repeatedly beating *me*. You have this all wrong." What happened next? Silence. Then the courtroom burst into laughter. The restraining order was ultimately granted, but this should give you an idea of what men are up against in the legal forum. (Silence Whispers, 2003, p. 1)

As this quote illustrates, abused husbands face an almost impossible task when they try to assert that their wives abuse them. It is no surprise, then, that prevention and intervention programs for male victims of spousal abuse are not as organized or systematic as those for abused women. In contrast to the well-organized efforts for battered women, there is no $3.7 billion Violence Against Men Act, no battered men's defense, and no legislation that empowers male victims of spousal abuse. Furthermore, although some resources exist for male victims of intimate violence, there seem to be even fewer resources for female abusers. Consider this woman who remarked on the recovery of her battering husband by stating, "[Now] he tries to understand my side of the argument. He talks to me rather than hits me. I still hit him, however. I would like to enroll in a class in anger management, but the [local] shelter for battered women does not help women with this problem" (Stacey, Hazlewood, & Shupe, 1994, p. 63).

Mandatory arrest policies have resulted in an exponential increase in the number of women prosecuted for spousal abuse offenses (Hamberger & Arnold, 1990), and at least some of these women are in need of services to assist them in resolving domestic encounters nonviolently. Although some services do exist in some locations (e.g., a 36-week-long intervention in Denver that is a modified version of a male batterers' program, and a 20-week-long anger management program at the University of Massachusetts Medical Center that has a strong cognitive-behavioral emphasis), there are no outcome studies published on the effectiveness of female batterers' programs (Dowd, 2001).

The dearth of resources for victims of husband abuse has been attributed to the feminist view of power relationships. According to feminist theory, men cannot be victimized by women in a society in which males are dominant (e.g., Dobash & Dobash, 1988). Because shelters and some batterers' programs were a result of the feminist movement against intimate violence, they are inherently unlikely to offer services to victims of husband abuse. Recently, several researchers (e.g., Stacey, Hazlewood, & Shupe, 1994) have realized that women can and do perpetrate aggression against their husbands and that not all of this aggression is in self-defense, and they have altered their treatment programs to address this problem. Furthermore, hotlines and Web sites have proliferated in recent years that address the issue of violence against men.

Some of these services are not sex-specific—that is, they are available to *all* victims of intimate violence. For instance, the National Domestic Violence Hotline (2003) accepts calls from male victims and refers them to local service providers and male victims groups ' . support. However, the majority, if not all, of these male victims groups are for male victims of sexual abuse, particularly those who were abused as children. Stop Abuse for Everyone (SAFE, 2003) is a program that offers services for all victims of intimate violence and is particularly sympathetic to male victims. It lists services that are available and responsive to male victims and provides training to law enforcement, health care providers, social services, and crisis lines on how to recognize husband abuse. It also offers a brochure that is specific to issues of husband abuse.

The first hotline in the United States specific to abused husbands, the Domestic Abuse Helpline for Men (2003), was developed in Maine in 2000. It offers a 24-hour hotline, referrals to mental health professionals who understand that intimate violence is not gender-specific, support groups, and advocacy assistance. Its personnel are in the process of setting up a shelter in Maine specifically for battered men because over the last three years they have had several requests for shelter assistance (Brown, J., personal communication, September 9, 2002). In addition, Valley Oasis in Lancaster, California offers shelter services for both men and women. The Oasis has 11 cottages, and men and women are kept separate unless the man's cottage becomes full and they need to combine the sexes. Furthermore, the state of Michigan required in 1996 that because the laws for intimate violence are gender-neutral, shelters that receive state funds must provide services to battered men. Since then, about one fourth of the shelters reported that they provided shelter to at least one abused man, with many others arranging for the men to be housed elsewhere. One problem is that most abused men probably do not know that they can utilize these programs because traditionally they have been geared toward women. Therefore, even though some abused men are now being served by these traditionally female-oriented programs, still many others do not reach out for the help that is available (Heinlein & Beaupre, 2002).

In addition, society is slow to accept the idea of an abused man. Many men do not seek help for fear of being ridiculed, and shame and embarrass-ment ensure that their abuse remains hidden in this society. Those victims who are willing to seek help either have trouble locating the few resources available or are unlikely to get the legal help they need to escape an abusive home. This situation is compounded when they have children: Many abused men state that they stayed with their abuser in order to protect the children from her violence. They knew that if they left their violent wives the legal system would most likely grant custody of the children to their violent wives, and that perhaps even their custody rights would be blocked by their wives as a continuation of the controlling behaviors she used during the marriage (McNeely et al., 2001).

Summary

The topic of abused husbands has been the subject of much controversy over the past three decades in this country. In this chapter, we attempted to move beyond the argument over whether or not men can be abused. There is ample evidence in the literature that many men are maltreated by their wives, both physically and emotionally, and that often this maltreatment is not in self-defense. Husbands who have experienced abuse have been shown to suf-fer many physical and psychological consequences, including injuries, fatal-ities, depression, alcoholism, stress, PTSD symptoms, fear, and distress. In addition, because the legal system and the public are slow to acknowledge their experiences, they may be doubly victimized—they may experience shame and humiliation if they choose to reveal their victimization. Many times, however, abused men choose to either stay in the abusive relationship to protect the children and/or keep silent about the abuse they experience, and, thus, they suffer in silence and isolation. If they do choose to seek help for their situation, they may be unaware of any of the few resources that are available to them. Many researchers have acknowledged for decades that men could be abused by their wives; however, because of the controversial nature of this topic, we are just beginning to understand its dynamics. There is little research on the predictors and consequences of this type of abuse, and only a few services have been established to help these victims.

9

Abuse in Gay/Lesbian/Bisexual/ Transgender Relationships

Ann is a lesbian with a Ph.D. who works at a major hospital in Massachusetts. Within two months of entering a relationship with Jane, Ann found that Jane was becoming very controlling, dictating where Ann could go and with whom she could spend time, and forbidding Ann to have contact with her former friends. Within four months, Jane started threatening Ann, screaming at her, and poking her in the chest if she did anything Jane did not like. The aggression escalated to pushing, shoving, and slapping. When Ann escaped the relationship, Jane started stalking and harassing her at home and work, threatening to beat or kill her. (Cabral & Coffey, 1999)

[Gary] could be a generous, sweet, charming, sensitive man . . . During our first year, Gary stayed sober . . . I finished coming out of the closet because I had finally found someone who meant so much to me that no one else's opinion mattered. A little over a year into the relationship, Gary started using drugs again . . . It was at this time that he was first violent with me . . . He immediately apologized and swore that it would never happen again . . . [A] friend of mine threw a party to which I wanted to go. Gary informed me that if I went without him he would wait outside and shoot me when I came out or crash the party and just start shooting . . . I did not go . . . [Another time when drunk] he told me he was going to kill me with his knife or maybe his gun. But first he was going to rape me . . . The next day, I packed what I could and left the house for good . . . Gary had allowed himself to almost kill me. (Rogers, 1999, pp. 11–14)

I had [male to female] sex reassignment surgery . . . I started dating Tom a week after arriving here . . . He was very gentle and loving and told me, "I will always love you for the woman you are." . . . A month after we'd known each other, I moved in with him. That's when the violence began. [He] began calling me "he-she," "it," and "boy." He also said that he would tell others that I was born male if I ever tried to break up with him. I was frightened of how my classmates and supervisor at work would treat me if they knew I was transgender . . . One evening . . . I came home to find Tom drunk and playing with a gun. He pointed the gun at me, yelled and berated me, and told me that he was going to kill me. I tried to leave the apartment but Tom chased me to the door, locked it, grabbed and choked me. I passed out. When I gained consciousness, he was raping me. (National Coalition of Anti-Violence Programs, 2001, pp. 13–14)

Conceptions of what it means to be a family have broadened considerably in recent decades in the United States. The nuclear family with a father, mother, and two or more children is no longer the majority family form in this country. Despite strong resistance from the religious right, one family form that is gaining increasing recognition and acceptance is the same-sex couple, with or without children. Laws are constantly changing, but same-sex civil unions have been legalized in Vermont, and in 2003 the Supreme Judicial Court of Massachusetts granted civil marriage rights to same-sex couples in *Goodridge et al. v. Department of Public Health.*

Little research has been conducted on intimate violence in same-sex relationships, primarily because intimate partner violence research has been greatly influenced by the feminist perspective. Feminists, as discussed in Chapter 1, place considerable responsibility on patriarchy in explaining intimate violence and emphasize the attempts of male perpetrators to exercise power and control over their victims. How well does that perspective apply to the three cases presented at the beginning of this chapter? Does a preoccupation with power and control seem to be at the core of these cases, or are other variables (e.g., alcohol abuse, stress) equally or more important? What are the implications of these cases for the feminist view of intimate violence as a gendered problem—that is, an attempt by males to retain power and control over their partners by any means? Do any of the other theories described in Chapter 1 throw additional light on the dynamics of these cases? Proponents of the view that violence is not a gendered problem, but a human problem, have pointed toward the little research conducted on violence in gay/lesbian relationships to highlight their view that male privilege is not at the heart of all intimate partner violence. Consider the "human problem" view—does violence in gay/lesbian intimate relationships challenge the feminist perspective, as they claim? In what ways?

Scope of the Problem

One of the problems in identifying the rates of maltreatment in non-heterosexual intimate relationships is a lack of consistency in the terms and definitions used. In this chapter, we will be using terms and definitions that have been accepted by social science organizations. Typically, the terms *gay* and *lesbian* have become familiar terms that apply to men who have sex with men and women who have sex with women. The term *bisexual,* referring to individuals who have sex with both men and women, is also familiar. In general, these are the preferred terms today, as compared to the older term, *homosexual,* with its negative connotations. A less familiar term, *transgender,* can be found in the titles of most non-heterosexual self-help and social service groups, as well as in much of today's literature on individuals who do not fit neatly under the heterosexual label. *Transgender* is sometimes restricted in its use to "individuals who have a persistent and distressing discomfort with their assigned gender" and who "were born anatomically as one biological sex but live their lives to varying degrees as the opposite sex" (Clements-Nolle, Marx, Guzman, & Katz, 2001, p. 917). More typically, it also includes *transvestites* (i.e., cross-dressers) and *intersexed* individuals (i.e., hermaphrodites, born with biological features of both sexes) (e.g., Lombardi, 2001). In this chapter, consistent with the language of organizations in the field, we use the term GLBT (gay/lesbian/bisexual/transgender) to refer to the large and diverse group of individuals who have sex or want to have sex with someone of the same sex.

Although these are the definitions we will be using in this chapter, it is important to note that different studies use different definitions of GLBT when studying family violence. In some studies, participants are asked to self-identify their sexual orientation. In others, participants are considered GLBT if they are currently in a same-sex relationship. In still others, participants only need to have been in a same-sex relationship at some point to be considered GLBT. Consider how these differing definitions can influence who would be considered GLBT in studies on intimate partner violence. Someone could currently be in a same-sex relationship but later consider it experimentation and eventually establish a long-term heterosexual relationship. Someone may self-identify as GLBT but may not yet have been involved in a sexual relationship. And someone else may self-identify as heterosexual, but several months or years later realize that they are gay/lesbian. Different studies, based on their own definitions of GLBT, may include or exclude individuals in each of these scenarios, and this determination can, in turn, influence our rates of intimate violence in this community. Thus, obtaining accurate data on the prevalence of intimate violence in GLBT individuals is a difficult process. In addition, as noted in Chapter 1, until very recently, members of the gay/lesbian community were reluctant to disclose any information about relationship violence because of fears of further stigmatization. Furthermore, as previously mentioned, mainstream family violence researchers typically did not investigate same-sex relationships—in part because of prevailing theoretical perspectives not accommodating the idea of women battering other women and because of the view that violence between male intimates is violence between equals, and therefore not abusive.

Given recent interest, there are a growing number of studies providing estimates of the prevalence of violence within same-sex relationships. All of these studies have serious methodological limitations, however, thereby compromising efforts to obtain precise estimates of GLBT intimate violence. For example, the only probability-based national study of intimate partner violence—the National Violence Against Women Survey (NVAW; Tjaden & Thoennes, 2000)—provided no direct evidence on the frequency of abuse experienced by individuals in same-sex relationships. Other national data sources are generally service agencies with GLBT clients or criminal justice system reports, and in both cases the samples are not representative of the GLBT community at large—for example, clients who voluntarily seek help may be more open about their gender identity than other GLBT individuals. Finally, most studies providing information concerning the frequency with which different kinds of abuse occur in GLBT relationships are relatively small and nonrepresentative. Nevertheless, taken as a whole, the studies provide clear documentation that intimate violence is just as pervasive in GLBT relationships as in other intimate partner relationships—and perhaps even higher.

National Studies

NVAW, as mentioned previously, only indirectly addressed the issue of violence in GLBT relationships. The survey contained an item where respondents indicated whether they had *ever* lived with a same-sex partner as part of a couple. Rates of intimate partner violence were then compared between respondents with a history of same-sex cohabitation and those reporting only opposite-sex cohabitation. These comparisons revealed significantly more intimate partner violence in the same-sex cohabitants than in the opposite-sex cohabitants. However, the abused women in same-sex intimate relationships were not typically being abused by their same-sex partners. Rather, intimate partner violence against both male and female partners was most commonly perpetrated by men (Tjaden & Thoennes, 2000). Consistent with these national data are the findings of smaller studies indicating that battered heterosexual women experience significantly higher levels of victimization from physical and nonphysical abuse than battered lesbians (Tuel & Russell, 1998) and that lesbians who report being involved in aggressive relationships have often been victimized by both male and female partners (Lie, Schilit, Bush, Montagne, & Reyes, 1991).

One important source of information on reported cases of intimate violence in GLBT relationships is the annual report of the National Coalition of Anti-Violence Programs (NCAVP). Based on information from 14 agencies serving GLBT and HIV-affected victims of intimate violence from 11 regions across the United States, the 2002 annual report documented over 5,000 cases of GLBT intimate violence (NCAVP, 2003). Most of the reported incidents came from agencies in coastal metropolitan regions (e.g., San Francisco, Los Angeles, New York, Boston). In addition, of the GLBT victims, 52% were male, 46% female, and 5% transgender. This study is valuable because it provides data on the magnitude and relative distribution of intimate violence in the GLBT community; however, the reported numbers are largely a function of evolving programming and organizational capacities and are not necessarily representative of GLBT intimate violence in the United States because the estimates are derived solely from GLBT individuals who use these services.

Overall, the national data indicate that women who have ever lived as part of a couple with another woman have experienced more intimate violence than women reporting only heterosexual relationships, but both groups of women have been victimized primarily by men. Approximately half of the clients of agencies dealing with intimate violence in same-sex relationships are men—perhaps because there are fewer shelters and other support services for abused men in general, or perhaps because half of the cases of intimate partner violence in the GLBT community are cases in which males are victimized. In the following sections, we focus on findings from more geographically limited studies, which typically have less representative samples but provide more specific information concerning the extent and nature of intimate violence in GLBT relationships.

Rates of Intimate Violence in GLBT Samples

There are four major types of evidence concerning the rates at which GLBT individuals experience intimate violence: (1) studies examining self-reported rates of intimate violence within samples consisting entirely of women who have sex with women; (2) studies comparing lesbians (and sometimes bisexual women) with heterosexual women; (3) studies comparing lesbians and bisexual women with gay and bisexual men; and (4) studies of gay and bisexual men, sometimes compared to heterosexual men.

Thus, the majority of the studies on this issue concentrate on violence in lesbian relationships. Of the available studies, some involve community samples, and some involve lesbians who were recruited precisely because they were victims or perpetrators of relationship aggression. Each study type has limitations, although in the aggregate the studies provide a useful supplement to the limited national probability sample and criminal report data. They also provide valuable information on the incidence and dynamics of several different types of aggression in GLBT intimate relationships, including physical, psychological, and sexual.

Rates of Physical and Nonphysical Aggression in Lesbian Relationships

Studies of physical and verbal aggression in lesbian samples yield a broad range of estimates of abuse. More than half of the lesbians recruited at music festivals and through other forms of convenience sampling say they have been abused, either physically and/or verbally, by a female partner (Lie & Gentlewarrier, 1991; Lockhart, White, Causby, & Isaac, 1994; McClennen, Summers, & Daley, 2002). Reports of psychological abuse are sometimes as high as 80% or 90% (Lockhart et al., 1994; Turell, 2000). In a nationwide self-selected sample of 100 battered lesbians, aggressive acts frequently or sometimes experienced included being pushed and shoved (75% of the women); hit with fists or open hands (65%); scratched or hit in the face, breasts, or genitals (48%); verbal threats (70%); demeaning comments in front of friends or relatives (64%); having eating or sleeping habits disrupted (63%); and having property damaged or destroyed (51%) (Renzetti, 1989).

Rates of Sexual Aggression in Lesbian and Gay Relationships

Although the general public has long recognized that men are capable of sexual assault, and that some men sexually assault other men as well as women, there has been considerable resistance to the idea that women can sexually assault their partners. Yet women are clearly capable of violent sexual coercion. Consider Maureen's story concerning her partner's jealousy over a male friend of Maureen's: "On several occasions her jealousy was out of control and she took it out on me sexually, by holding me down on the bed, grabbing my breasts, and trying to force something into my vagina, insisting that 'this is what you want from him' or words to that effect" (Girshick, 2002, pp. 1502–1503). The notion that women are incapable of sexual assault persists even within the lesbian and bisexual community: "After Deirdre's partner sexually assaulted her by fisting (inserting the whole hand into the vagina), she recounted: "Later that evening when she came back to say something to me, I looked at her and said, 'You raped me.' And she looked at me and said, 'I gave you exactly what you wanted and exactly what you deserved' " (Girshick, 2002, pp. 1507–1508).

Studies of small nonclinical samples indicate that rates of sexual aggression within lesbian relationships may be as high as, or higher than, rates of sexual aggression in heterosexual relationships. In a study of sexual assault victimization among college students, 30.6% of lesbians, as compared to 17.8% of heterosexual women, reported having been forced to have sex by a partner against their will (Duncan, 1990). In other samples, between 12% and 30% of lesbians have reported being victims of forced sex, or sexual aggression, by their current or most recent partners (Turell, 2000; Waterman, Dawson, & Bologna, 1989). In a San Francisco sample (Brand & Kidd, 1986), many lesbians reported having been victimized both physically and sexually by intimate male as well as female partners. Males were the sexual and physical aggressors in intimate relationships significantly more often than women were; however, within committed relationships in which abuse occurred, lesbians were receiving as much abuse from their female partners as heterosexual females were receiving from their male partners.

There appears to be considerably less scholarly research on levels of sexual violence in male same-sex than in female same-sex relationships—which is consistent with the feminist movement's general concern with victimization of women. The limited findings suggest that gay men are subjected to forced sex by their partners more often than heterosexual men are. For examples, among college students, approximately 12% of gay men, as compared to 3.6% of heterosexual men, reported having been forced to have sex by a partner against their will (Duncan, 1990; Waterman et al., 1989). Other estimates of forced sex in gay male relationships range from 12% in either past or present intimate relationships (Turell, 2000) to more than 33% over a lifetime (Kalichman et al., 2001). Moreover, gay men who experience sexual aggression are more likely to have been physically assaulted by a male partner and to report being afraid to request male partners to use condoms (Kalichman et al., 2001).

Some forms of sexual aggression, such as forced unprotected (particularly anal) sex, can have abusive side effects that are particularly problematic in the gay community; that is, these practices can result in the spreading of AIDS. Some gay men who use coercive sexual tactics already have AIDS, yet insist that their partners submit to unprotected sex, despite the risks. Consider the dilemma of this gay man: "I am HIV positive because Alex insisted that we have sex without a condom. He told me, 'If you love me, you'll do this,' even though he knew he could easily infect me. I've been going to a counselor for about six months and I've thought about leaving him but I feel trapped because I am HIV positive" (NCAVP, 2000, p. 26).

Comparisons of Abuse Rates in Lesbians and Gay Men

There is some evidence that within intimate relationships lesbians are abused at higher rates than gay men. For example, among a large, diverse GLBT sample, lesbians reported more physical abuse, coercion, threats, shaming, and use of children for control than gay men, and bisexuals reported less abuse than gay men or lesbians (Turell, 2000). In addition to self-report data on abusive experiences, the criminal justice system provides some information that can be used to make estimates concerning the relative incidence of intimate violence in lesbian and gay relationships. For example, of all non-impounded abuse prevention orders issued in one Massachusetts district court in the year 1997, reported psychological aggression and physical assault were higher in female same-sex relationships than in male same-sex relationships. By contrast, in heterosexual assault cases, the reported assault rates of male and female perpetrators were nearly identical (Basile, 2004).

Findings on the relative rates of sexual aggression in lesbian versus gay male relationships are not as clear, however. For example, some studies show that sexual aggression, including unwanted fondling and penetration, is higher in lesbian (30%) than in gay male (12%) relationships (e.g., Waterman et al., 1989), whereas others show equal rates in both communities (e.g., Turell, 2000; Waldner-Haugrud & Gratch, 1997). In these studies, rates of sexual aggression reported by both lesbians and gay men ranged from 12% who reported sexual aggression (Turell, 2000), to approximately 33% who reported unwanted fondling, to over 50% who reported unwanted penetration (Waldner-Haugrud & Gratch, 1997).

Intimate Violence and Transgendered Individuals

There is little data on intimate aggression in transgendered relationships—perhaps in part because the number of transgendered individuals is relatively small. According to one source, "There are no actual statistics on the numbers of transvestites, transsexuals, or intersexuals in the world at large. Due to the nature of transgenderism, accurate counts cannot be made. The estimated ratio of [male to female] transsexuals to genetic males is between 1:2,000 and 1:80,000. The estimated ratio of [female to male] transsexuals to genetic females is between 1:2,000 and 1:125,000. Researchers estimate that the percentage of men who have cross-dressed is quite high, perhaps as high as 50 percent. Female cross-dressers are comparatively uncommon" (Rohde, 2002, section 1.15). Consistent with their low numbers, transgendered individuals constitute only a very small share of the cases of GLBT intimate violence. In 2002, transgender cases made up only 4% of the cases of GLBT victimization seen at agencies serving GLBT and HIV-affected intimate violence victims (NCAVP, 2003). The forms of abuse to which these individuals are subjected may, however, be fairly extreme. For example, data from the Portland, Oregon Survivor Project's 1998 Gender, Violence, and Resource Access Survey (NCAVP, 2003) indicated that 50% of their transgender and intersex respondents had been raped or assaulted by a romantic partner. Interestingly, when asked if they were survivors of intimate violence, only 62% of these victims replied in the affirmative.

Special Issue: Child Abuse in GLBT Families

As has been demonstrated, there is a growing body of research on intimate violence in GLBT couples. As discussed in a later section, there is also a small but growing set of studies investigating the extent to which GLBT adults were victims of child abuse. To our knowledge, however, there are no studies examining the relative rates of child abuse in children growing up with at least one GLBT parent as compared to children growing up with one or two heterosexual parents. This is a noteworthy gap in the literature, given estimates that between 6 and 14 million children nationwide live with at least one gay parent (American Civil Liberties Union, 1999). Although at least one state (Florida) has laws expressly barring lesbians and gay men from adopting children, more than 21 states have granted second-parent adoptions to lesbian and gay couples. There is substantial evidence that growing up with at least one gay or lesbian parent does not have any ill effects on the child's development (Patterson, 2003). Given the widespread acceptance of corporal punishment in families with heterosexual parents, as well as the prevalence of child abuse in this country, it seems important that there be some investigation of the level of coercive and/or abusive parenting being used by GLBT parents.

Predictors and Correlates

Explanations of abuse in GLBT relationships place considerable emphasis on macrosystem and exosystem homophobia and homonegativity. Attention has also been given to characteristics of victims that enhance their vulnerability to abuse within the microsystem as well as to individual/developmental characteristics of abusers. Most available studies have serious methodological limitations, such as omitting a clear definition of abuse, relying on retrospective reports of abuse within the lifetime, not obtaining representative samples, and not controlling for potential biases; nevertheless, their findings are useful in identifying pathways for future research and intervention. For a thoughtful discussion of the limitations of the relevant literature, see Burke and Follingstad (1999).

Macrosystem

As we have demonstrated throughout this book, the American macrosystem is quite violent. Children, women, the elderly, and men may all be physically, psychologically, and/or sexually victimized within their closest relationships. However, it is also true that members of some groups within the macrosystem may be victimized more than others, and this victimization may lead to higher levels of violence within their families. Among the groups at heightened risk of societal victimization are immigrants, ethnic minority group members (discussed at length in Malley-Morrison & Hines, 2004), and individuals socially identified as homosexual. Although there is no solid empirical evidence linking macrosystem attitudinal variables with GLBT intimate violence, there is an extensive theoretical and clinical literature postulating this link.

Macrosystem level values in the United States have long included *homophobia* (fear, disgust, anger, hatred, and intolerance in regard to homosexuality), *homonegativity* (negative attitudes toward lesbians and gay men), and *heterosexism* (the normalizing and privileging of heterosexuality). These values seem consistent with a strongly gendered version of a society that sanctions aggression toward individuals who violate gender roles. Among the ways in which such fears and attitudes are expressed are hate crimes against individuals presumed to be gay, lesbian, bisexual, or transgendered. Violence motivated by homophobia and heterosexism has been described as the most visible, violent, and culturally legitimated type of hate crime in the United States (Jenness & Broad, 1994). In one survey of New York high school students, hostility toward lesbians and gay men was greater than hostility toward ethnic and racial minorities (Herek, 1989). Such hostility has a long history in this country: Men were executed for sodomy in the colonies as early as 1624, and lesbians and gay men have been subjected for centuries to various forms of institutional violence, including castration and clitoridectomy, forced psychiatric treatment, and dishonorable discharge from the armed services (Herek, 1989).

There is substantial evidence that macrosystem anti-gay attitudes are associated with pervasive harassment and victimization of GLBT individuals, as well as with discrimination (Herek, Gillis, Cogan, & Glunt, 1997). An analysis of more than 20 national, state, and local studies of anti-gay violence and victimization revealed that 52%–87% of the respondents had been subjected to verbal harassment; 21%–27% had been pelted with objects; 13%–38% had been stalked or chased; 10%–20% had been vandalized; 9%–24% had been physically assaulted; and 4%–10% had been assaulted with a weapon (Berrill, 1990). In general, perpetrators were young White males, acted in groups, and were not affiliated with organized hate groups but probably were influenced by the rhetoric of those groups. In another victimized GLBT sample, approximately 74% of the victims were male, 22% were female, and 4% were transgendered (Kuehnle & Sullivan, 2001). Almost 45% of the respondents had been victimized by a stranger; about 10% by a family member, ex-lover, acquaintance, or roommate; about 20% by a neighbor, landlord, or tenant; about 12% by law enforcement or security personnel; and about 6% by service personnel. Nearly 50% of the offenses were serious personal offenses. About 18% required outpatient medical treatment, 3% required hospitalization, and just over 1% resulted in death. There is also strong evidence that the AIDS epidemic has contributed to a trend of increasing violence toward members of the GLBT community (Berrill, 1990).

What does macrosystem anti-gay hostility have to do with abuse within GLBT relationships? Intimate violence in GLBT relationships is often attributed directly or indirectly to these macrosystem attitudes (e.g., Allen & Leventhal, 1999; Island & Letellier, 1991; Renzetti, 1998; Russo, 1999). Three main arguments linking macrosystem homophobia/homonegativity with violence in GLBT relationships are: (1) macrosystem homophobia/homonegativity is a weapon used by batterers to intimidate their same-sex partners into enduring their abuse and staying within the abusive relationship (Allen & Leventhal, 1999); (2) prior experience with homophobia and homonegativity leads GLBT victims to feel there is no alternative to the violence they are enduring (Island & Letellier, 1991); and (3) macrosystem homophobia leads to internalized homophobia, which leads to partner abuse within GLBT relationships (Renzetti, 1998).

This issue of *internalized homophobia* (or *internalized heterosexism*), which represents GLBT individuals' internalization of negative attitudes and assumptions concerning homosexuality, is a frequently cited cause of violence in GLBT relationships (Margolies, Becker, & Jackson-Brewer, 1987; Szymanski & Chung, 2003). Research with lesbians and bisexual women has indicated that the higher their level of internalized homophobia, the greater their depression (Herek et al., 1997; Szymanski, Chung, & Balsam, 2001); distress (McGregor, Carver, Antoni, Weiss, Yount, & Ironson, 2001);

alcohol consumption (DiPlacido, 1998); dissatisfaction with social support (McGregor et al., 2001; Szymanski et al., 2001); and conflict concerning sexual orientation (Szymanski et al., 2001), symptoms that are risk factors for intimate violence (Szymanski & Chung, 2003).

Although noting that there has been no direct empirical research on the relationship between these societal values and GLBT intimate violence, Renzetti (1998) asserts that, "societal homophobia (a social-structural variable) produces internalized homophobia (a psychological variable), which in turn may generate, among other outcomes, partner abuse in homosexual relationships" (p. 123). Informal support for this position comes from clinicians and social service personnel working with victims. According to Allen and Leventhal (1999), "The threat of or actual verbal, physical, and sexual violence we face from strangers on the street or family members or acquaintances can also leave us feeling vulnerable for abuse in our relationships. Because violence against us is sanctioned, it may be difficult for those in our communities who are battered not to internalize blame for it . . . When a batterer exploits the very vulnerabilities created by oppression, the victim may not be able to find respite from abuse" (p. 79).

Box 9.1   Norma and Tammy

I'm a 55-year-old gay lesbian woman living in rural Southern Ohio . . . My new girlfriend Tammy, who's only 35, moved into my trailer very quickly after we got together. Over several months, Tammy took over all of our finances, including my credit cards and monthly disability check . . . Once, after a big fight during which she hit and kicked me, we made up by going out and getting a puppy that I named Sammy.

During our last fight, Tammy was drunk, she got out her gun, loaded it in front of me, and started calling for the dog. After an hour of begging her to stay away from the dog she pointed the gun at me. I didn't call the police because I don't think they'd know how to handle it. I'm a butch lesbian, I worked in a factory most of my life. Tammy is a tiny little Avon saleswoman. Sammy and I finally got out and went to a friend's house. When we went back to the trailer the next day, Tammy was gone. I'm afraid things will get worse when she comes back.

SOURCE: National Coalition of Anti-Violence Programs, 2003, p. 7. Reprinted with permission from the National Coalition of Anti-Violence Programs.

Box 9.1 describes the case of Norma and Tammy. What is the role of homonegativity—both in the macrosystem and as internalized by Norma—in this abusive relationship? How many forms of abuse has Norma experienced? In what ways are these forms of abuse similar to those in other family relationships? What contributing factors appear to be operating? To what extent are they different from other cases of intimate violence? Might some of the circumstances of the case change if Norma and Tammy were living in an urban area where homosexuality is more widespread and accepted?

Exosystem

Although victimization of GLBT individuals may be related to macrosystem values, the actual acts of victimization tend to take place within the individual's exosystems and microsystems, such as schools or jobs. For example, among older GLBT individuals from GLBT support groups and agencies around the United States, 71% of males and 29% of females reported experiences of abuse related to their sexual orientation, including verbal and physical attacks or discrimination in housing or employment (D'Augelli & Grossman, 2001). Many GLBT college students experience considerable hostility: On one campus, more than 75% of gay, lesbian, and bisexual students had been verbally harassed, usually by other students but sometimes by faculty (D'Augelli, 1992). Almost every respondent had heard derogatory anti-gay/anti-lesbian comments on campus, and approximately 75% said they sometimes feared for their personal safety because of their sexual orientation.

Being gay also appears to be associated with discrimination in the workplace. For example, data from the 1989–1991 national General Social Survey showed that gay and bisexual men earned almost 25% less than their heterosexual counterparts, even after controlling for race, sex, age, education, and area of residence (Badgett, 1995). Although it is unknown to what extent these trends still exist today, it appears that homonegativity in the workplace, like homonegativity in the macrosystem, may also operate to keep GLBTs in abusive relationships. As one 36-year-old battered gay man explained: "Alan is threatening to out me to my parents and my employer. I know I'll be fired if my employer finds out that I'm gay. I feel very isolated" (NCAVP, 2000, p. 12).

Microsystem

Characteristics of the Victim

Lesbian Relationships. A childhood history of abuse appears to be a predictor of the victimization of aggression in women's intimate same-sex relationships. In one national sample of lesbians and bisexual women who had been sexually abused by their partners, 71% were incest survivors (Girshick, 2002), and among lesbian respondents recruited at a music festival, a history of child abuse in the victims of partner abuse was associated with greater verbal abuse in the current relationship (Lockhart et al., 1994). By contrast, Renzetti (1992) did not find a high prevalence of child abuse in the histories of the battered lesbians she studied.

Gay Male Relationships. Age, HIV, substance use, and personality have been investigated as possible victim-level risk factors for intimate violence among gay men. A probability sample of urban gay men revealed that gay men aged 40 and younger were at greater risk of intimate aggression than gay men aged 60 and over, and gay men who had tested positive for HIV were more likely to be victims of partner battering than gay men testing negative for HIV (Greenwood, Relf, Huang, Pollack, Canchola, & Cataria, 2002). In addition, in a nationally representative sample of HIV-infected adults, 11% of the gay and bisexual men reported physical harm from a partner or another person close to them since their diagnosis (Zierler et al., 2000). More importantly, among the respondents reporting harm, nearly 45% believed it was their HIV-seropositive status that prompted the physical aggression. Finally, among 700 homosexual men in the New York City area, ecstasy users were more likely than others to be victims of intimate violence (Klitzman, Greenberg, Pollack, & Dolezal, 2002), and borderline personality traits and insecure attachment are also characteristic of gay male victims of partner physical abuse (Landolt & Dutton, 1997).

Characteristics of the Relationship

Dependency and independency issues appear particularly important relationship predictors in violent lesbian relationships (Miller, Greene, Causby, White, & Lockhart, 2001). Among maltreated lesbians, the greater their wish to be independent and the greater their abusive partner's dependency, the more frequent and varied the abuse (Renzetti, 1992). Intimate violence in lesbian couples is also associated with relationship conflicts, including disagreements over who controls or has the most responsibility for the couple's finances, jealousy in both partners, the victim's unemployment, and both the perpetrator's and victim's drug/alcohol abuse (Lockhart et al., 1994).

As is true of other forms of intimate violence, a major predictor of one partner's violence in GLBT relationships is the other partner's violence. For example, among 350 lesbians in Arizona, aggression in intimate relationships was typically mutual; only 30% of the women who had used aggression against a female partner described it as self-defense (Lie et al., 1991). In gay male relationships, the more psychologically and physically abusive one man is, the more abusive his partner is (Landolt & Dutton, 1997).

Special Issue: Mutuality of Abuse

As just mentioned, one of the strongest predictors of violence in GLBT relationships is the other partner's violence. In addition, as shown in regard to intimate violence in heterosexual relationships, it is often difficult to determine whether a perpetrator and victim can be clearly differentiated or if the violence in the relationship is mutual. The principal position of the GLBT professional community is that mutual abuse in GLBT relationships is a myth (e.g., Asherah, 2003; Hart, 1986; Merrill, 1998; Peterman & Dixon 2003; Renzetti, 1992). The main components of the argument are theoretical: (1) calling GLBT abuse mutual minimizes its significance; (2) men are trained to be aggressive and consequently can never be victims; and (3) lesbian women may fight back against their aggressors more often than heterosexual women, but their aggression is self-defense or an expression of rage resulting from earlier victimization. It has also been argued that because court personnel are still generally unfamiliar with the nature of same-sex battering, they often issue mutual restraining orders even in cases where one partner is the sole batterer (Fray-Witzer, 1999, p. 25). According to Merrill (1998), "[When] a victim happens to be a man or a perpetrator happens to be a woman, providers, including criminal justice and mental health professionals, are almost always confused about how to proceed. For example, sometimes battered gay and bisexual men who call the police for assistance are mistakenly arrested as perpetrators, and sometimes lesbians who batter are mistakenly referred to battered women's shelters" (p. 130). To the extent that both parties are seen as equally guilty for the violence without adequate verification, the message is conveyed that what has occurred is "fighting" rather than one party's abuse of the other. Moreover, Fray-Witzer noted, once mutual restraining orders have been issued, if a victim pushes to escape from the abuse, the victim becomes vulnerable to the abuser's threats of criminal prosecution.

Resistance to the idea of mutual violence is reflected in an effort to reconceptualize the potentially different roles of intimate aggression in same-sex relationships. In a study of 62 lesbians identifying themselves as either victims or perpetrators of partner abuse, Marrujo and Kreger (1996) distinguished among primary aggressors, primary victims, and participants. From their perspective, lesbian *primary aggressors* (27% of the sample) are similar to heterosexual perpetrators in that they are pathologically jealous, controlling, manipulative, entitled, and have problems with anger. Lesbian *primary victims* (39% of the sample) are clinically similar to victims of heterosexual intimate violence in experiencing low self-esteem and depression. By contrast, *participants* (34% of the sample) represent a different group, characterized by a repeated pattern of "fighting back" against the primary aggressor with the intention of hurting, injuring, or getting even with that aggressive partner—a response that the authors appear to assume is not itself abusive. The authors also appear to assume that in this last type of relationship, it is always possible to identify a primary aggressor. What do you think?

We have noted in previous chapters that there is strong resistance to the notion that women can be abusive of their male partners, yet there is considerable empirical evidence that some heterosexual women do behave abusively (Hines & Malley-Morrison, 2001). As yet, there does not appear to be definitive empirical evidence establishing the extent to which women may also be mutually abusive against each other within a same-sex relationship, or the extent to which men may use their aggression against each other in a same-sex relationship.

Box 9.2   Blake and Trudy

Blake called the Advocates for Abused and Battered Lesbians (AABL) and explained that a court domestic violence advocate had told her to call because she was a lesbian who had been in a fight with her girl-friend over the weekend and had been arrested. Blake explained that she had been with her partner for five months. At first, things had been good between them, but an issue developed because Blake's lover, Trudy, decided that she should make a real woman out of Blake, whom she considered too masculine. Arguments over what Blake should wear when they went out became increasingly loud and usually ended with Blake giving in to Trudy's demands. Blake said she was big-ger and heavier than Trudy, and knew she could overpower her, but was nevertheless afraid of Trudy, especially when Trudy started throw-ing things. In the incident leading to Blake's arrest, Blake had stopped for coffee with a friend after work, and when she got home Trudy threw a mug at her, yelled at her, and accused her of being unfaithful. Blake said she responded by pushing Trudy into the couch, but imme-diately apologized. Irate, Trudy started punching Blake, who grabbed her wrists and yelled at her to stop. Trudy broke free and started breaking things, then became calm, walked to the phone, said she was going to call someone for support, but instead called the police, who came and arrested Blake.

When the AABL staff called Trudy to get her side of the story, Trudy described several forms of her own abuse of Blake, but put responsibility for her abuse onto Blake, saying she should be more flexible. Trudy admitted that she insists on sex after a fight because it is a way of making up, but also admitted that Blake does not want to have sex at those times because she is tense. Based on their discussions with Trudy, the staff decided that she is the real abuser, not Blake, and they encouraged Blake to join a support group for battered lesbians.

SOURCE: Goddard and Hardy, 1999.

Box 9.2 presents a case provided by Advocates for Abused and Battered Lesbians (AABL) to illustrate the challenge of identifying the victim in cases of lesbian violence. The counselors at AABL believe that identifying the victim is important even in cases where the violence may appear to be mutual; moreover, counselors cannot assume that the individual arrested in such a case is necessarily the perpetrator. In what ways is this philosophy relevant to some of the other cases you have read? From our perspective, the argument that GLBT violence cannot be mutual has not been established empirically, and based upon the empirical research of heterosexual relationships it is likely that much of GLBT intimate violence is indeed mutual.

Individual/Developmental

Family History

Macrosystem homophobia may play itself out within the family microsystem of sexually nonconforming boys or girls. In one sample of homeless gay and lesbian youth, several of the respondents reported that their parents had expelled them from home after learning about their sexual orientation; others left because of physical child abuse (Mallon, 1998). According to one child welfare advocate, "We just keep seeing kids getting beaten up and thrown out of their houses, kids getting beaten up by their fathers for being gay, or young lesbians getting sexually abused by male relatives trying to change them so they won't be gay"(Mallon, 1998, p. 98).

In one small sample of gay and heterosexual college students (Harry, 1989), gay males reported significantly more physical abuse by their parents than did heterosexual males.

The strongest data on the relative rates of childhood maltreatment of GLBT individuals versus heterosexual individuals come from the National Survey of Midlife Development in the United States (MIDUS) (Corliss, Cochran, & Mays, 2002). Compared to heterosexual respondents, gay and bisexual men reported having experienced more childhood emotional and physical maltreatment, and lesbian and bisexual women reported having experienced more severe physical maltreatment. The researchers postulated that there were at least three different pathways by which children with a minority sexual orientation became subject to higher levels of parental abuse: (1) they may be punished for directly disclosing their sexual orientation; (2) they may be proportionally more likely to display atypical gender behaviors that anger their parents; and (3) they may be engaging in undesired behaviors such as substance abuse, which could lead to greater conflict with their parents. Whatever the reason, given that childhood maltreatment has often been identified as a major risk factor for later intimate violence, it is possible that the higher levels of childhood maltreatment that GLBT youths experience put them at risk for either perpetrating or sustaining intimate partner aggression in later relationships.

Most of the available evidence suggests that childhood maltreatment is a risk factor for later violence in lesbian relationships. Among lesbian batterers from both urban and rural communities, every batterer had a family history of violence (Margolies & Leeder, 1995). Approximately 70% of these batterers had been sexually abused as children, 65% had been physically and/or verbally abused, and nearly all had witnessed their mothers being abused by their fathers or stepfathers. A history of extensive child abuse was also found in a Tucson sample of lesbians who perpetrated aggression in their relationships: Over 80% reported having experienced parental emotional aggression, nearly 60% had been subjected to physical aggression, and more than 25% reported at least one act of sexual aggression (Lie et al., 1991).

The link between childhood maltreatment and current intimate violence has been studied only rarely in gay and bisexual men. In one probability sample of nearly 3,000 gay and bisexual men, gay/bisexual men sexually abused in childhood were more likely to participate in sadomasochism/bondage and discipline sexual practices than men without such a history. The findings suggest that for these gay and bisexual men, childhood sexual abuse led to an increased likelihood of engaging in submissive or aggressive sexual roles in adult intimate relationships (Paul, Catania, Pollack, & Stall, 2001).

Alcohol and Substance Use/Abuse

As shown in Chapters 7 and 8, alcohol and substance abuse are risk factors for intimate violence, and there is considerable evidence that compared to heterosexuals, GLBT individuals abuse alcohol and other substances at higher rates. For example, four population-based studies conducted in the year 2000 indicated that compared to heterosexual women, lesbians and bisexual women were more likely to drink heavily (Gruskin, Hart, Gordon, & Ackerson, 2001). Some investigators have reported that 20%–35% of the gay/lesbian population are alcoholic or at least have a drinking problem—a much higher percentage than in the population at large (Renzetti, 1992). Furthermore, compared to heterosexuals, gay men use significantly more "poppers," sedatives, hallucinogens, tranquilizers, and stimulants (Woody et al., 2001), and transgendered individuals are at higher risk for substance abuse (Lombardi & van Servellen, 2000).

Although there is considerable evidence that GLBT individuals are at risk for elevated levels of alcohol and drug use, there is little systematic empirical research examining alcohol or substance abuse as predictors of violence in GLBT relationships. However, there seem to be three major themes in the relevant literature: (1) substance abuse is not a cause of GLBT intimate violence but may be used as an excuse for being abusive (e.g., Island & Letellier, 1991; Renzetti, 1992); (2) lesbians and gay men may drink or use drugs in order to feel more powerful, and, based on their expectations concerning the effects of alcohol or drugs, may behave more aggressively (e.g., Renzetti, 1992); and (3) perpetrators may use drugs or alcohol as part of a coercive pattern; for example, in a sample of almost 300 gay males and lesbians from 14 states, nearly 20% of both the men and women reported that their partners had gotten them drugged or stoned as a sexually coercive tactic (Waldner-Haugrud & Gratch, 1997).

Alcohol/Substance Abuse and Violence in Lesbian Relationships. In one sample of battered lesbians, 39% said that either they or their partners had been under the influence of drugs or alcohol at the time of the battering incident, and an additional 28% said that both had been under the influence (Renzetti, 1998). Renzetti speculated that lesbians may be motivated to drink or use drugs because they believe alcohol or drugs make them more powerful and assertive (which they act out by becoming abusive toward their partners) or because the substance abuse serves as a basis for excusing the abuse (Renzetti, 1998). In addition, *frequency* of alcohol use is positively associated with the number of different types (physical, emotional, sexual) of aggression both perpetrated and sustained by lesbian participants (Schilit, Lie, & Montagne, 1990). In this study, over 60% said that they and/or their partners had used alcohol or drugs prior to an abusive incident. On the other hand, among lesbian batterers from urban and rural communities, there was no direct relationship between drug/alcohol abuse and violence (Margolies & Leeder, 1995).

Personality Characteristics

A number of personality characteristics appear to be predictive of physical aggression in lesbian relationships. These include a high need for control and fusion with their partners (Miller et al., 2001); low self-esteem (Margolies & Leeder, 1995); the batterer's dependency (Renzetti, 1992); and the batterer's jealousy (Renzetti, 1992). In gay male couples, perpetration of both psychological and physical abuse appears to be associated with an "abusive personality"—i.e., with borderline personality disorder, trait anger, insecure attachment, and recollections of poor parent-child relationships during childhood (Landolt & Dutton, 1997).

Applying the Ecological Model

Overall, most of the predictors of maltreatment in GLBT relationships appear to be largely similar to predictors of maltreatment in other family and intimate relationships: A family history of abuse, substance and alcohol use, certain personality traits, and certain types of relationship conflicts are all predictive of aggression in GLBT relationships. However, several researchers argue that there are unique stressors that fall at the macrosystem and exosystem levels that may also influence levels of aggression in GLBT relationships. These stressors include homophobia, homonegativity, and heterosexism. In addition, the proliferation of HIV in this community provides for an additional type of abuse and an additional type of disability that may increase one's vulnerability to abuse. Consider now the case in Box 9.3. What forms of abuse has Jason experienced? What role do macrosystem, exosystem, and microsystem factors play in the abuse Jason has undergone? What individual/developmental factors can you identify that have been shown to be risk factors in other forms of family violence? Are there any features of this case that are unique to GLBT relationships?

Box 9.3   Jason, a 16-Year-Old Gay Male

I left home at 16 when my parents found out I was gay. My mother cried a lot and my father called me a "freak of nature," beat me up pretty badly, then told me he no longer considered me to be his son . . . I couldn't find a job because of my age but some friends . . . told me how easy it was to turn tricks and make some easy money . . . Julio picked me up one night . . . He seemed different than all of the others. He took me to dinner, told me that I was beautiful, asked me questions about myself and seemed really interested in me . . . Julio asked me to move in with him two months after we met. I knew I was in love with him and couldn't wait to start our life together. He wouldn't give me the keys to his house because . . . for "security reasons" he didn't want his keys "floating around." When I was home alone though, I felt trapped because the security door would automatically lock behind me when I left. I was also not allowed to use the telephone because he didn't want me to keep in touch with "the other trash from the streets."

Things were going pretty well until Julio and I had our first disagreement. He'd been drinking and was angry at me for breaking a plate when I was washing it. I tried to apologize but he just kept screaming about how inconsiderate and selfish I was. When he told me he was going out to meet his friends, I started crying and told him that I didn't want to be locked in the house alone. He hit me with the door and pushed me away. I grabbed him by the back of his shirt and, when it ripped, he fell forward onto the sidewalk. He hit his head and it started to bleed . . . [He] pushed me away and said he was going to call the police. I got scared and ran away. I went back later that night and things seemed calm. Julio told me everything was going to be okay and I fell asleep in his arms. I was awakened a few hours later by the police who took me away in handcuffs. I was charged with domestic violence, sentenced to three years probation, and mandated to attend a batterers' group for 52 weeks.

SOURCE: National Coalition of Anti-Violence Programs, 2001, pp. 17–19. Reprinted with permission from the National Coalition of Anti-Violence Programs.

Consequences

Outcomes of intimate violence also appear similar across same-sex and opposite-sex intimate relationships. As with other forms of family violence, the most severe outcome of violence in GLBT relationships is death. A U.S. Department of Health and Human Services report on intimate homicides from 1981 to 1998 (Paulozzi, Saltzman, Thompson, & Holmgreen, 2001) revealed that 6.2% of the intimate partner homicides of men were by same-sex partners and 0.5% of the intimate partner homicides of women were by same-sex partners. The report did not provide information on the relative proportion of same-sex intimate partnerships in the country during this period, although according to a 1980 estimate, 4%–10% of the population is GLBT (Marmor, 1980, in Lockhart et al., 1994). Thus, the rate of intimate homicide in gay and lesbian relationships is not disproportionately high relative to their representation in the population.

Some information on nonfatal injuries resulting from intimate violence in GLBT individuals is available from the NCAVP report on data from the year 1999 (NCAVP, 2000). Intake information from 1,175 clients where injury information was recorded revealed that in 60% of the cases there had been no physical injury. Of the remainder, 29% had received minor physical injuries, and 10% had sustained serious injuries. Among the clients with injuries, 57% did not require medical attention; 12% needed it but did not receive it; 13% received outpatient treatment; and 4% were hospitalized. Much higher levels of injury were reported by a sample of 52 gay and bisexual men recruited from domestic violence programs and AIDS agencies. In this sample, 79% of the men had been physically injured by their partners at least once, and 85% had suffered significant property or financial loss as a direct result of their partner's abuse (Merrill & Wolfe, 2000).

Research concerning the psychosocial correlates of intimate aggression generally cannot establish clearly which correlates are risk factors for violence and which are outcomes. Moreover, it is difficult to separate the outcomes of abuse in GLBT relationships from the effects of other factors, such as society's homonegativity and internalized homophobia. There is some evidence that aggression in GLBT relationships, like aggression in heterosexual relationships, is predictive of mental health problems. In a sample of both lesbians and heterosexual women, nonphysical abuse was predictive of lower self-esteem, and physical abuse was predictive of depression; the gender of the batterer was not a significant predictor of either depression or self-esteem (Tuel & Russell, 1998). Among gay and bisexual men at a gay pride festival, men who were victimized by sexual aggression reported higher levels of dissociation symptoms, trauma-related anxiety symptoms, and borderline personality characteristics. Also, in comparison to gay and bisexual men without a history of unwanted sexual contact in adulthood, those with such a history were significantly more likely to report crack cocaine and nitrite inhalant use, and treatment for substance abuse problems (Kalichman et al., 2001).

Prevention and Intervention

A 1998 report from the National Institute of Justice acknowledges that batterer intervention programs for gays and lesbians are still in their infancy—in part because of a lack of theoretical consensus concerning the causes of maltreatment in GLBT relationships and in part because of the reluctance of members of these communities to expose their problems and possibly contribute further to negative stereotypes (Healey, Smith, & O'Sullivan, 1998). Prevention programs are even more limited. Much of the responsibility for education efforts stems from grassroots organizations.

The Social Service System Response: Shelters

Although there has been a general increase in the number of shelters available for battered women in the United States, these shelters are not equally available to all women victimized by violence. Sometimes, women who differ from agency personnel (e.g., lesbians, women of color) are seen as inappropriate clients by shelter staff (Loeske, 1992, in Donnelly, Cook, & Wilson, 1999). For example, among personnel from state-funded women's shelters in three southern states, the average client was a young white heterosexual woman with a poor or working class background (Donnelly et al., 1999). Personnel from some shelters indicated that they rarely direct outreach efforts at abused lesbians, or even deliberately exclude them to avoid alienating contributors, important board members, or other residents. One respondent from Alabama reported, "Well, we had a situation where a woman came into the shelter, and she didn't identify herself as a lesbian at first. After a few days we found out that her 'partner' was female. The other women, they weren't too happy. The woman that she shared a bedroom with was so uncomfortable that she wound up leaving. From now on, if this happens, we'll give them their own rooms" (Donnelly et al., 1999, p. 730). Another program director explained, "Our shelter is pretty much limited in who we serve. We lean toward a feminist philosophy but, of course, we are in the Bible Belt in the South, so it's not any kind of radical feminist approach. We would not be able to operate in the community. Our family violence philosophy pretty much limits those we serve to family members. We do not consider same-sex partners or unmarried partners" (Donnelly et al., 1999, p. 733).

The Criminal Justice System Response

The nature of the criminal justice system response to intimate violence in GLBT relationships varies considerably by state and probably even within states. As of 2002, 29 states and the District of Columbia had gender-neutral domestic violence laws that included household members and dating partners (National Gay and Lesbian Task Force, 2002). An additional 18 states had domestic violence laws that were gender neutral but applied only to members of a common household. Three states (Delaware, Montana, and South Carolina) had domestic violence laws that explicitly excluded same-sex victims of intimate violence.

By 1995, there were 11 states with no provisions for same-sex, nonrelated cohabitants to obtain a restraining order against an abusive partner (Griffin, 1995, in Peterman & Dixon, 2003). Although 34 states had laws designed to provide equity in response to intimate violence in same-sex relationships, courts often do not treat same-sex intimate violence offenders the same way they do heterosexual offenders. For example, in same-sex intimate violence situations the perpetrator often can bail out more easily, receive lesser penalties, and not be required to attend counseling like other intimate violence offenders.

The Role of Police

There are mixed views on the role of the police as a resource for victims of intimate violence in GLBT relationships. According to one perspective, the police are indifferent to lesbian battering, either because they just do not care whether one lesbian beats up another, because they assume beating is part of the lesbian culture, or because they just cannot take intimate violence seriously when there is no man involved (Russo, 1999). In her study of 100 battered lesbians, Renzetti (1989) found that only 19 had ever called the police for help, and 15 of these victims found the police response to be only a little helpful or not helpful at all. On the other hand, one battered gay man reports that, "Overall, my experience with the San Francisco Police as an openly gay male victim of intimate violence was very positive. Only once did I experience blatant homohatred, when two policewomen refused to file an incident report about a restraining order violation and referred to me as 'she' and 'this woman' to each other and to other officers in my presence" (Island & Letellier, 1991, pp. 21–22).

Box 9.4   Criminal Justice Experiences

When I was struggling to get free, I bit George's finger and arm. I pushed him away and called the police. When the officers came, they arrested both of us and put us in the same jail cell. I'm still not sure what we're being charged with. The last time that George assaulted me, I got a temporary restraining order but the judge refused to grant a permanent order. (Gay white male) (NCAVP, 2001, p. 23)

Sheila, my partner of 3 years, was arrested for assaulting me. When the police arrived they didn't recognize us as a couple and treated it as if we were roommates in a mutual fight. My saving grace was the fact that I didn't fight back and so Sheila was eventually arrested after I had to explain our "situation" several times . . . While Sheila was in jail, I contacted a shelter so I would have a safe place after Sheila got released. I felt relieved to have a safe space but soon I realized that I feel like an "outcast." (38-year-old white lesbian) (NCAVP, 2003, p. 14)

My treatments often made me vomit several times a day. One time when I was in the bathroom throwing up Beth came in to the bathroom and started screaming at me, she called me a faker and started slamming my head on the bathtub until I lost consciousness. The next day when I was lying on my mattress on the floor, I asked Beth to bring me some water. She ran at me and kicked me several times. Later that evening I called the police. After they arrived and heard both sides of the story they said that they "don't understand lesbian relationships." They told Beth to go take a walk to cool down. (39-year-old white lesbian, cancer survivor) (NCAVP, 2003, p. 16)

He knocked me down and repeatedly kicked my knees and ribs. I attempted to call 911, but he ripped the phone out of the wall. They called back and he told them everything was OK, but they dispatched a car anyway . . . The police lieutenant said, "You guys need to get your shit together or someone is going to jail tonight." After the police left, I . . . quietly left the house. I called 911 from my cell phone and got a gay operator, who was sympathetic to my needs. He convinced me to go home and wait for the police to arrive . . . The police interviewed me outside the house and checked my injuries. They went upstairs, and not finding any injuries on my boyfriend, arrested him for domestic violence assault. They asked me to sign the arrest warrant. That was the hardest thing I have ever done. I endured several weeks of legal proceedings. Although I got an "Order of Protection," when it came down to it, I dropped the charges. By that time, I had already moved and just wanted it to be over. (40-year-old gay white male) (NCAVP, 2003, p. 21)

For the next two years, Robert beat me up on several occasions and finally broke my jaw. I got a restraining order against him . . . He called to apologize three days after it had been served. He was being so nice that I let him back into

(Continued)

(Continued)

> the house and . . . he became abusive again. He broke the dishes and called me a "faggot spic." I called the police and they arrested him. Later, Robert called me from the police station and said that since I got him arrested, I should bail him out. I did bail him out but I didn't let him come home with me. Several days later, I returned to court to request a year's extension on the restraining order . . . [The] judge told Robert that he would be arrested again if he came near me . . . (34-year-old gay Latino male) (NCAVP, 2001, pp. 10–11)

SOURCE: National Coalition of Anti-Violence Programs. Reprinted with permission from the National Coalition of Anti-Violence Programs.

Box 9.4 presents the stories of several GLBT survivors describing their experiences with the criminal justice system. As can be seen, their experiences are quite varied. What factors do you think might account for the variations in their experiences? To what extent do you think that the experiences of victims of intimate violence in heterosexual relationships may also vary widely? What is the solution to such variation?

Summary

The prevalence and incidence of intimate partner violence in GLBT relationships seems to be at least as great as in heterosexual relationships. Similarly, some of the risk factors are the same as those found in other family relationships—for example, family history of abuse, substance abuse, and dependency. Others appear to be more unique to GLBT relationships—for example, societal homonegativity, internalized homonegativity, and AIDS— yet these factors can be conceptualized under the general category of "stressors," which, as shown previously, are influential risk factors for other types of family violence. There is very little research on outcomes specific to intimate partner violence in GLBT relationships, although death, physical injuries, and some psychological injuries are possible consequences. The development of intervention and prevention programs addressing GLBT intimate partner violence is in its infancy, and currently many shelters and state laws do not have provisions for intimate violence in GLBT relationships. However, many states have gender-neutral language in their domestic violence laws, and, although not always the case, the criminal justice response to incidents of violence in GLBT relationships has been shown to be helpful in at least some situations. As you read Chapter 12 about the response of the criminal justice system to other forms of family violence, consider the extent to which particular adjustments need to be made to the system to accommodate problems associated with intimate partner violence in GLBT communities.

10

Elder Abuse

The 70-year-old woman ran to her bedroom, slammed the door, and called 911. Even though he was shouting, "Get out of my face, or I'll kill you!" the drunken man chased her, kicked in the door, grabbed her, threw her against a dresser, and wrung her neck with her necklace until the necklace snapped. Then he grabbed her again, dragged her, and yelled, "Come out and see the damage I did to your house, bitch!" She told him that she had called the police; in response, he told her that she better leave fast. When the police came, they found her outside, and then took the man—her 37-year-old son—into custody. (Doege, 2002a)

The above case is an example of one of the many forms that elder abuse can take. In this case, an older woman was physically abused by the son who lived with her. Other elders have reported abuse by other family members, including their spouses and daughters. Furthermore, as with other types of abuse, elder abuse can be of many types: In addition to physical abuse, elders can suffer from neglect, sexual abuse, psychological abuse, abandonment, and financial exploitation. Determining when an elder is the victim of maltreatment is not always as clear-cut as in the above scenario, however. In this chapter, we consider definitional issues in elder maltreatment, as well as research addressing its prevalence, predictors, consequences, and appropriate prevention/intervention efforts.

The "discovery" of elder abuse occurred in the 1970s. After Suzanne Steinmetz, a well-known elder abuse researcher, introduced the concept during a congressional hearing in 1978, more systematic research into its prevalence and dynamics began (Anetzberger, 1987). As is also the case with research on many of the other types of abuse discussed in this book, elder abuse research has been plagued with problems. Although researchers agree that family members sometimes maltreat older members, there have been numerous debates over definitions and dynamics of elder abuse (e.g., Pillemer, 1993; Steinmetz, 1993); consequently, research on the effects of elder maltreatment and ways to prevent or intervene in it is still quite limited.

Scope of the Problem

Definitional problems in the area of domestic elder abuse include issues related to neglect of the elderly and whether financial exploitation is abusive. For instance, although neglect is considered a common form of *child* maltreatment, is it possible to neglect an *adult* in ways that should be considered abusive? Adults are usually self-sufficient and self-determining unless they are in some ways impaired. Therefore, how can they be "neglected"? Or, how impaired must they be in order for us to apply the label "neglect" meaningfully? How about the label "self-neglect," which can be found in many studies on elder abuse, and refers to cases in which older people neglect to eat properly, fail to practice good hygiene, etc.? If an older person neglects his or her own well-being, should we say that person is being maltreated? Or should we reserve such labels for acts committed (or omitted) by someone other than the individuals themselves? Finally, many researchers argue over whether financial exploitation should be considered abusive. An example of financial exploitation is when family members use the elderly person's income for their own benefit. The controversy surrounding this type of abuse concerns cultural differences in perceptions of family obligations. That is, many cultures would not consider this type of financial exploitation to be abusive because family members have an obligation to share their resources with one another.

For the purposes of this chapter, we discuss the various types of abuse identified by the National Center on Elder Abuse, which conducted the first National Elder Abuse Incidence Study (NEAIS) in 1996. Similar to the National Incidence Studies on Child Maltreatment (see Chapter 4), the design for NEAIS employed a sample of 20 nationally representative counties. Moreover, the data for this study included information from two types of sources: (1) reports of domestic elder abuse to protective service agencies; in this case, to adult protective service (APS) agencies, and (2) *sentinel agencies* (i.e., agencies who regularly work with the elderly), including financial institutions, law enforcement agencies, hospitals, and elder care providers; these agencies are sometimes mandated (depending on the state) to report cases of elder abuse to APS, but not all cases of elder abuse recognized by these agencies will necessarily be reported to APS. Data from the NEAIS, similar to data on other forms of family violence from mandated reporters, are likely to significantly underestimate the actual incidence of elder abuse in the United States because the study counts only cases recognized by certain professional agencies.

To arrive at the various subtypes of elder abuse, the investigators analyzed state definitions of elder abuse (many of which differed); convened roundtables of professionals dealing with elder abuse; reviewed definitions offered by elder abuse experts; and pilot tested their definitions with the APS agencies and sentinels that would be part of the study. According to the NEAIS, elder abuse perpetrated by family members can be of the following types: (1) *Physical abuse,* which "is the use of physical force that may result in bodily injury, physical pain, or impairment . . . The unwarranted administration of drugs and physical restraints, force-feeding, and physical punishment of any kind are . . . examples of physical abuse"; (2) *Sexual abuse,* which "is nonconsensual sexual contact of any kind with an elderly person"; (3) *Emotional or psychological abuse,* which "is the infliction of anguish, emotional pain, or distress"; (4) *Neglect,* which "is the refusal or failure to fulfill any part of a person's obligations or duties to an elder . . . Neglect typically means the refusal or failure to provide an elderly person with such life necessities as food, water, clothing, shelter, personal hygiene, medicine, comfort, personal safety, and other essentials included as a responsibility or agreement"; (5) *Abandonment,* which "is the desertion of an elderly person by an individual who has assumed responsibility for providing care or by a person with physical custody of an elder"; (6) *Financial or material exploitation,* which "is the illegal or improper use of an elder's funds, property, or assets"; and (7) *Self-neglect,* which "is characterized as the behaviors of an elderly person that threaten his/her own health or safety" (National Center on Elder Abuse, 1998).

According to the NEAIS, in 1996 approximately 450,000 of the 44 million people over the age of 60 in this country at that time were abused and/or neglected (excluding self-neglect) in a domestic setting, which is equivalent to a rate of 11.3 per 1,000 elders. Of these abuse incidents, 16% were reported to APS for investigation. Therefore, there were five times as many abusive incidents that went unreported as were reported. When self-neglect is added to the definition of elder abuse, over 551,000 elders were abused in 1996, 21% were reported to APS agencies for investigation, and four times as many cases went unreported as were reported. Self-neglect was the type of abuse most commonly reported to APS. Why might this be? What kinds of dynamics could be involved here?

Of the types of elder abuse that were substantiated by APS agencies, the largest percentage were for neglect (48.7% of all cases); followed by emotional or psychological abuse (35.4%); financial exploitation (30.2%); physical abuse (25.6%); abandonment (3.6%); sexual abuse (0.3%); and other types (1.4%). Of all cases reported to APS, physical abuse was the most likely to be substantiated (61.9% of reported cases were substantiated); followed by abandonment (56%); emotional or psychological abuse (54.1%); financial exploitation (44.5%); neglect (41%); and sexual abuse (7.4%).

The NEAIS is probably the most comprehensive study of elder abuse to date; however, it does suffer from methodological problems (e.g., its reliance on reports from sentinels), and its results should be interpreted with caution. Other studies have been conducted on representative samples of elders and have also established preliminary rates of elder abuse. For instance, of over 2,000 elders in the Boston metropolitan area in the mid-1980s, 32 out of every 1,000 elders were the recipients of abuse (physical, emotional, and/or neglect) (Pillemer & Finkelhor, 1988). If these results could be generalized nationwide, between 701,000 and 1,093,560 elders would have been abused that year. In a random sample of 342 elders in New Jersey, five were abused, which is a rate of 15 per 1,000 (Gioglio & Blakemore, 1982), considerably smaller than that established by the Boston study but somewhat similar to the NEAIS estimates. Each of these studies must also be interpreted with caution because of narrow definitions of what constitutes abuse (Pillemer & Finkelhor, 1988) or because of vague definitions of elder abuse (Gioglio & Blakemore, 1982). Can we determine the exact rates of elder abuse then? As with other types of abuse, it is difficult to establish a reliable estimate, but according to a report issued by the House Select Committee on Aging (1991), between 1 and 2 million elders are abused each year in the United States.

Consider the cases presented in Box 10.1. Each one represents at least one of the types of elder abuse just discussed. What are the differing dynamics in each of these cases? Do you think that the predictors and correlates of each type of abuse may be different? If so, what are the different factors operating in each case? Are there also likely to be differences in the possible consequences of each of the forms of maltreatment? How might prevention and intervention efforts differ depending on the situation?

Box 10.1   Cases of Elder Abuse

Mary, age 78, has lived with her daughter, Rose, age 51, for the past 10 years since Rose's husband died. Rose was a waitress until two years prior to these incidents and has since been unemployed. They both live off of Mary's Social Security check, which makes it tough to get by.

Rose spends most of her days in front of the television and drinks more than 10 alcoholic beverages a day. Mary gets increasingly frustrated with Rose because she feels Rose should be out looking for a job.

Mary and Rose used to be constant companions and used to enjoy one another's company. But since Rose began drinking, there have been a number of conflicts between her and her mother, several of which have ended in abuse. For instance, she pushed her mother against a wall 6 months ago but apologized profusely afterwards. When conflicts over Rose's working and drinking problems grew worse, Rose got upset and told her mother that she was a terrible mother and wife and blamed Mary for Mary's husband leaving when Rose was a child. When Mary told Rose to get out after that comment, Rose slapped Mary across the face. In addition, Mary found out that Rose had been withdrawing a lot of money from their joint checking account, most likely to pay for alcohol. When she confronted Rose about this, Rose pushed her mother, who fell to the floor. Mary twisted her ankle in this incident and could barely walk for days. Rose didn't help her mother at all during this time period, and Mary could barely get to the bathroom, nor could she prepare meals for herself (Baron & Welty, 1996).

An older frail woman who was caring for her husband with Alzheimer's disease remarked, "He would beat me pretty bad, choke me. He grabbed me and said 'I'll kill you.'" (Pillemer, 1985, p. 154).

Mr. W. lives with his adult daughter, who is schizophrenic. He is dependent upon her for all his activities of daily living. When she is not suffering from the effects of schizophrenia, she provides for his care; however, when she does have psychotic episodes, she neglects his care (Greenberg, McKibben, & Raymond, 1990).

Agnes, age 85, was widowed last year. She moved in with her 55-year-old daughter Emily because Agnes has problems with arthritis and congestive heart failure. Emily is often incredibly frustrated and worried about her mother, her child, and her husband who may be forced into early retirement. One time, Emily found herself calling her mother names and accusing her of ruining her life. Another time, she slapped her mother (American Psychological Association, 2003).

Another major problem facing researchers in elder abuse involves identifying the dynamics of this type of abuse. In child abuse, a dependent child is maltreated by a parent, and in spousal abuse spouses either abuse each other or one spouse is the primary victim of abuse. In the case of elder abuse, however, there is a wider range of possibilities for types of abusive relationships. For example, (1) *spousal elder abuse* can occur in which long-standing spousal abuse situations continue into old age; (2) elder abuse can be the result of *caretaker stress,* in which a family member (e.g., an independent adult child) who is caring for an impaired elder becomes frustrated at the elder and becomes abusive; or (3) a well-functioning elder can be taken advantage of by an adult family member who is dependent upon the elder; in other words, an otherwise healthy elder is abused by a *dependent adult child.* Each of these types of elder abuse is discussed in the following sections. Before reading on, though, consider again the cases presented in Box 10.1. Which cases do you think are examples of spousal elder abuse? Caretaker stress? Abuse by an adult dependent child?

Spousal Elder Abuse

According to the NEAIS data, abuse by spouses is more likely to go unreported than reported; only 19% of cases reported to APS as compared to 30.3% of unreported cases are committed by spouses. In the random sample survey of elder abuse in Boston, *nearly three fifths* of the perpetrators were spouses (Pillemer & Finkelhor, 1988). Yet even though a substantial percentage of elder abusers are spouses, this type of abuse has been the least researched. Perhaps, as Pillemer and Finkelhor suggested, the reason for this lack of attention is because elderly spouses hitting each other does not conjure up the same abhorrent images as does the situation of a stronger, healthier, middle-aged adult hitting an elder. Moreover, although many people probably assume that the consequences of spousal elder abuse are not as severe as other forms of elder abuse, Pillemer and Finkehor's data show otherwise: According to their study, there were no differences in the level of violence committed, the number of injuries sustained, or the degree of upset engendered in the victim between cases where the abuser was a spouse versus those where the abuser was an adult child. Because the elderly are much more likely to live with their spouses only than with their adult children only (Pillemer & Finkelhor, 1988), this area of spousal elder abuse needs more attention.

The rate of abuse by elderly spouses may be as much as 41 per 1,000 elders (Pillemer & Finkelhor, 1988). Other studies have shown that between 4% and 6% of elders are involved in physically abusive spousal relationships

(Harris, 1996; Mouton, Rovi, Furniss, & Lasser, 1999). Injuries have been shown to occur in as many as 6% of physically abused elderly husbands and 57% of physically abused elderly wives (Pillemer & Finkelhor, 1988). However, even though these rates are quite substantial, little effort has been made to protect these elderly victims of spousal abuse, probably because neither the spousal abuse nor the elder abuse system is specifically designed to deal with this form of maltreatment. Spousal abuse programs are designed with young couples in mind, and elder abuse programs are designed around the notion that caretaking relatives are abusing the elders. Therefore, even though the elderly victims of spousal abuse should conceivably have two systems to turn to for help or to investigate their plights, they often fall through the cracks (Brandl & Cook-Daniels, 2002).

Elder Abuse by Caregiving Relatives: Caregiver Stress Hypothesis

In contrast to spousal elder abuse, the abuse of elders by adult relatives who are entrusted with their care is the most researched and most widely recognized type of elder abuse. Statistics compiled by APS, as seen in the NEAIS, indicate that frail, confused, elderly women over the age of 80 are the most likely group to experience elder abuse, primarily at the hands of family members. According to work by Steinmetz (1988), these elderly reside in the homes of their middle-aged children, who must adjust their lifestyles to accommodate to this new responsibility. Steinmetz (1993) noted that, "the stress, frustration, and feelings of burden experienced by caregivers who are caring for dependent elders can result in abusive and neglectful treatment" (p. 223). In support of this statement, it has been shown that the more stressful the caregiving is *perceived* to be by middle-aged adults, the more likely they are to resort to using abusive behaviors against the elder (Steinmetz, 1988). Furthermore, Fulmer and O'Malley (1987) found that in comparison to 147 elders who were not referred for elder abuse, those 107 who were referred were more dependent on their caregivers.

Elder Abuse by an Adult Dependent Child: Adult Dependent Child Hypothesis

Proponents of the view that elder abuse is primarily committed by adult dependent children relying on their aging parents for help rather than by adult caregivers taking care of elderly parents (e.g., Pillemer, 1993) point to the fact that the majority of case control studies on this issue (e.g., Bristowe & Collins, 1989; Godkin, Wolf, & Pillemer, 1989; Homer & Gilleard,

1990; Phillips, 1983; Pillemer 1985, 1986; Pillemer & Finkelhor, 1989; Pillemer & Suitor, 1992) have found that victims of elder abuse do *not* differ from nonvictims in their health status and level of dependency. In addition, although women tend to be the primary caretakers of the elderly, males are more likely than females to perpetrate elder abuse (National Center on Elder Abuse, 1998), which points towards a weakness in the caretaker stress hypothesis. Pillemer (1993) argued that the popular assumption that elders are abused by their middle-aged, stronger caregivers stems from media attention and the attention of public officials. Indeed, we seem to be swayed morally by the notion that frail elders should be protected from the frustrations of their caregiving relatives. As shown in a later section, the majority of elder abuse intervention programs seem to be designed to reduce the stress of caregivers (e.g., Scoggin et al., 1989). But, as Pillemer (1993) argued, the more rigorous studies of elder abuse clearly show that we need to look at the characteristics of the abuser, not the characteristics of the victim, if we are to clearly understand elder abuse.

Research on characteristics of the abusers shows that typically they are the ones who are dependent. For example, in a survey of community agencies in Massachusetts, two thirds of the elder abuse cases were characterized by the adult child being financially dependent upon the abused elder (Wolf, Strugnell, & Godkin, 1982). Of reported cases of elder abuse in Wisconsin, 75% of the abusers were living in the elder's home, and 25% were financially dependent upon the elder (Greenberg, McKibben, & Raymond, 1990). Pillemer's own research also supports his supposition that elder abuse is committed by adult dependent children. In a series of studies, he found that in comparison to nonvictimized elders, those elders who were abused were more likely to have their adult children dependent upon them in several areas, including housing, household repair, financial assistance, transportation, and cooking and cleaning (Pillemer, 1985, 1986; Pillemer & Finkelhor, 1989).

Predictors and Correlates

The NEAIS of 1996 reported predictors and correlates of elder abuse separately for cases reported by APS and by the sentinels. The APS reports contained data on the income level, ethnicity, gender, age, and physical and mental frailty of the victims, and the sentinel reports had data on all of those variables except ethnicity and income. Furthermore, APS and sentinel reports contained data concerning the perpetrator's gender, age, race/ethnicity, and relationship to the victim. Many of the smaller-scale studies on elder abuse also provide data on predictors and correlates. Following is a summary of the findings of these reports and their similarities and differences at most levels of the ecological model.

Macrosystem

Abuse reported by APS in the NEAIS was most likely to have been perpetrated against older individuals earning under $15,000 per year. Specifically, in 91.5% of neglect cases, 75% of emotional/psychological abuse cases, 75.5% of physical abuse cases, 77.6% of financial exploitation cases, and 100% of abandonment cases, the victims made under $15,000. Also according to APS reports, the majority of the victims were White (84%), followed by Blacks (8.3%), Hispanics (5.1%), Asians (2.1%), and Native American Indians (0.4%). Whites were overrepresented in every category of abuse, with the exception of abandonment, in which a slight majority of victims (57.3%) were Black. The ethnicity of the perpetrators reflected that of the victims: The majority were White, according to both APS (77%) and sentinel (63.5%) reports (National Center on Elder Abuse, 1998).

Microsystem

Frailty of the Victim

Researchers who assert that elder abuse is committed primarily by over-stressed caretaking relatives tend to focus on the characteristics of the victims that predispose them to be at risk for abuse. For example, in a Wisconsin sample of 204 elder abuse cases, 51% of the victims were frail, 12% had Alzheimer's, and 20% were homebound (Greenberg, McKibben, & Raymond, 1990). In comparison to the Wisconsin elderly population in general, in which 20% shared a household with someone other than a spouse, 73% of the abused elders lived with an adult child (75%, however, lived in the elder's home, not the child's home). In these cases, the frailty and incapacities of the elder are viewed as the factors that put them at risk for abuse by their caregiver relatives.

The NEAIS also provides evidence that frailty in elderly people may put them at risk for abuse. According to APS reports, the majority of elder abuse victims are unable (47.9%) or somewhat unable (28.7%) to care for themselves and sometimes confused (27.9%) or very confused and disoriented (31.6%). However, these statistics stand in stark contrast to those characterizing cases reported by sentinels: Only 18.8% were unable to care for themselves, 33.1% were somewhat able to, 7.5% were very confused and disoriented, and 37.9% were somewhat confused. Thus, the cases that were not necessarily reported to APS were ones where the elders showed higher functioning (National Center on Elder Abuse, 1998). What are the implications of these findings for the caregiver stress hypothesis? For the dependent adult child hypothesis?

Gender and Age of the Victim

According to both APS and sentinel reports of the NEAIS, elderly females were more likely to be abused than elderly males, even after accounting for elderly females' overrepresentation in the population. This was the case for every type of abuse reported to APS except for abandonment, in which males were more likely to be victimized. According to APS reports, the majority of reported elder abuse incidents were perpetrated against the oldest old—those who were aged 80 or older—and continually decreased as the age of the victim decreased. This pattern held true for elder abuse overall and for neglect, emotional abuse, physical abuse, and financial exploitation. (Sexual abuse and abandonment occurred too infrequently for exact rates by age to be estimated.) A different picture emerged for elder abuse cases reported by sentinels: Those who were aged 60–70 were the most likely to suffer from physical abuse, emotional abuse, neglect, and self-neglect. Those aged 71–80 were the most likely to suffer from abandonment, and those over age 80 were the most likely to be neglected.

Individual/Developmental

Researchers who stress that it is the dependency of the abuser on the elder—not the other way around—that contributes most to elder abuse tend to concentrate most on characteristics of the abuser (i.e., on individual/developmental level factors) that predict elder abuse. These predictors seem to mirror the most powerful predictors found for child and spousal abuse. For example, the abusers may have been subjected to abuse themselves when they were children, often by the elderly parent they are now abusing (e.g., Steinmetz, 1978). Abusers are also likely to have problems with drugs and alcohol and may also have chronic psychological problems (Bristowe & Collins, 1989; Greenberg, McKibben, & Raymond, 1990). Furthermore, abusers are likely to have mental and emotional problems, be hospitalized for those problems, and be violent and possibly arrested in other situations (Pillemer & Finkelhor, 1989; Pillemer & Wolf, 1986). Finally, in comparison to families not characterized by elder abuse, those experiencing elder abuse are more likely to have a family member with emotional problems and interpersonal difficulties and to be more socially isolated (Godkin, Wolf, & Pillemer, 1989).

The NEAIS also provides data on individual/developmental level characteristics of the abusers. According to both APS and sentinel reports, the majority of abusers were male (53% and 63.1%, respectively). According to APS reports, males were more likely to commit every type of abuse except neglect. (Sentinel reports were not broken down by abuse type.) Also according to both APS and sentinel reports, the majority of abusers were under the age of 60 (66% and 60.7%, respectively), and, according to APS reports, individuals under 60 committed the majority of all types of abuse. Finally, according to both APS and sentinel reports, children were the most likely perpetrators of elder abuse (47% and 30.8%, respectively), followed by spouses (19% and 30.3%). Other perpetrators included parents, grandchildren, siblings, friends and neighbors, and other relatives, the relative distribution of which depended upon the agency reporting (National Center on Elder Abuse, 1998).

Special Issue: Abuse of Caretakers

Although not a well-researched area, several studies have shown that elders can be very abusive with their caretakers, usually their spouses or children. For instance, between 29% and 33% of elders are physically abusive towards their caretakers (Coyne, Reichman, & Berbig, 1993; Phillips, Torres de Ardon, & Briones, 2000), and 16% of elders may be *severely* violent toward their caretakers (Paveza et al., 1992). In addition, 34%–51% of elders have been shown to be verbally abusive of their caretakers, 18%–34% physically abusive, and 7% sexually abusive (Hamel et al., 1990; Steinmetz, 1988). When comparing these statistics to those of caretaker abuse toward the elders, one wonders which problem is more severe. Hamel et al. (1990) reported that abuse of caretakers seemed to be a function of long-standing interaction patterns between the elder and caretaker. This result was confirmed by Phillips et al. (2000), who found that many caretakers did not view the behaviors as abusive; rather, the behaviors were seen as normal and merely the way the elder had behaved for years. In addition, four variables seemed to predict caretaker abuse: younger age of the elder; greater difference between the caretaker's past and present images of the elder; a perception of a power imbalance in favor of the elder; and greater interpersonal conflict (Phillips et al., 2000).

Consequences

Although little research has been done to assess the psychosocial or physical consequences of elder abuse, a few effects have been identified. Physically abused elderly are at an increased risk for injuries, and neglected elderly are at an increased risk for malnutrition and dehydration. One study that followed elderly residents of New Haven, Connecticut for 13 years (Lachs, Williams, O'Brien, Pillemer, & Charlson, 1998) found that those who were abused at any time during the follow-up period had a poorer survival rate (9%) than either those who were self-neglecting (17%) or those who were never abused (40%). In addition, victims of elder abuse were significantly more distressed than nonvictims; however, victims who had more social support, a higher sense of mastery, more feelings of self-efficacy, and active reaction patterns were less distressed than their counterparts (Comijs, Penninx, Knipscheer, & van Tilburg, 1999). In addition, depression, suicidal ideations, shame, and guilt have been shown to be possible consequences of elder abuse, although it is unclear whether the abuse or the symptoms came first (Anetzberger, 1998; Pillemer & Finkelhor, 1988; Reis & Nahmiash, 1997).

Finally, death resulting directly from the abuse is a potential outcome of elder maltreatment. An example of this outcome is presented in Box 10.2. As you read this case, consider what types of abusive behaviors the daughter may have perpetrated against her father in the past. Also consider which of the three dynamics of elder abuse this case represents: Is it a case of spousal abuse, elder abuse perpetrated by a caregiving child, or elder abuse perpetrated by an adult dependent child? How common do you think such a scenario may be? That is, how often do you think elder abuse results in the death of the elder? A discussion of these particular issues of elder homicide can be found in Box 10.3.

Box 10.2   The Murder of Handy Morrow

Handy Morrow was an 85-year-old man who lived in Detroit with his daughter Edna. Handy liked to tape record religious sermons from the radio; however, one day while he tried to record a sermon, his daughter was playing her television set too loudly. Handy asked Edna to turn her television down, but she refused. After the verbal conflict that followed, Handy went down into the cellar and turned off the electricity to Edna's room. Edna became so enraged by Handy's act that she started yelling at him. The entire argument to this point, and the incidents that follow, were all recorded on Handy's tape recorder. While Edna was yelling at Handy, "thwacks" could be heard in the background, and Handy was heard groaning. Edna then yelled, "Shut your face! I don't want to hear you! I don't want to see you!" Apparently, the "thwacks" that were heard were blows from a tire iron that Edna used to beat her father. Handy was found four days later by his grandson. By that time, Handy was covered in his own waste and hideously bruised, with a broken arm. In addition, he had not been fed or given anything to drink since the incident with Edna. Handy died 24 days later.

SOURCE: Bachman, 1993.

Box 10.3   Homicide of the Elderly

The rates and dynamics of homicide against the elderly are difficult to ascertain, probably because so few elderly are murdered. Therefore, subdividing the murders that were perpetrated against the elderly into the types of offender-victim relationships results in unreliable numbers. A few national and state studies have been undertaken, though, to determine who kills the elderly and why.

Rates of murder against the elderly by a family member tend to stay constant at about 1 per 100,000 people aged 65 and over (Bachman, 1993). In contrast to those aged 64 or younger, who are most likely to be murdered by acquaintances, the elderly are equally likely to be murdered by acquaintances, strangers, or relatives. In addition, although younger people are most likely to be murdered during some sort of conflict, the elderly are most likely to be murdered during the commission of a felony (Bachman, 1993).

According to some of the most recent statistics released from the National Crime Victimization Survey (NCVS), between 1992 and 1997, 1,000 murders were committed against elders. During that time period, the average population of elders was 31.3 million; therefore, the murder rate for the elderly was 3.2 per 100,000. Relatives, intimates, or others known to the victim committed half of these murders, and relatives and intimates committed 25% (or 260) of all murders against the elderly in those years. Finally, elderly victims of murder were twice as likely to have been killed by a relative or intimate than younger victims (Klaus, 2000).

Although these data from the NCVS provide an indication as to how often the elderly are murdered by their family members, they do not give information concerning which family member is committing

(Continued)

(Continued)

the homicide. Recall that elder abuse can be committed by spouses, children, or other relatives, and the dynamics of each of these types of abuse could be very different. One source of data on the identity of the relative who killed an elder comes from the Bureau of Justice Statistics report on age and intimate partner homicide. According to the FBI's Supplementary Homicide Reports, between the years 1993 and 1999, the rate of intimate partner homicide against females ages 65 and over was between 0.5 and 0.7 per 1,000 females, and for males it was between 0.1 and 0.2 per 1,000 males (Rennison, 2001). Furthermore, according to the State of New Jersey's reports on domestic violence and homicide, between the years 1985 and 1997 there were 32 male and 64 female elderly victims of spousal homicide. Overall, 0.4% of elder male victims of spousal abuse and 0.4% of female elder victims of spousal abuse were killed by their partners (Rosenthal, 1999).

In a minority of these elder spousal homicide cases, homicide-suicide was perpetrated. That is, in addition to killing a spouse, the homicide perpetrator, usually male, also killed himself. The rates of homicide-suicide among the elderly tend to be about 0.4 to 0.9 per 100,000, or approximately 12% of all elder homicides (Cohen, Llorente, & Eisdorfer, 1998). The perpetrators of homicide-suicides are males, their victims are their female partners, and there is usually some sort of depression or psychological disturbance in the male perpetrator. Based on her research on elder homicide-suicide in Florida, Cohen (2000) debunked the myth that homicide-suicides are mercy killings, altruistic events, or suicide pacts. Rather, homicide-suicides, she says, are usually acts of desperation and depression. Guns are the most common weapon, and homicide-suicides tend to fall under three typologies: dependent/protective, aggressive, and symbiotic.

1. *Dependent/protective homicide-suicides* comprise almost half of all spousal homicide-suicides. The husband, who may or may not be ill himself, is caring for a chronically ill spouse. There is evidence of serious depression in the husband, which most likely has gone untreated, and his homicide-suicide is not an impulsive act; however, preceding the act, there is usually either a decline in the perpetrator's physical health and/or talk of entering a nursing home/assisted living facility. The wife in this case is not a willing or knowing participant in the homicide-suicide—the decision is solely made by the man.

2. *Symbiotic homicide-suicides* comprise 20% of homicide-suicides in the elderly. In this case, both are sick, and both have talked about wanting to die or being better off dead, although the male is the perpetrator of the act.

3. *Aggressive homicide-suicides* comprise 33% of all cases. In these instances, there is a history of verbal and/or physical conflict and a history of partner violence. The male perpetrators are usually nine to ten years older than their female partners, and neither has a physical illness. The trigger in this instance is typically talk of separation or divorce by the victim. Rather than being planned, the homicide-suicide in this case is a surprise, and very violent (Cohen, 2000).

Prevention and Intervention

To start this chapter on elder abuse, we introduced the case of a 70-year-old woman who was being physically abused by her 37-year-old son. During that incident, the woman called the police, and this was the third time she had called police in the previous few years. The first time happened two years prior—during that incident her son had thrown things around the house, spit in her face, and threatened to kill her. When a judge questioned her about the incident, she stated that she was deeply hurt by what he had done, especially because he was the one who used to protect her from her husband's abuse when he was alive. Even though she was hurt, she stated that she needed him around the house because he did all of the heavy work for her, including shoveling and yard work. Her son received 18 months probation because of this incident (Doege, 2002a).

While still on probation, her son acted out in a drunken rage again—he smashed her bedroom mirror, destroyed a television set, and grabbed his mother's arm so tightly he ripped off a piece of her skin. This time, her son served 140 days in prison, partly because of the pleas that she gave for his freedom. In the incident that introduced this chapter, which occurred nine months after he was released from prison, he was charged with aggravated battery and obstructing an officer. The mother stopped cooperating with prosecutors, and the son was then charged with six counts of intimidation of a witness. Although the son will still be put on trial, the fact that the mother will probably not appear makes efforts to stop the violence that much more difficult (Doege, 2002a).

This situation, in which the victim of elder abuse is unwilling to be proactive in eliminating the violence in his/her life, is not uncommon. Researchers in the field of elder abuse point to several reasons why elders are sometimes unwilling to take the necessary steps to free themselves from abuse. For instance, in the case of spousal abuse, they may have just given up on the idea of living a life that is free of maltreatment (Vinton, 2002). There may also be a belief, which is particularly strong among the elderly, that it is inappropriate to share the secrets of the abusive behavior with strangers or even friends (Vinton, 2002). There are also concerns for elderly victims similar to the ones that the 70-year-old woman in the above story voiced: If their abusers were removed from the home, then there is the possibility that the elders would be isolated and no one would be there to help with transportation, finances, or chores (Doege, 2002a). Finally, there is a reluctance to turn in a loved one, a reluctance that seems to be more pronounced when the abuser is a child rather than a spouse (Vinton, 1991).

By 1991, all states had elder abuse statutes in their laws and/or had amended their existing APS laws to bring elders under APS protection. Currently, every state has an intervention agency, usually APS, that handles cases of elder abuse; however, there is little to no uniformity in the laws concerning, definitions of, and provisions for elder abuse across the states. For instance, 46 states have mandatory reporting laws, such that professionals who care for elders must report suspected cases of abuse; failure to do so usually can result in some kind of misdemeanor charge. The other four states (Colorado, New York, Wisconsin, and Illinois) have voluntary reporting. Of those states with mandatory reporting, 15 not only require mandatory reporting from professionals but also from *anyone* who suspects elder abuse. Furthermore, in many cases of elder abuse, domestic violence laws (for cases of spousal elder abuse) or guardianship laws (for cases in which a frail elder needs to be removed from an abusive home) can be relied upon in certain cases of elder abuse (Griffin, 1999b; Jones, 1994; Williams & Griffin, 1996).

Even though most elders go to the doctor on a regular basis, some investigators have found that in states in which physicians are mandated to report elder abuse, fewer than 10% of referrals to APS are made by health care professionals (Lachs, 2003). Some researchers surmise that this reluctance by physicians to report results from a conflict between doctor-patient confidentiality and the law. Furthermore, an elderly patient is an adult who has certain rights, such as the right to self-determination. If an elder understands that he/she is being abused but wishes to remain in that situation, a physician may believe that there is little that can be done, or that removal from the situation into a nursing home is the most appropriate action (e.g., Gottlich, 1994; Macolini, 1995).

In order to assist physicians in dealing with cases of potential elder abuse, Lachs and Pillemer (1995) provided a guide. First, they argue, the physician must ask two crucial questions: Does the patient accept intervention, and if not, does the patient have the capacity to accept or refuse intervention? If the patient accepts intervention, the physician can implement a safety plan, educate the patient about elder abuse, provide assistance with the possible causes of the abuse, and refer the case to the appropriate agency. If the patient does not accept intervention and is capable of refusing intervention, the physician should educate the patient about elder abuse, provide the patient with information about assistance services and referrals, develop a safety plan, and develop a follow-up plan. Finally, if the patient does not accept intervention but does not have the capacity to reject intervention, the physician should call APS and discuss with them the following options: assistance with financial management, conservatorship, guardianship, and court proceedings for orders of protection.

It is not only physicians who have a poor record of reporting elder abuse to APS. As mentioned previously, the NEAIS showed that most cases do not get reported to APS, and many elder abuse victims are not getting the intervention they may need. To deal with this problem, several professionals have suggested that multidisciplinary teams be formed to investigate and diagnose elder abuse. These teams should include a doctor, nurse, social worker, and an APS specialist. Other team members may include police, a sheriff, district attorney, forensic pathologist, DNA expert, neurologist, psychiatrist, and a member of the clergy (Lachs, 2003).

Recently, a multidisciplinary strategy was attempted in New York City's housing projects. In this intervention, after the police were called for an incident of elder abuse, in half of the cases a police officer and a domestic violence counselor returned to the home several days later to discuss elder abuse, legal options, police procedures, and social services with the victims. In addition, if the abuser was present, the police officers would make it clear that the household would be monitored regularly by the police. This project had surprising results: The households receiving the multidisciplinary intervention actually had an increase in the number of calls to police during the 12-month follow-up and reported significantly higher levels of physical abuse than households not receiving the intervention (Davis & Medina-Ariza, 2001). Interpretation of these results is difficult—did the abuser get angry at the police report and intervention and subsequently increase the level of physical abuse? Did the intervention make the victim more sensitive to the nature of abuse and/or more willing to call the police, even though the amount of abuse may have decreased or remained the same? It is difficult to answer these questions based on the data gathered, and obviously more research is needed on the effectiveness of multidisciplinary teams in ameliorating the problem of elder abuse.

Advocates argue that in order for elder abuse to be dealt with effectively, we must educate the public about the problem, increase the availability of respite care, increase the supports available for families caring for elders, and encourage counseling and treatment for the problems that contribute to elder abuse (American Psychological Association, 2003). The available interventions seem to concentrate mostly on the type of elder abuse in which the stressed adult caregiver abuses the frail elder. As mentioned previously, this type of elder abuse is the most widely known type, and, consequently, it receives the most intervention efforts, even if it is not the only, or even the most prevalent, form of elder abuse. In a study of intervention programs for elder abuse, the most accepted strategies were those providing nursing and other medical care and homemaking assistance. The programs that had been most accepted and successful for the abuser were those providing supportive counseling to reduce anxiety, stress, and depression, along with education and training (Nahmiash & Reis, 2000). Campbell Reay and Browne (2002) found that an education and anger management intervention program for people who abuse and/or neglect the elderly proved successful in reducing the amount of strain, depression, and anxiety on the abuser. In addition, physically abusive caretakers reported declines in their use of physical aggression, although their average rate of using physically aggressive tactics was still quite high. These declines in psychological and behavioral problems were maintained at follow-up.

In addition to programs that aim to help the abuser deal with his/her stress, strain, and abusive reactions to problems, several programs have been devised to aid the elderly victim. For instance, some shelters have been established, but these are very few and have little funding. Moreover, it seems as if elders are reluctant to use them (Lachs, 2003). One team of advocates in Wisconsin has teamed up with shelters for victims of wife battering to try to provide shelter services to abused elderly women. This Older Abused Women's Program, which was established by the Milwaukee Women's Center in 1992, has served over 500 women ages 50 and older (Doege, 2002b). Analysis of potential clientele revealed that younger older women physically abused by their spouses or adult children were the least likely to accept services for elder abuse. By teaming up with the domestic violence advocates, members of this group hoped to find ways to help these women help themselves. Results on the effectiveness of this partnership are not yet available (Raymond, 2002). In addition to this project, advocates in Florida, a state that has a high percentage of elderly in its population, report that most domestic violence shelters in the state have housed elderly women, participated in joint activities with elder groups, and/or targeted elderly women for special programming (Vinton, 2002). Although the use of domestic violence shelters is a viable, and perhaps attractive, alternative for abused elderly women, there are no such efforts to aid elderly men, who are also at risk for abuse.

In addition to the possible use of shelters for intervention, strategies fostering empowerment of the elder victim, such as support groups, information about rights and resources, and volunteer buddy/advocates are currently the most successful victim-oriented strategies, whereas referrals to community activities and programs are the least successful (Nahmiash & Reis, 2000). Several intervention programs to prevent financial abuse have begun to appear. These types of programs, which may be run by social service agencies, assist the elderly in writing checks for their monthly expenses and in preparing budgets to help them manage their expenses. They also offer the elders a safe place to hold their valuables and to cash their checks (Baron & Welty, 1996). Finally, the National Center on Elder Abuse has a Web site with resources for people who want to learn more about or report elder abuse, including hotline numbers in all states, fact sheets on elder abuse, and other resources for families, caregivers, and professionals dealing with elder abuse (http://www.elderabusecenter.org).

Summary

As this chapter has shown, elder abuse is a complex, multidimensional problem that is plagued by definitional issues. Several types of elder abuse have been identified, including physical abuse, neglect, sexual abuse, abandonment, psychological abuse, and financial exploitation. The issues of physical abuse and neglect seem to be the most well researched. In addition, three dynamics of elder abuse have been found: spousal elder abuse, elder abuse by a stressed adult caregiver, and elder abuse by an adult dependent child. Although researchers argue about which dynamic is the most prevalent, there is evidence that each of these types exists. However, the majority of the intervention and prevention programs are aimed at the dynamic in which the elder is abused by a stressed adult caregiver. There are some resources for female victims of elderly spousal abuse, but few resources are currently available for either male victims of elder abuse or elders being victimized by adult dependent children. Research into the predictors of elder abuse shows that variables such as low income, isolation, family patterns of abuse, substance abuse, psychopathology, and victim frailty may all play roles in the various forms of elder abuse. Furthermore, the victims seem to suffer greatly—they can become psychologically distressed and physically injured by the maltreatment. In addition, they have a shorter life expectancy than those elders who are abuse-free and may even be killed as a direct result of the maltreatment. Because of these sometimes tragic consequences of elder abuse, prevention and intervention programs are needed that address all the possible types of elder abuse. However, instituting such programs will not guarantee that they will be used by the elder victims, many of whom do not reach out for help. Therefore, campaigns also need to be established that educate the public about issues concerning elder abuse and the ways in which elders can seek help for it.

11

Hidden Types
of Family Violence

Abuse of Siblings, Parents,
and People With Disabilities

As we got older, it got worse. I would have knives pulled on me. Then I would turn around and pull a gun on . . . my sister or younger brother. I became very violent, especially toward my sister (Wiehe, 1997, p. 30). [My brother] took my pet frog and stabbed it to death in front of me while I begged him not to. Then he just laughed! (p. 54). My brother threatened to kill me if I told our parents about him molesting me. I was 3 or 4 years of age at the time; he was about 18. He showed me the butcher block we kept in the cellar with the ax and blood. He said he'd kill me there if I told (p. 67).

Adolescent male held knife to parent to get ride to local shopping center. . . . Argument about adolescent female's threat to run away from home; adolescent kicks and pushes parent. . . . Adolescent male argues with stepmother about being away from home too much; adolescent throws dog at stepmother (Evans & Warren-Sohlberg, 1988, pp. 207–208).

My mother wasn't around much, and I always felt in my sisters' way, like I held them back from things they wanted because they had to help care for me. My sisters would slap me and shut me in my room (32-year-old woman with congenital osteogenesis imperfecta; Center for Research on Women with Disabilities, 1999, Chapter VII, paragraph 1). After my child was born, my husband became jealous and didn't want me to get up and take care of her. He would take my chair away from me and tied me up when I pulled myself out of bed. I left him the first chance I had (49-year-old woman with spinal cord injury since age 17; Center for Research on Women with Disabilities, 1999, Chapter VII, paragraph 2).

These quotes are from studies assessing what we deem "hidden forms of family violence." The first set of quotes is from a qualitative study on sibling abuse (Wiehe, 1997). Typically, the general public does not view aggression between siblings as "abuse" but as "normal sibling rivalry." Do you think the opening quotes describe nonabusive interactions? Does it make a difference who the perpetrator is? Would you consider any of the examples mere "sibling rivalry"? Does the sibling have to be older than his/her victim (i.e., there is a power relationship) for the act to be considered abusive?

This last question also pertains to the second set of quotes, which are police reports from a study of "parent abuse" episodes reported to the police in a northwestern city (Evans & Warren-Sohlberg, 1988). However, is it meaningful to say that a child is abusing his/her parents when the parents are the ones in a position of authority and power? Would you consider any of the examples given abusive? Who is ultimately at fault in these situations?

The final set of quotes is from people with disabilities who were victims of abuse from family members. Although these quotes refer more obviously to abusive behaviors than any of the other examples do, there are many questions as to how frequently people with disabilities are abused, especially in comparison to people without disabilities. We have shown in previous chapters that disabilities are possible *consequences* of family violence, but can they also be risk factors?

Sibling Abuse

Scope of the Problem

Consider the three cases described in Box 11.1. Would you consider these three scenarios to be abusive? If so, what types of abusive behaviors are occurring? What actions, if any, would you take if you found out that these interactions were occurring in your neighbor's family? Should they be referred to the authorities? Should they get counseling? Does the fact that perpetrators and victims are siblings affect your judgments at all?

Box 11.1   Three Cases of Possible Abuse

---

1. I would usually be playing by myself somewhere. He would barge in mad or drunk or both and he'd want me to either get something for him or cook for him. When I would say "No," he'd get extremely angry and start hitting me, cursing at me, kicking me, pushing me around in a total frenzy of violence. It started out by slapping, pushing, cursing. The older [I] got, the more severe it became. I have suffered from a broken nose and collarbone, countless bruises and scratches. I still have a BB in my leg where he shot at me with a BB gun. He kicked me with steel-toed boots on my upper arm, and it was red and purple for weeks. I thought he had broken it. (Wiehe, 1997, pp. 17–18)

2. [When I was] about age 3, [he] started fondling me, which progressed to full sexual intercourse over the next years, starting when I was about 9 or 10 and continuing to age 15, when I ran away and became a hooker. (Wiehe, 1997, p. 64)

3. I was being constantly told how ugly, dumb, unwanted I was. Already about 2 years of age I was told, "No one wants you around. I wish you were dead . . . your real parents didn't want you either so they dumped you with us." I grew up feeling if my own family doesn't like me, who will? I believed everything . . . that I was ugly, dumb, homely, stupid, fat—even though I always was average in weight. I felt no one would ever love me. When you're little, you believe everything you're told—it can last a lifetime. (Wiehe, 1997, p. 44)

Sibling abuse is considered the most underresearched and underrecognized form of family violence. Do you think the causes, effects, and correlates of maltreatment are necessarily different if the perpetrator is a sibling rather than a parent? Because there are implicit norms that siblings fight and tease each other, the public at large usually does not recognize that behaviors between siblings can cross the line from simple sibling rivalry to sibling abuse. As one mother stated, "They fight all the time. Anything can be a problem . . . it's just constant, but I understand that this is normal. I talk to other people, their kids are the same way" (Steinmetz, 1977, p. 43).

Physical Abuse

Although underresearched, physical violence between siblings appears to be the most prevalent form of family violence. In the 1975 National Family Violence Survey (NFVS), 800 out of every 1,000 children committed an act of violence against their siblings in the previous year; thus, on a national level, 50.4 million children were violent with their siblings (Straus & Gelles, 1990a). Many of these acts were incidents of minor violence, such as slapping and pushing; however, 530 out of every 1,000 (53%) children committed an act of severe violence (e.g., kicking, punching, biting, choking, attacks with a knife or gun). Although initially surprised by the amount of violence children perpetrated, Straus and Gelles argued that the results should have been expected, considering that children tend to imitate and exaggerate their parents' behaviors and that there are implicit norms that permit violence between siblings. After all, "kids will be kids" and "it's just normal sibling rivalry."

Additional analyses of the NFVS provided prevalence rates for different types of violence perpetrated on siblings; specifically, 74% of children pushed or shoved a sibling, 42% kicked, bit, or punched a sibling, and 16% beat up a sibling. The finding that 530/1,000 youngsters were severely violent against their siblings means that there were 19 million sibling attacks nationwide that would be considered assault by legal definitions. Moreover, it was estimated that 109,000 children used a knife or gun on a sibling in that year and that 1.5 million children had used a knife or gun on a sibling at least once in their lifetime (Straus & Gelles, 1990a). In addition, the statistics are likely to be underestimates for several reasons: Parents do not know about every fight their children have; most fights are probably forgotten because they are considered normal and commonplace; and the staff interviewed only parents from two-parent homes.

These data also give an idea of the ages and genders of children aggressing against siblings. Violence steadily declined as the children got older; 90% of 3- and 4-year-olds versus 64% of 15- to 17-year-olds committed an act of violence that year—an important decrease. Nevertheless, nearly two thirds of late adolescents had committed an average of 19 acts of violence against their siblings in the previous year. Moreover, although there seemed to be more boys who were violent against siblings than girls (83% versus 74%), it is noteworthy that nearly 75% of girls were violent. At the youngest ages, girls and boys were nearly equal in the amount of violence perpetrated, and the difference between the sexes steadily increased with age. Severe violence was more prevalent in all-boy families (67%), followed by mixed-sex families (52%), and all-girl families (40%) (Straus, Gelles, & Steinmetz, 1980).

Although some of these incidents can certainly be deemed "sibling rivalry," there is no reason to believe that all cases of violence between siblings are "normal," especially when 1.5 million children have actually wielded a knife or gun against their siblings. In the first study to specifically address the problem of sibling abuse, Vernon Wiehe (1997) gathered qualitative data on the experiences of 150 people who reported abuse by their siblings. The opening quotes to this chapter were from his study. Here are some more of their voices:

> Once my sister was ironing. She was a teenager. I was between 4 and 5. I was curious as to what she was doing. I put my hands flat up on the ironing board and she immediately put the hot iron down on my hand. She laughed and told me to get lost. (p. 21)

> I was deliberately shot in the face with a BB gun by my brother . . . I lost an eye as a result. (p. 21)

> My brother discovered that hitting in the solar plexus caused one to black out. So he would hit me and watch me pass out. (p. 22)

> My brother made several serious attempts to drown me . . . and later laughed about it. (p. 24)

Do you think that any of these cases are simply "sibling rivalry"? If not, what should be done with the perpetrator of these acts? There are cases in which the physically abusive behavior of siblings turns into siblicide—that is, one sibling kills another, as discussed in Box 11.2. Obviously, these acts of murder cannot be dismissed as "normal sibling rivalry." What should be done with the perpetrators of siblicide?

Box 11.2   Siblicide

In accordance with other researchers in the area (e.g., Gebo, 2002), we use the term "siblicide" rather than "fratricide" to refer to children who murder their siblings. We feel that "fratricide" has too much of a connotation of brothers murdering brothers, and, as statistics show, this situation is not always the case. According to the most recent statistics released in the Uniform Crime Reports from the FBI, there were

(Continued)

(Continued)

73 brothers and 26 sisters murdered in 2001. This represents just over 5% of murders committed within the family and over 0.5% of all murders that year. However, we do not know the age of the perpetrator or victim of the siblicide (i.e., child, adolescent, or adult), and there could be important variations in the characteristics of siblicide depending upon the age of the siblings.

Generally, siblicide is more common among adults than among juveniles, and children under 12 usually do not kill their siblings (e.g., Gebo, 2002; Underwood & Patch, 1999). In fact, among all sibling victims of homicide between the years 1976 and 1994, 78% were adults killed by their adult siblings, and only 9% were juveniles killing their juvenile siblings (Gebo, 2002). In juvenile relationships, older siblings generally are the perpetrators of homicide against their younger siblings (Daly, Wilson, Salmon, Hiraiwa-Hasegawa, & Hasegawa, 2001; Gebo, 2002), whereas the opposite is true in adult relationships: Younger siblings kill their older siblings (Gebo, 2002). Furthermore, there is generally less than a five-year age differential between the perpetrators and victims of siblicide (Gebo, 2002). Siblicide is also more common among males than among females. The most common type of siblicide is brothers killing brothers, followed by brothers killing sisters, sisters killing brothers, and finally sisters killing sisters (Daly et al., 2001; Gebo, 2002).

Sexual Abuse

Although some people argue that sexual abuse between siblings may merely be acts of sexual curiosity, most acknowledge that sibling sexual abuse can happen, and some studies have addressed this issue. For example, Finkelhor (1980) found that among college students, 15% of females and 10% of males reported some type of sexual experience with a sibling. Two additional studies revealed that a substantial percentage of juveniles in treatment for sexual abuse had committed sexually abusive acts against a sibling: Johnson (1988) found that 46% of 47 sexually abusive boys victimized a sibling, and Pierce and Pierce (1987) found that of the 59 sexual offenses committed by 37 adolescents, 40% were against sisters and 20% were against brothers. In fact, some argue that sibling incest may be as much as five times more common than parent/child incest (e.g., Finkelhor, 1980; Smith & Israel, 1987).

In the cases of sibling sexual abuse in Wiehe's (1997) qualitative study, the majority of the perpetrators were boys who were three to ten years older than their victims, and the majority of the victims were their sisters (79%). There was usually more than just one incident. When the abuse continued, the types of abuse often became more serious and differentiated, as in the scenario presented in Box 11.1. Most of the abusive situations occurred when the parents were out of the house and had left the older sibling in charge.

Sibling incest may be one of the more severe and frequent types of sexual abuse. Although most quantitative studies involve small samples, their findings indicate that 46%–89% of sibling incest cases involved penile penetration, and the majority of offenders committed more than 16 acts of abuse, significantly greater than father/daughter incest, stepfather/step-daughter incest, or any other type of sexual abuse (Adler & Schutz, 1995; Cyr, Wright, McDuff, & Perron, 2002; O'Brien, 1991). Furthermore, many abusive situations occur over a four- to nine-year time span (Cole, 1982; Finkelhor, 1980; Laviola, 1992; Russell, 1986), and physical force is used in more cases of sibling incest than in father/daughter incest (Rudd & Herzberger, 1999; Russell, 1986). For such reasons, many researchers (e.g., Jones, 2002) have argued that the current definition of sexual abuse in the literature—i.e., that there must be a five-year age discrepancy between the perpetrator and victim for sexual behavior to be considered abusive—provides serious underestimates of the extent of intrafamilial sexual abuse.

Emotional Abuse

Another form of sibling abuse—emotional abuse—may be the most excused type. After all, we were all teased at some point in time, often by our siblings, so children must learn how to handle that, right? But the kind of "teasing" that many of Wiehe's (1997) respondents experienced—the name-calling, ridicule, degradation, fear arousal, destruction of personal possessions, and torture or destruction of a pet—went far beyond simple "teasing," and the effects can last a lifetime. Consider the following examples provided by Wiehe:

> My brothers and a cousin tied me to a stake and were prepping the ground around me to set it on fire. They were stopped and built a dummy of me instead and burnt that. (p. 46)

> If my brothers found out I cared about something—for example, toys—they were taken and destroyed in front of me. (p. 52)

My second oldest brother shot my little dog that I loved dearly. It loved me—only me. I cried by its grave for several days. Twenty years passed before I could care for another dog. (p. 54)

If you had a sibling who did any of these things to you, what would your reactions be? Could you simply ignore the experiences? Would you consider them abuse? Why or why not?

Predictors and Correlates

The best and most consistent predictor of any type of sibling abuse is the presence of another type. That is, if children are physically or sexually abused by a sibling, they are likely to be emotionally abused by that same sibling—and if they are being sexually abused by a sibling, they are probably also experiencing physical abuse from the sibling (Johnson, 1988; Wiehe, 1997). Moreover, children abused by their siblings typically come from families characterized by parental physical, emotional, and/or sexual abuse of the children (Wiehe, 1997). In addition to this microsystem-level predictor of all types of sibling abuse, predictors of physical and sexual sibling abuse have been found at the microsystem and individual/developmental levels of the ecological model. Following is a summary of the scant literature to date.

Physical Abuse

Microsystem. Some characteristics of the victim have been associated with sibling physical abuse. For example, they may be withdrawn and lack assertiveness; however, it is difficult to say whether these characteristics are predictors or results of the abuse. The most extensively studied microsystem-level predictors of sibling physical abuse are characteristics of the family. Physically abusive siblings are often reproducing parental behaviors; because the parents maltreat each other, they tend to ignore and/or minimize their children's violence (Wiehe, 1997). Not surprisingly, families in which children physically abuse each other appear to be characterized by chaos, disorganization, and such problems as parental drug and alcohol abuse, parental mental illness, and/or marital difficulties (Wiehe, 1997). The parents may not only be violent but also lack warmth and positive affect. The balance of care in these families and the attention given to the children is often inappropriate. For example, the perpetrating child is often asked to take care of his/her younger sibling when the older child is still not mature enough or capable of being a caretaker (Green, 1984; Wiehe, 1997).

Individual/Developmental. Perpetrators of sibling physical abuse are frequently impulsive and quick-tempered children who often feel powerless or inferior for some reason. Usually they are older children who have a problem with a new child coming into the family. Their feelings of inferiority or powerlessness may stem in part from the lack of attention they perceive from their parents, and they may act aggressively to express hostility toward their parents, gain attention, or master their victimization status by becoming a perpetrator (Green, 1984; Wiehe, 1997). Or their feelings of inferiority and powerlessness may stem from a learning disability, some type of organic dysfunction, or some other physical or psychological impairment (Green, 1984).

Sexual Abuse

Microsystem. Families characterized by sibling incest have been described as distant and inaccessible (Smith & Israel, 1987) and dysfunctional in child-rearing practices and rules (Laviola, 1992). There is often marital discord, parental rejection and/or physical abuse of the perpetrator, a negative and argumentative family atmosphere, and an all-around dissatisfaction with family relations (Adler & Schutz, 1995; Rudd & Herzberger, 1999; Worling, 1995). Many of these families are large (Cyr et al., 2002; Finkelhor, 1980; Rudd & Herzberger, 1999; Russell, 1986). Sometimes either the father is not present or there is a general problem with alcoholism or mental illness in the family (Cyr et al., 2002; Meiselman, 1978; Rudd & Herzberger, 1999). Indeed, in one study nearly 34% of sibling incest cases came from single-parent homes and 26% from reconstructed families (O'Brien, 1991). Moreover, a substantial minority of sibling incest offenders may have observed sexual activity between their parents or between their parent and another adult (Smith & Israel, 1987). On the other hand, a minority of these families may be characterized by rigid, puritanical sex values where sexual expression in the children is discouraged (O'Brien, 1991). Victimization seems to run in families characterized by sibling incest, as many mothers had been either sexually or physically abused themselves (Adler & Schutz, 1995). In 22% of families in one study (O'Brien, 1991), there was another type of incestuous behavior occurring, such as father/daughter incest (Smith & Israel, 1987).

Individual/Developmental. Often, sibling perpetrators have themselves been sexually abused; of clinical samples of sibling sexual abusers, 23%–67% reported a history of sexual abuse (Becker, Kaplan, Cunningham-Rathner, & Kavoussi, 1986; O'Brien, 1991; Smith & Israel, 1987). In addition, as many as 92% of sibling incest offenders may have also suffered physical abuse in their families (Adler & Schutz, 1995). Sibling incest perpetrators may also suffer from a variety of psychological problems: In one study (Adler & Schutz, 1995), 75% of the offenders had sought mental health intervention on a previous occasion for problems such as conduct disorders, attention deficit disorder, drug abuse, adjustment disorder, social phobia, dysthymia, post-traumatic stress disorder (PTSD), depression, and learning disabilities (Adler & Schutz, 1995; Becker et al., 1986). Many incest perpetrators have also had other types of sexual disorders (e.g., exhibitionism, voyeurism) that were not directed toward a sibling, and they may have been arrested for a nonsexual crime (Becker et al., 1986; O'Brien, 1991).

Consequences

Although few empirical studies have been conducted on the consequences of sibling abuse, some research shows that a large percentage of sibling abuse survivors experience some type of clinically significant psychological distress (e.g., Cyr et al., 2002). As with other types of abuse, the frequency, duration, and intensity of sibling abuse mediate the consequences, as does the parents' reaction to the abuse. Many of the victims described by Wiehe (1997) came from homes in which abuse was endemic. The parents did not always intervene in the sibling abuse and even acted in ways that further victimized the child. For example, many parents denied the suffering of the victim and minimized or ignored any evidence of abuse or its consequences. Often, the parents would blame the victims, saying they must have done something to deserve it. More commonly, they would punish both the victim and perpetrator, usually physically. Other common, but unfortunate, reactions to sibling violence are indifference because the parents have their own problems to worry about or, even worse, joining in on the abuse (Wiehe, 1997).

Wiehe (1997) reported that the parents of the victims in his study were more aware of sibling physical and emotional abuse than of sexual abuse. For several reasons, victims did not always tell their parents a sibling was sexually abusing them. First, they may have been too developmentally imma-ture to understand what was happening to them. Often, they were threat-ened with retaliation if they told, or the perpetrators, especially if they noticed an autonomic pleasure reaction from the victims, would tell them that they "wanted it," and then the victims would blame themselves for the abuse. In addition, the overall climate of the home may have made it impos-sible to talk about the abuse. Either sex was never discussed, or the victims realized that telling would effect no change. Even if the victim was able to tell the parents about the sexual abuse, the parents would often deny what was happening. As reported in Chapter 5, the closer the perpetrator is to the mother, the less likely she is to believe the victim; in fact, mothers are least likely to believe the child if the sexual abuser is a brother (Cyr et al., 2002). Consider the effects that disclosure had on this victim:

> When I tried to tell my father about it, he called my mother and brother into the room, told them my accusations and asked him if it was true. Naturally, he said I was lying and my mother stood there supporting him. Nothing happened except that I got beaten later by my mother for daring to say anything and for "lying." My brother then knew that from then on, there was nothing he couldn't do to me. He was immune from punishment. Never again did I say a word since to do so would have only meant more abuse from the both of them. (Wiehe, 1997, p. 102)

The consequences of experiencing sibling abuse in such atmospheres are usually devastating to the victims, who often live in constant fear of further abuse. They learn not to cry for help because it often leads only to further abuse (Wiehe, 1997). Even as adults, victims of sibling physical, emotional, and sexual abuse experience problems in a wide variety of areas (Russell, 1986; Wiehe, 1997). They tend to suffer from poor self-esteem and feelings of absolute worthlessness and have problems in their relationships with members of the opposite sex (Laviola, 1992; Russell, 1986; Wiehe, 1997). The experience of sibling abuse has been shown to predict the use and experience of relationship violence during adulthood (e.g., Russell, 1986; Simonelli, Mullis, Elliott, & Pierce, 2002). If a female is abused by her brother, she may have distrustful, suspicious, fearful, and even hateful feelings toward men, and many do not marry because of these feelings (Russell, 1986; Wiehe, 1997). If they do get involved in relationships, they often replay the victim role because they think it is normal. Some report that if they are married, they want to have only one child to save their own child from the experiences they had (Wiehe, 1997). Adult victims may still experience extreme anger toward their perpetrators, anger that they sometimes take out on inappropriate targets (Wiehe, 1997). Drug and alcohol problems, depression and suicidal ideation, anxiety, eating disorders, and post-traumatic stress symptoms are frequently cited as problems for sibling abuse survivors (Rudd & Herzberger, 1999; Wiehe, 1997). Many have sexual problems as adults (Laviola, 1992; Wiehe, 1997)—either becoming promiscuous (Rudd & Herzberger, 1999) or withdrawing from and avoiding sexual activity, even in marriage, possibly because of flashbacks of the sexual abuse during sex (Wiehe, 1997). Finally, sibling sexual abuse victims may experience recurrent failures in the workplace and social settings, and thus have a general dissatisfaction with life (Abrahams & Hoey, 1994).

Prevention and Intervention

Currently, there is a dearth of formal programs for preventing or intervening in sibling abuse, probably because it has received so little recognition as a problem. Incest is always illegal, whether committed by an adult or a child, but siblings who sexually abuse a sibling may do so for the duration of their childhoods. Although the responsibility for preventing and intervening in sibling abuse should lie with the parents, are there steps that can be taken when those parents choose to ignore, minimize, or even join in on the abuse? If the parents are too wrapped up in their own problems to deal with what is happening between their children, what can be done to stop or prevent sibling abuse?

One step is to bring these behaviors to public attention. Then, if a teacher sees signs of maltreatment in a child, he or she will not excuse the abusive behaviors simply because they came from a sibling and not a parent. Child protective services (CPS) must then take these reports seriously and intervene, even if only for the reason that in families where sibling abuse is occurring, there are often other types of abuse as well. Intervention services should then focus on the family as a whole to try to help them figure out why one child is abusing the other and what other types of abuse are going on in the family. The child victim will also need therapy to deal with the abuse experienced, but these steps will not be taken if the public is not educated about this type of abuse.

Parent Abuse

Scope of the Problem

Parent abuse—that is, the abuse of young or middle-aged adults by their teenage or younger children—has been the least researched type of family violence. One reason may be people's reluctance to stretch the term "abused" to refer to parents whose own children are aggressive toward them. If the normal parent-child relationship is one in which a parent has power over the child, is it possible for the child to assume power and abuse the parent? If so, whose fault is it? If the parents are responsible for the child's behavior and the child aggresses toward the parent, can the parent conceivably be "abused" by the very child whose behavior they are responsible for?

Whether "parent abuse" is the most appropriate term, there is considerable evidence that parents can be the victims of their children's aggression. Presumed rates differ depending on the nature of the study. For instance, in the 1975 NFVS, the rate of child-to-parent violence was very similar to the rate of interparental violence (Straus & Gelles, 1990a). Specifically, according to the parents, 180 of every 1,000 children aged 3 to 17 committed a violent act toward them in the previous year, and 90 of every 1,000 children aged 3 to 17 committed a severely violent act (i.e., an act with a high likelihood of causing an injury). Teenagers aged 15 to 17 are especially likely, in comparison to younger children, to inflict an injury on their parents. The NFVS showed that, in the survey year, 100 per 1,000 teenagers perpetrated a violent act, and 35 per 1,000 teenagers a severely violent act, against their parents. When the rates are projected to the U.S. population as a whole, 9.7 million parents were victims of violence from their children, and 4.8 million were victims of severe violence (Straus & Gelles, 1990a).

In another analysis of the 1975 NFVS, using only families with children aged 10 to 17, 9% of parents were victimized by at least one act of violence from their adolescents in the preceding year, translating into 2.5 million parents nationwide (Cornell & Gelles, 1982). Further, 3% of parents were kicked, punched, beaten up, and/or had a knife or gun used on them by their adolescents that year, which equaled 900,000 parents nationwide. These numbers from the NFVS are considered to be underestimates of the true rate of child and adolescent violence toward parents because (1) the NFVS did not include families headed by single or divorced parents, and some studies show that "parent abuse" is more common in these types of families (e.g., Evans & Warren-Sohlberg, 1988; Livingston, 1986), and (2) most likely, the families with the most severe violence problems were not included (Ulman & Straus, 2003). However, other surveys of high school boys show similar rates of "parent abuse" to the NFVS: Between 7% and 11% reported using violence against their parents within the previous one to three years (Brezina, 1999; Peek et al., 1985). Higher rates of abuse were found in one survey of 18-year-old, low-income youth: Over the course of a lifetime, 32% of boys and 29.5% of girls reported hitting a parent (Langhinrichsen-Rohling & Neidig, 1995).

An analysis of official police reports of adolescent-to-parent abuse in a Washington city between 1984 and 1987 (Evans & Warren-Sohlberg, 1988) provides useful information about disputes that can lead to an adolescent striking a parent. The 73 cases of "parent abuse" investigated by police during the three years of the study comprised approximately 5.2% of all family violence cases. In 80% of the cases, the adolescents instigated the argument with their parents, typically over aspects of the home life (e.g., responsibilities around the house, privileges, problems with siblings; 34% of cases), money issues (e.g., allowances and spending patterns; 22% of cases), and alcohol use by the adolescent (19% of cases). Seven cases, confined solely to adolescent daughter/mother disputes, involved arguments concerning sexuality.

Predictors and Correlates

Macrosystem

Two macrosystem-level variables have been investigated as possible contributors to parent abuse—namely, ethnicity and attitudes toward aggression. Adolescents who assault their parents are more likely than non-assaultive adolescents to approve of delinquency and violence and believe that they will not be formally punished for their aggressive behavior (Agnew & Huguley, 1989). Furthermore, although most adolescents who assault their parents tend to be White (Agnew & Huguley, 1989; Charles, 1986), an analysis of police records showed that minority adolescents were overrepresented, relative to their distribution in the population, in cases of parent abuse (Evans & Warren-Sohlberg, 1988).

Microsystem

Characteristics of the Victim. The most well-studied characteristic of victimized parents is the parent's gender. Studies consistently show that mothers are much more likely to be victimized by nonlethal aggression than fathers (Cornell & Gelles, 1982; Livingston, 1986; Ulman & Straus, 2003). Among cases reported to police, 75% of adolescent male violence and over 95% of adolescent female violence was directed toward mothers; overall, 80% of the victims were mothers (Evans & Warren-Sohlberg, 1988). When *fathers* are victimized, it is usually by their older male children (Cornell & Gelles, 1982). One possible reason why mothers are victimized more often than fathers is that fathers typically spend less time with children and use less corporal punishment on children than mothers (Ulman & Straus, 2003).

Relationships With Peers. Two studies have examined the association of parent abuse with peer group characteristics. One set of researchers (Agnew & Huguley, 1989) found that adolescents who assaulted their parents were more likely than non-assaultive adolescents to have friends who assaulted their parents, and the other found that parent abuse was associated with belonging to a delinquent peer group (Kratcoski, 1985).

Family Dynamics. Several family-level variables are possible contributors to parent abuse. For example, low family integration (Kratcoski, 1985), weak attachments between parents and children (Agnew & Huguley, 1989), marital conflict, and parental minimization of children's hitting (Charles, 1986) have all been shown to contribute to children hitting their parents. Single mothers seem especially prone to being hit by their children: In cases reported to police, 46% involved single-mother households, in contrast to 17% of households headed by single mothers nationwide at that time (Evans & Warren-Sohlberg, 1988). It has also been shown that approximately 25% of single mothers with teenagers are hit (Livingston, 1986).

Other Types of Family Violence. The most consistent finding is that children who assault their parents were victimized by violence from the very parents they are now aggressing against (Browne & Hamilton, 1998; Evans & Warren-Sohlberg, 1988; Herbert, 1987; Langhinrichsen-Rohling & Neidig, 1995). In fact, the frequency of parent abuse is strongly related to the frequency of other violence in the home (Herbert, 1987). The more frequently children witness or experience violence from their parents, the more likely they are to hit their parents (Herbert, 1987). Indeed, four fifths of one sample of children who had hit their parents had recently been victims of their parents' violence, and extrafamilial exposure to violence did not add to the children's tendencies to abuse their parents (Browne & Hamilton, 1998).

Several different types of family violence have been implicated as contributors to children and adolescents hitting their parents. In one study, 29% of physically abused, 24% of emotionally abused, and 29% of sexually abused children were violent toward their parents (Browne & Hamilton, 1998). Both corporal punishment (Brezina, 1999; Larzelere, 1986; Mahoney & Donnelly, 2000; Ulman & Straus, 2003) and more severe physical abuse of children (Hotaling, Straus, & Lincoln, 1990; Kratcoski, 1985; Mahoney & Donnelly, 2000; Meredith, Abbott, & Adams, 1986; Ulman & Straus, 2003) have been shown to predict children's aggression against their parents, as has sibling violence (Kratcoski, 1985). Interparental violence also predicts children's violence toward parents (Hotaling et al., 1990; Kratcoski, 1985; Langhinrichsen-Rohlin & Neidig, 1995; Meredith et al., 1986), particularly in families in which the mother is the sole aggressor toward the father or both parents are violent toward each other (Ulman & Straus, 2003). Although being the victim of violence by parents is more strongly related to children hitting their parents than witnessing interparental violence (Ulman & Straus, 2003), parent abuse was most likely when both types of parental violence were present (Straus & Hotaling, 1980). Thus, there seems to be a cumulative effect of both parent-to-child and interparental violence on the child's violence toward the parent (Ulman & Straus, 2003).

Consider these results now in regard to the issue of labeling adolescent aggression toward a parent as "abusive." If the parents somehow contributed to their own victimization by using violence and teaching their children the appropriateness of violence, should they still be called "victims of parent abuse"? Why or why not?

Individual/Developmental

Gender. Although the NFVS showed no significant gender differences in the use of aggression toward parents (Ulman & Straus, 2003), most studies show that boys are more likely to use violence than girls (e.g., Charles, 1986; Cornell & Gelles, 1982; Evans & Warren-Sohlberg, 1988). Interestingly, one study showed that girls inflicted more injuries because they used an object during their aggressive incidents (Charles, 1986).

Age. Analyses of the 1975 NFVS showed that young children have the highest rates of hitting parents: 33% of three- to five-year-olds hit their parents in the previous year, compared to 10% of children ages 14–17 (Ulman & Straus, 2003). These results were explained by the fact that as children age, they have more control over their behavior and experience less corporal punishment from their parents; therefore, they are less likely to aggress. The majority of acts that the young children used would usually not be considered "abusive" by certain standards because they do not cause any harm or injury to the parents. The majority of children who injure parents seem to be teenagers (Harbin & Madden, 1979; Straus & Gelles, 1990a), although there are accounts of ten-year-olds injuring their parents (Harbin & Madden, 1979). Of parent abuse cases reported to the police, 60% were committed by adolescents between the ages of 15 and 17 (Evans & Warren-Sohlberg, 1988). Finally, there is evidence that for boys, severe aggression against parents increases with age (e.g., Agnew & Huguley, 1989; Cornell & Gelles, 1982), whereas daughters' severe aggression decreases with age (Agnew & Huguley, 1989).

Delinquent Behavior. Other individual/developmental variables predictive of aggression against parents abuse include adolescent psychosocial problems. A substantial percentage of adolescents who assault their parents have prior police records and/or contact with a social service agency (Evans & Warren-Sohlberg, 1988), are delinquent at school (Cornell & Gelles, 1982), and have a history of substance abuse (Pelletier & Coutu, 1992; Schiff & Cavaiola, 1993).

Consequences

The only well-researched consequence of parent abuse is the gravest possible consequence, the death of the parent. Although we do not know the percentage of child-to-parent aggressors who kill their parents, there is clear evidence that it happens. Consider the case in Box 11.3. What factors do you think prompted Terry to kill the parents who supposedly cared for him? Consider that Terry's father attempted to stab Terry once and pulled a gun on him once. How culpable do you think Terry is for his actions considering the situation he was in?

Box 11.3   Kids Who Kill Parents

Terry Adams and his two older sisters had long been physically and emotionally abused by their parents, both alcoholics. . . . Once [Terry's sisters] were gone, Terry became the sole target of the abuse. At the age of 16, he decided there was no alternative but to leave . . . and make it on his own. Terry had to go through his parents' bedroom to get out of the house. While he was attempting to escape this way during the early morning hours, Mr. Adams woke up and confronted his son. Terry told his dad that he was tired of the way he was living, that he was old enough to make it on his own, and that he knew what he was doing . . . Maybe he could not leave home legally, Terry told his dad, but he didn't care: He was leaving anyway. Mr. Adams slugged his son, knocking him down. When the youth got back up, his father pushed him over, and he fell into a closet where several guns were kept—including a .22 caliber rifle. Terry grabbed the rifle and fired at his father. He remembered his father screaming, "Oh my God!" Terry's mother, who was in bed when her husband was shot, woke up when the gun went off. Terry could not remember actually shooting his parents, particularly his mother. All he could remember was her face "when she sat up in bed . . . the agony within the terror. . . . The rest of it is more or less, sort of hazed out for me. I remember waking up completely. Standing there looking at two dead bodies. Two people. What have I done now,

(Continued)

(Continued)

> you know. Like it was a dream." . . . Court records showed that four petitions alleging neglect, abuse, and physical abuse in the Adams home had been filed and investigated by the state social services agency. Three of the petitions named Terry as the victim and had been filed in the two years preceding the homicides. The latest referral was made when Terry was 15. . . . Ten months after the state agency terminated supervision, Mr. and Mrs. Adams were dead.

SOURCE: Heide (1995), pp. xiii–xiv.

Adolescent Parricide

The most recent statistics from the FBI's Uniform Crime Reports show that in 2001, 94 mothers and 110 fathers were killed by their children, which represents 1.5% of all homicides and 11.3% of all family homicides. Approximate half of these homicides occurred during arguments. Neither the age of the offender nor the age of the victim was recorded in these data, so it is impossible to know how many of these parents were victims of elder abuse versus how many were victims of parent abuse. However, one extensive study of homicide data between the years 1977 and 1986 may provide a clearer picture (Heide, 1995). Each year throughout this period, over 300 parents were killed; 15% of the mothers, 25% of the fathers, 30% of the stepmothers, and 34% of the stepfathers were killed by adolescents under the age of 18.

Who Kills and How? In Heide's (1995) analysis, sons were responsible for approximately 85% of parricides, which is proportional to the gender distribution of homicide offenders in general. In the majority of cases, the child who killed his parent was a White, non-Hispanic male. The proportion of Black youth who killed their parents (~30%) exceeded their representation in the population (12%) at that time but was less than their representation as homicide offenders in general (~49%). Guns are typically the weapon of choice for adolescent parricide offenders: 82% of the fathers, 75% of the stepfathers, 65% of the mothers, and 56% of the stepmothers were killed with a gun.

Who Was Killed? Parents and stepparents who were murdered were typically White, non-Hispanic. Murdered stepparents were typically younger than murdered biological parents, and murdered male parents were on average younger than murdered female parents. Finally, mothers were significantly more likely to be White than fathers were (Heide, 1995).

Why Did the Adolescents Kill? Heide (1995) identified three types of adolescent parricide offenders: (1) the severely abused child; (2) the severely mentally ill child; and (3) the dangerously antisocial child. The *severely abused child* is the most frequently encountered adolescent parricide offender: Over 90% of adolescent parricide offenders were abused by their parents, could no longer tolerate the abuse, and thus perceived the homicide as an act of desperation. *Severely mentally ill children* are psychotic individuals with disorganized personalities; they tend to have distorted perceptions and disjointed communication skills. When more than one family member is killed and there is extreme violence or dismemberment of the bodies, most likely the child was severely mentally ill. *Dangerously antisocial children* have a sociopathic personality but do not experience delusions or hallucinations as severely mentally ill children might. However, they do display irrational behavior, poor judgment, and shallow emotions. They do not learn from experience, nor do they experience anxiety or guilt over their violations of the rights of others. In many cases, these typologies are not mutually exclusive—any given adolescent parricide offender could fall into one, two, or all three of these categories (Heide, 1995).

For the most part, adolescent parricide offenders tend to suffer from severe maltreatment and have few alternatives to remove themselves from the maltreatment. They typically suffer both abuse and neglect, and often experience injuries and threats of serious injury or death. The family is typically characterized by severe spousal abuse and alcoholism/heavy drinking, both of which predate the child abuse. Many children who experience this type of home environment do not kill their parents; however, typically, adolescent parricide offenders had made unsuccessful attempts to get help and, around the time of the homicide, had been increasingly considering either suicide or flight as a means of escape. The homicide was usually a reaction to a perceived or real imminent threat from their parents. Generally, adolescent parricide offenders do not have an extensive delinquent history, if any. However, a gun is usually easily accessible in the house. During the homicide, parricide offenders typically report a dissociative state, and the parent's death is usually seen as a relief for both the adolescent and other family members (Heide, 1995).

There seem to be important differences between homes in which the mother was killed by a son and those in which the father was killed by a son. In homes where matricide occurred, the mother was typically psychologically and sexually abusive of her son. The fathers were passive and pacified their wives, while being physically and emotionally distant and nonsupportive of their sons (Cormier, Angliker, Gagne, & Markus, 1978). In homes in which patricide occurred, the fathers were usually severely physically abusive, especially of the mothers, who were typically weak, helpless, and passive. The sons watched this abuse and were often called upon by their mothers to protect them (Cormier et al., 1978). When daughters killed their parents, it was usually the case that the parents were abusive (Cormier et al., 1978).

What Do We Do With the Offenders? Reactions of the judicial system to the adolescent parricide offender are becoming more compassionate, and the battered child defense has been successfully used in trials of parricide offenders. With proper intervention, many adolescent parricide offenders can function normally within society and be productive. Treatment is usually consonant with that of the severely abused child—for example, the offenders must understand the family dynamics that led them to murder (Heide, 1995). Experts stress that adolescent parricide offenders should not go to prison because good therapy is usually not available there (Heide, 1995). However, they should be institutionalized for a few years, until they are able to obtain therapy and gain self-control (Duncan & Duncan, 1971; Russell, 1984).

Abuse of Family Members With Disabilities

Scope of the Problem

The relationship between disability and abuse is complex. A physical or mental disability in either the victim or the perpetrator can be a risk factor for abuse, and physical or mental disabilities in victims can also be an *outcome* of maltreatment. In previous chapters, we have given considerable attention to disabilities (particularly psychiatric) as causes and outcomes of maltreatment. For example, Denver's Domestic Violence Initiative found that 40% of women with disabilities who sought their services had been disabled as a direct result of domestic violence (National Center on Elder Abuse, 1999). Similarly, it has been estimated that 37% of children who experience injuries from abuse develop some form of disability as a result (Crosse, Kaye, & Ratnofsky, 1993). In this section, however, we focus on victims of maltreatment whose disabilities appear to be *antecedent* to the maltreatment.

Child Maltreatment

According to the American Association of University Affiliated Programs for Persons with Developmental Disabilities (2001), "Children with disabilities are, on average, 3.4 times more likely to be maltreated. Broken down by form of maltreatment, children with disabilities are 3.88 times more likely to experience emotional abuse, 3.79 times more likely to be physically abused, 3.76 times more likely to be victims of neglect, and 3.14 times more likely to be sexually assaulted than children without disabilities" (paragraph 4). Information on cases of child maltreatment reported to CPS confirms that the proportion of children with disabilities reported for child abuse exceeds the proportion of children with disabilities in the population. According to the National Center on Child Abuse and Neglect (NCCAN), 17.2% of all abused children had disabilities, and 15.2% of children who had been sexually abused had disabilities (Crosse et al., 1993). These percentages are far greater than the proportion of children with disabilities in the population, which is 2% for children under the age of three, 3.4% for children between three and five years, and just over 11% for children between 6 and 14 years of age (U.S. Census Bureau, 1997).

Higher rates of abuse of children with disabilities were found by the Boys Town National Research Hospital: 31% of school-aged children with disabilities compared to 9% of children without disabilities were maltreated (Fried, 2001). Among children with mental retardation, 30% had a documented history of maltreatment, and almost two thirds had experienced multiple forms of abuse. Among the children with mental retardation, more boys (56.5%) than girls (43.5%) had experienced maltreatment.

A maltreatment prevalence rate of approximately 30% in children with disabilities has also been found in more localized studies. Of 49 consecutive outpatient cases seen in the Abuse Referral Clinic for Children with Disabilities in Philadelphia, 18% of children with disabilities showed evidence of child maltreatment, and another 13% were at high risk for maltreatment, resulting in an overall rate of 31% of children with disabilities experiencing or likely to experience abuse (Giardino, Hudson, & Marsh, 2003). Similarly, a Nebraska population-based epidemiological study of all public and parochial school children, plus all children ages zero to five in special needs and early intervention programs, revealed a maltreatment prevalence rate of 31% for children with disabilities compared to 9% for children without disabilities (Sullivan & Knutson, 2000a). For every disability except autism, maltreatment rates for the children with disabilities exceeded those for the children without disabilities. Moreover, significantly more of the maltreated children with disabilities (71%) than the maltreated children without disabilities (60.6%) experienced multiple episodes of maltreatment. Across all disability types, neglect was the most common form of maltreatment, followed by physical, then emotional, and finally sexual abuse. A further comparison of the maltreated and non-maltreated runaways in Nebraska revealed a high level of disabilities among children who had run away from home (Sullivan & Knutson, 2000b). Moreover, a majority of the runaways who were diagnosed as mentally retarded, visually impaired, or attention deficit disordered without conduct disorder were also victims of family maltreatment.

The type of maltreatment experienced by children with disabilities may vary in relation to the type of disability they have. In New York State, pregnancy or birth complications were associated with later physical abuse of the child; low IQ in the child was associated with neglect; and a handicapping condition in the child was associated with sexual abuse (Brown, Cohen, Johnson, & Salzinger, 1998).

Wife Abuse

According to Fiduccia and Wolfe (1999), independent of age, race, ethnicity, social class, or sexual orientation, women with disabilities are, in general, raped, assaulted, and abused at double the rate of women without disabilities. In a sample of 100 women with disabilities, 92% of the participants rated "abuse and violence," in general, as one of their top five concerns (Freeman, Strong, Barker, & Haight-Liotta, 1996). Although these studies refer to violence in general, not necessarily violence in the family, the dependence of women with severe disabilities on caretakers can make them vulnerable to unique forms of abuse, including denial of medications, oversedation, disconnection of a wheelchair's power supply, and putting something dangerous in the path of a blind woman (Berkeley Planning Associates, 1997). One woman with a disability explains the problem well: "You finally say, 'OK, this is it, I'm going to do whatever I can to change this marriage. And by the way, can you bring my scooter to me, so I can leave you?'" (Saxton, Curry, Powers, Maley, Eckels, & Gross, 2001, p. 402).

The evidence is mixed, however, as to whether women with disabilities are at heightened risk for maltreatment as compared with women without disabilities. In a national study of women with and without disabilities, 62% of each group had experienced some form of abuse at some point in their lives, most typically from intimate partners, and 13% of the women with physical disabilities reported having experienced physical or sexual abuse from an intimate within the previous year (Young, Nosek, Howland, Chanpong, & Rintala, 1997). Overall, the women with and without physical disabilities did not differ from each other, or from other national survey samples, in rates of emotional, physical, or sexual abuse. Similarly, there is no evidence that people who use special equipment or require assistance are more likely than individuals not in need of assistive devises to be recipients of abuse from partners, family members, or others (Horner-Johnson, Drum & Pobutsky, 2002). However, women with physical disabilities may experience a longer duration of physical and sexual abuse than women without physical disabilities (Young et al., 1997).

Elder Abuse

There is much less research on the role of disability in elder abuse than in the abuse of women and children. As noted in Chapter 10, there is only limited support for the assumptions that the development of mental and physical problems and disabilities makes older people more burdensome and a greater source of stress for family caretakers, and that it is their dependency and the stress it produces that put them at risk for abuse. For example, in a study of nearly 3,000 elderly adults in New Haven, Connecticut, self-reported chronic conditions, particularly a history of stroke and hip fracture, increased the risk for referral to a state protective services agency (Lachs, Williams, O'Brien, Hurst, & Horwitz, 1996); cognitive impairment and depressive symptoms were also strongly associated with reports of suspected elder maltreatment. However, none of these impairments made an independent contribution to the prediction of maltreatment reports after controlling for the role of sociodemographic variables (e.g., race, income, age).

There is also some debate over the extent to which dementia in older people is a risk factor for abuse. Although elders with dementia may be at risk for being abused, there is evidence that these elders may be even more abusive than their caretakers. For example, among caregivers who called a telephone helpline specializing in dementia, 12% said they had used at least one physically abusive behavior (e.g., pinching, biting, kicking, striking) toward the elder with dementia in their care, but one third reported that the elder had been abusive toward them (Coyne, Reichman, & Berbig, 1993). Caregivers who had been abused by their elderly relative with dementia were more likely than caregivers who had not been abused to use abusive behaviors themselves. This finding of reciprocal violence between elderly adults with dementia and their caretakers has also been found in other studies (e.g., Paveza et al., 1992).

Predictors and Correlates

Box 11.4 describes a boy with multiple handicaps abandoned by his father and subjected to severe physical punishment by his mother. How many risk factors for abuse appear in his case? Might he have been abused if he did not have disabilities? What role do you think his disabilities play in his maltreatment? How should this case be handled?

Box 11.4   Peter

Peter is a six-year-old boy who lives in subsidized housing with his 24-year-old mother (Ms. R), who receives public assistance, and his three-year-old brother. Peter was referred to a hospital center for children with multiple disabilities because of severe problems with over-activity, impulsivity, inattentiveness, and oppositional behavior. He is also visually impaired, has a seizure disorder and language and com-munication disorders, and was diagnosed as mildly mentally retarded at age three. His mother reports that he is aggressive and reckless, and she is distraught over her inability to handle him. She makes frequent use of moderate to severe corporal punishment, including slapping him in the face and hitting him on the legs and buttocks with objects such as pad-dles and belts. Several times before Peter's admission to the hospital, she had lost control to the point that her beatings left bruises on his but-tocks that lasted for several days. Ms. R says that she does not use such techniques with Peter's younger brother because he doesn't need them.

Ms. R reports that her father was an alcoholic who battered her mother and who punished his daughters with belts, paddles, and hair-brushes. Ms. R began using drugs during high school, quit school, moved in with her boyfriend, and gave birth to Peter when she was 20. Peter was born six weeks prematurely, had serious respiratory prob-lems, and was a jittery, hard-to-soothe baby. While Peter's father still lived with the family, he frequently got drunk and beat Ms. R in front of the children, and beat the children.

SOURCE: Ammerman, Lubetsky, & Stubenbort, 2000.

Macrosystem

Macrosystem risk factors identified for abuse of people with disabilities are societal views devaluing people with disabilities, race and poverty, archi-tectural barriers to escape, and lack of accessibility in shelters (Fried, 2001; National Coalition Against Domestic Violence, 1996; Nosek & Howland,

1998; Saxton et al., 2001). Moreover, women with disabilities are more likely than women without disabilities to have low incomes and be unemployed, factors that increase their vulnerability to repeat victimization (Crossmaker, 1991).

Children and adults with physical disabilities are often negatively stereotyped because they do not live up to popular social images and the cultural valuing of independent, self-sufficient, attractive, and well-functioning people (e.g., Sobsey, Randall, & Parilla, 1997; Curry, Hassouneh-Phillips, & Johnston-Silverberg, 2001). Because of these stereotypes, perpetrators—and often victims—appear to believe that maltreatment is acceptable. As one woman explained, "I wasn't able to say 'knock it off' to my family who was doing my personal care. I thought it was normal to be tossed around in my chair. To have a comb dragged through my hair so it comes out. To be left on a toilet for an hour . . ." (Saxton et al., 2001, p. 403).

In the Boys Town National Research Hospital study (Fried, 2001), both race and poverty were risk factors for maltreatment of children with disabilities. Specifically, African-American and Native American children with disabilities were at increased risk for maltreatment compared to children with disabilities from other ethnic groups, and 14% of the maltreated children with mental retardation had family incomes below the poverty line, compared to less than 1% of their non-maltreated peers. How do the risk factors in this study compare with what you have learned about risk factors for other forms of family violence?

Traditionally, architectural barriers have served to keep children and adults with disabilities in positions of greater dependence and isolation than their disabling conditions require. Since the Americans with Disabilities Act of 1990, this situation has gradually improved, although there continues to be considerable resistance to the goal of making all facilities just as accessible to individuals with disabilities as to the citizenry at large. This continued problem is perhaps best illustrated by the lack of wheelchair-accessible shelters (National Coalition Against Domestic Violence, 1996). As one abusive man told his partner who had a physical disability, "No one is going to help you because you're handicapped. They aren't going to let you into those shelters. If you don't shut up, you aren't going to get your medication or food" (Russell, 1995, p. 18, as cited in Curry et al., 2001, p. 62).

Exosystem

Exosystem-level risk factors identified as contributing to the abuse of family members with disabilities include social isolation, a lack of social support, and family stress. Women with disabilities are more likely than women without disabilities to be socially isolated, which increases their vulnerability to repeat victimization (Crossmaker, 1991). It has also been reported that children with craniofacial abnormalities have problems forming positive attachments to their caregivers, and these difficulties are associated with family stress, isolation, and inadequate social support, which have all been identified as risk factors for abuse (Wald & Knutson, 2000).

Microsystem

Victim Characteristics. Among reported cases of abuse, the risk for maltreatment may be heightened particularly for boys with disabilities. Of substantiated cases of child abuse reported to CPS, 21% of boys, compared to 8% of girls, had disabilities (Crosse et al., 1993). Whereas boys and girls without disabilities were about equally often victims of substantiated abuse, children with disabilities were more than twice as likely to be boys than girls; furthermore, the number of abused boys with disabilities was higher and the number of abused girls with disabilities was lower than would be expected based on the distribution of children without disabilities (Sobsey et al., 1997). Moreover, in the sample of children with disabilities, boys were found in all of the maltreatment categories, except emotional abuse, at rates significantly higher than what would be expected based on the frequencies in the sample without disabilities. This was especially true for sexual abuse, where twice as many boys were victims of sexual abuse as would be expected. Nevertheless, even among children with disabilities, girls were more frequently victims of substantiated sexual abuse.

Family Characteristics. Several family risk factors for maltreatment of children with disabilities have been found, including interparental violence, having a parent with a history of childhood maltreatment, having a parent with an illness or disability, and living in a single-parent home (Fried, 2001). Also, dependence on the perpetrator for daily living assistance has been shown to be a risk factor for wife abuse among women with disabilities (Nosek, 1996).

Consequences

As in other populations, maltreatment of individuals with disabilities can lead to additional and more serious disabilities, and death. In one of the few studies to address these issues, national records showed that adolescent girls with preexisting psychosocial or medical conditions appeared particularly vulnerable to serious injury (Moskowitz, Griffith, DiScala, & Sege, 2001).

In addition, childhood physical and sexual abuse are associated with suicidality in children and adolescents with mental retardation (Walters, Barrett, Knapp, & Borden, 1995).

Prevention and Intervention

There appear to be few specific programs designed to prevent abuse of individuals with disabilities. There are, however, a number of initiatives designed to educate clinical, social service, and criminal justice personnel to the importance of investigating the possible role of maltreatment in physical and psychological symptoms presented by individuals with disabilities (e.g., Kemp & Malkinrodt, 1996). In addition, programs have been developed to assist maltreated individuals with special needs (e.g., Berger, 2001; Finn & Stalans, 2002; Nosek & Howland, 1998; Orelove, Hollahan, & Myles, 2000; Tyiska, 2001). Perhaps such initiatives will reduce the frequency of events such as the following:

> A 36-year-old physically disabled woman survived three years of violence by her boyfriend; she has agoraphobia and therefore finds it difficult to travel without experiencing panic, fear, and anxiety attacks. Because she could not travel back and forth to court with ease, the District Attorney dropped the case against the boyfriend, stating that "the defendant has a right to face his accuser." (Fiduccia & Wolfe, 1999, p. 3)

Several studies have been conducted to determine what kinds of services are needed by people with disabilities. For example, the Center for Research on Women with Disabilities (1999) surveyed programs delivering abuse-related services nationwide and found that the average number of women with a physical, mental, or sensory disability served during the past year was 20; however, the number of female clients with disabilities varied broadly across programs, ranging from 0 to 12,000. Of women with disabilities receiving services, the most common type of disability was mental illness, and the least common was a visual or hearing impairment. On average, 10% of the women served by each program had a physical impairment; 7% were mentally retarded or otherwise developmentally disabled; 21% were mentally ill; and 5% had a visual or hearing impairment. For nearly half of the programs, fewer than 1% of the clients served within the past year had any physical impairment. The most common service provided was accessible shelter or referral to an accessible safe house or hotel room (83%). Although a substantial percentage of the responding agencies provided some assistance to women with disabilities, less than 25% of the services responded to the survey, and it is likely that there were fewer accommodations to the needs of clients with disabilities in the nonrespondents.

Summary

This chapter reviewed the scant literature on three hidden forms of family violence, specifically violence against siblings, parents, and people with disabilities. Although some argue that sibling abuse is merely sibling rivalry, many of the statistics and case examples argue otherwise; that is, many boys and girls are the victims of behaviors that could be labeled physical, sexual, and/or emotional abuse at the hands of their siblings. Furthermore, because sibling abuse is not always recognized as a form of maltreatment, these children often have few resources to help them: Their parents do not always believe they are victimized, sometimes parental behaviors further victimize them, and they cannot always tell their stories without ridicule, even as adults, because of disbelief that siblings can abuse each other. The second hidden form of maltreatment discussed, parent abuse, may not be considered abuse primarily because the children are behaving similarly to their parents, and oftentimes their violent behavior is a reaction to the severe abuse they receive from their parents. Can children abuse their parents, then? Is there ever a situation in which innocent parents are subject to extreme violence from their children? The final hidden form of abuse discussed is not subject to these same theoretical questions. Few would argue that aggression against a child or adult family member with a disability is not abusive. However, the extent to which children and adults with disabilities are at increased risk of family violence is subject to debate, and more research is needed to settle this issue. Children with disabilities certainly seem to be at heightened risk for abuse; however, studies are inconsistent as to whether women and elders with disabilities are at greater risk. Furthermore, there is no research as to whether adult men with disabilities are at greater risk for abuse, an important research question considering that adult males are more likely than adult females to become disabled as a result of an accident.

12

Responding Effectively
to Family Violence

Tommy (4 months old) is the child of Frank (age 21) and Wendy (age 29). The family is known to the social services department because of concerns . . . at the time of Tommy's birth Frank's heroin addiction might affect his parenting capacity, but the health professionals agreed to offer support and to refer back to social services if necessary . . . Last night, the police arrested Frank attempting to break into a shop. Frank then told the police that he was concerned for his son, whom he had left at home on his own. The police . . . [found] Tommy in a cot in front of an electric fire, very overheated and screaming. In the room around him were syringes and other signs of drug use. A piece of clothing hanging on the fireguard had started to singe and smoke . . . A woman police officer . . . waited at the house for Wendy's return. She reported that Wendy seemed drunk . . . but seemed to be capable of caring for Tommy . . . Police records show that both Frank and Wendy have been charged before with minor drug-related offenses . . . On one occasion a year ago, the police were called out to a domestic dispute. Wendy alleged that Frank had punched her and she had a black eye coming up, but when the police arrived she seemed to want to shrug it off and did not want to press charges. (Beckett, 2003, pp. 159–160)

Pamela married David when she was fifteen; after fourteen years of an abusive marriage, Pamela left David. David sought her out at her job, assaulted her and tried to abduct her. David kept threatening to kill Pamela. She got a restraining order against David. When Pamela called the police about David violating the protection order she was told, "Well, that's a civil matter. You have to handle that with your own lawyers." Later, David gunned Pamela down in the parking lot of a restaurant. (Swisher, 1996, in Baker, 2002, p. 1)

'He was drunk and I was drunk . . . He says you ain't going anywhere, and I said I am. I pushed him and he pushed me to the floor. I got up and I just tore into him. I scratched him right on the neck and he punched me in the mouth and cut inside my lip. Next thing I know I'm standing in the hallway, there's cops handcuffing me. (Christina, an interviewee, describing the incident that got her arrested.) (Dasgupta, 1999, p. 204)

A woman called adult protective services reporting that she had just been contacted by her elderly neighbor, Mrs. M, who resided with her adult son . . . Mrs. M [had said] that her son had been drinking and earlier in the day had struck her about the face and arms when she refused to give him her social security check . . . Mrs. M had communicated her fear that her son [who had recently been released from prison] would again assault her . . . In this and many emergency situations, a joint home visit by a protective service worker and a law enforcement officer is indicated. (Ramsey-Klawsnik, 1995, pp. 45–46)

The first case above illustrates the kinds of family problems that the criminal justice system addresses every day. Who is (or are) the victim(s) in this case? What should be done to address the problems besetting this family? Should one or both parents be arrested? Put in jail? What should be done with Tommy? Up until this time, social service reports indicated that Tommy was being adequately cared for and that his father doted on him. Is this sufficient reason to allow Tommy to remain in the home with his parents? If not, what is the best alternative?

The second case portrays an all-too-common situation—one where a woman has tried to make use of the criminal justice system to protect her from assaults by her spouse, but the system has failed. Although the criminal justice system has undergone tremendous change in the past 20–30 years in the way it responds to intimate violence cases, there is considerable debate over the extent to which both civil and criminal statutes, and procedures developed to prevent or reduce intimate violence, have had positive results. The third case demonstrates one of the ways women become involved with the criminal justice system in intimate violence situations. The laws are gender neutral. It is illegal for spouses to assault each other, and oftentimes when police are called to intimate violence situations they arrest both parties because both may show signs of injury. Does the legal standard of gender equity seem reasonable to you? Do you think justice is served by Christina being arrested? Why or why not? The final case illustrates many of the themes discussed in Chapter 10, including financial exploitation and physical abuse of a well-functioning, but vulnerable, older parent by a less well-functioning adult child. Note that the preferred intervention here, as in many cases of family violence involving criminal behavior, is a coordinated response by social service as well as criminal justice personnel. Do you agree with this method of intervening? What other methods do you suggest?

Responding to Child Maltreatment

In considering the community response to child maltreatment, we begin by discussing the laws that led to these programs being implemented in the first place. We also discuss how child protective services (CPS) generally work and give expanded consideration to the treatment of child sexual abuse cases. We then consider some of the special issues related to the criminalization of child maltreatment.

Child Abuse Laws and CPS Procedures

In 1962, following publication of the landmark work by Kempe and colleagues on the battered child, the U.S. Children's Bureau adopted the first laws mandating physicians to report any known cases of child abuse to social service agencies. The logic behind the laws was this: Because children are often not able to speak for or defend themselves, professionals who deal with children have the obligation to protect them from abuse. If physicians report any child maltreatment cases they see to social service agencies, appropriate services and protection could be offered to these children and their families. Within five years, all states had passed child abuse reporting laws for physicians, and care was taken to ensure that any professionals reporting such cases would be exempt from civil or criminal liability (Davidson, 1988; Nelson, 1984; Zellman & Fair, 2002).

As research into child abuse expanded, an increased understanding of the problem led to a broader definition of acts constituting child abuse; by the 1970s, sexual abuse, emotional maltreatment, and neglect were also included in the definition. In addition, it became evident that several other types of professionals were sometimes in a position to detect child maltreatment. By 1986, virtually every state required that nurses, social workers, other mental health professionals, teachers, and school staff, as well as physicians, be mandated reporters of child abuse (Fraser, 1986; Zellman & Fair, 2002).

Perhaps the most influential child abuse law ever passed in this country was the Child Abuse Prevention and Treatment Act (CAPTA; P.L. 93–247) in 1974. This law freed millions of federal dollars to support state child protection agencies; however, in order to receive this federal funding, the states had to conform their child abuse reporting laws to the federal standards. The states also had to create policies and procedures for reporting and investigating alleged child abuse and had to offer treatment services for these families. CAPTA also gave states the power to remove children from homes if the children were deemed to be in danger (Myers, 2002; Zellman & Fair, 2002).

In 1980, the federal government passed the Adoption Assistance and Child Welfare Act, which required that states do everything they could (*reasonable effort*) to prevent the removal of children from their families, and to do all they could to reunite children with their families in the event of a removal. In 1997, the government clarified the "reasonable effort" phrase in the Adoption and Safe Families Act: "in making reasonable efforts, the child's health and safety shall be of paramount concern" (42 U.S. Code § 671(a)(15)(A)) (Myers, 2002, p. 306). In February 2001, President George W. Bush called for a $200 million increase in funding for the Promoting Safe and Stable Families program. This funding was to be used for helping states to ensure the safety of abused children by enabling them to be quickly adopted into safe, stable homes when efforts to keep the family together had failed (Carter, 2001). Congress reauthorized the Promoting Safe and Stable Families program (P.L. 107–133), including the additional money requested by the president, later in 2001 (Child Welfare League of America, n.d.).

Under the federal regulations, the job of CPS is to receive incoming calls concerning potential child abuse cases, investigate these cases, deem if abuse has occurred, and then take appropriate action. CPS workers tend to collaborate with local law enforcement and other child professionals in medicine, mental health, education, and law to determine what the appropriate action should be. Most cases do not go to court unless the abuse is very serious, probably because, as the U.S. Supreme Court declared in 1987, "child abuse is one of the most difficult crimes to detect and prosecute, in large part because there often are no witnesses except the victim" (*Pennsylvania v. Ritchie*, 1987, p. 60). Therefore, the majority of cases are referred for tertiary intervention (see Chapter 4) (Myers, 2002).

The Processing of Child Abuse Cases

Although all states have laws concerning the reporting of child maltreatment, there are no common guidelines for assessment, substantiation, and processing of these cases; however, in general, reports that are substantiated by the state's CPS agency are referred to juvenile rather than criminal courts.

The hearings are dependency hearings designed to determine the extent to which the child is safe in the family home rather than to determine the guilt of a parent. In rare cases, typically sexual abuse cases, where there is also a goal of punishing the perpetrator rather than merely protecting the child, the case may go to criminal court. To illustrate a modal example of the processing of a substantiated case, we present a summary of the typical procedure in Massachusetts (Bishop et al., 2001). In Massachusetts, in 1994, approximately 33,850 cases of child maltreatment were reported; of these, approximately 15,550 were substantiated, and 2,600 were deemed serious enough by the Massachusetts Department of Social Services (DSS; the state's CPS agency) to warrant the filing of a care and protection (C & P) petition in juvenile or district court. The purpose of a C & P is to seek removal of the child from the home. In 1994, the Boston Juvenile Court, which hears approximately 20% of all C & P cases in Massachusetts each year, heard 526 cases—an average of more than one a day for every weekday of the year. Following the filing of the C & P, an arraignment is held at which there is typically sufficient evidence of risk to warrant the judge to grant temporary custody of the child to DSS. Generally, DSS places the child in foster care until the initial hearing on the case, which, according to the law, must take place within 90 days of the arraignment. At the hearing, the judge has several options, including dismissing the case and returning the child to the home, permanently removing the child from parental custody, or continuing (postponing) the case to allow for gathering additional evidence or providing services to the parents. If, at the initial or subsequent hearings, the child is removed from parental custody, it is the responsibility of the state to find a permanent home for the child through adoption or legal guardianship.

Unfortunately, the social service and criminal justice systems, as you probably already know, are very flawed. When the child abuse laws were first enacted back in the 1960s, it was thought that maltreatment of children was very rare and that prevention would be a low-cost venture (Nelson, 1984). However, it soon became clear that the vast numbers of child abuse referrals were overwhelming for CPS agencies short on funds, staff, and time. In reaction to the unmanageable caseloads, CPS agencies have been forced to narrow their definitions of abuse so that only the most serious cases get attention. Unfortunately, this process means that many families needing services fall through the cracks of the system because the abuse is just not severe enough to receive CPS attention (Zellman & Fair, 2002).

Because CPS is so overwhelmed in their caseloads, many "mandated" reporters have chosen not to report certain cases of child abuse. This fact is evident in the third National Incidence Study (NIS-3), which collected data from both CPS agencies and professionals who dealt with children. The

NIS-3 revealed that in 1993, only 28% of abused children under the Harm Standard and 33% of abused children under the Endangerment Standard received CPS attention. (See Chapter 4 for a more detailed description of the NIS studies.) Even more alarming is the finding that only about 25% of the children who were seriously harmed or injured by abuse or neglect received CPS attention. In addition, these percentages represented a decrease from the previous NIS study; thus, as time went on, fewer and fewer professionals were reporting cases of child abuse and neglect to the designated authorities (Sedlak & Broadhurst, 1996).

To understand why mandated reporters were not reporting all the cases they were mandated to report, Zellman and her colleagues (Zellman & Fair, 2002) conducted a study to determine how many professionals failed to report cases and why. They found that 40% of mandated reporters decided not to report a case of suspected child abuse at some point in their careers. Child psychiatrists were the professionals most likely *not* to report a case of suspected child abuse. Their main reason was their belief that reporting would not help the child or family because it would not help them see the seriousness of the problem or stop the maltreatment. Professionals who failed to report were also concerned that a report would disrupt the treatment of their client and/or family, and some believed they could help the family better than the CPS agency. Many respondents believed that in some cases reports to CPS agencies would only hurt matters, not help them—especially the more mild cases of abuse that would probably be screened out by CPS agencies anyway. Moreover, reporting such cases to CPS would carry a risk that the family would terminate treatment, and, therefore, even the child professional would not be able to monitor the child's safety (Zellman & Fair, 2002).

The study by Zellman and Fair adds to the considerable evidence that the CPS system is overburdened and unable to carry out the goals imposed by law. Child professionals realize this fact and are therefore less likely to report certain types of maltreatment to the authorities, even when they are legally mandated to do so. One implication of this practice is that if child abuse is not reported to the authorities, then appropriate statistics on the extent of this problem cannot be obtained, and funding may consequently be cut (e.g., Finkelhor, 1993). However, overreporting of abuse may also be a problem. According Besharov (1993), requiring professionals and encouraging lay-people to report "suspected" cases of abuse is part of what overburdens the system; too much time is spent investigating cases of unfounded maltreatment, and victims who truly need help are being ignored because there are simply not enough resources to find and help them. What do you think? Does it seem likely that overreporting is what interferes most with the process of protecting children from abuse, or is underreporting the more likely problem? Or could both underreporting and overreporting be problems? What are the implications of these problems? What can be done to solve them? We have already discussed cultural and societal influences on physical abuse in this country—for example, the acceptance of violence in general, the acceptance of physical aggression as appropriate child discipline, and the belief that parents should be able to raise their children as they see fit. Consider your solutions to problems in the CPS system. Would they be readily accepted in this country?

Child Sexual Abuse Cases in the Criminal Justice System

As previously mentioned, most of the maltreatment cases that go to court are heard in a juvenile or sometimes district court. However, in some severe cases, especially of sexual abuse, the case may be heard in a criminal court. In a meta-analysis of 24 studies with samples involved in criminal prosecution of child maltreatment, 79% involved only sexual abuse cases and the remaining 21% involved a combination of physical and sexual abuse cases (Cross, Walsh, Simone, & Jones, 2003). The great majority of the child abuse cases referred to prosecutors (approximately 72%) were carried forward without dismissal. Among the cases carried forward, plea rates averaged 82% and conviction rates 94%. In comparison to national data on felony dispositions, child abuse was less likely to lead to the filing of charges and to incarceration than most other felonies but more likely to be carried forward without dismissal. In addition, rates for diversion, guilty pleas, and trial and conviction for the child abuse cases were essentially consistent with those for all violent crimes (Cross et al., 2003). Although perpetrators of child sexual abuse seem to be treated similarly to other felons in criminal court, the profile of the abusers is quite different. Compared to other felons, child sexual abusers are more likely to be employed, to have been married, and to be mostly non-Hispanic Whites—characteristics that are also in sharp contrast to the child maltreatment cases heard in juvenile court (Cullen et al., 2000).

Most of the studies of the adjudication of child sexual abuse cases in criminal court do not separate intrafamilial from extrafamilial abuse. However, in one study focusing specifically on the prosecution of child sexual abuse cases where the alleged perpetrators were family members, Martone, Jaudes, and Cavins (1996) examined state attorney office records addressing 451 allegations of intrafamilial sexual abuse between 1986 and 1988. They found that 72% of the cases showed probable cause of sexual abuse. The complaints filed in these cases included 77 felony charges, 29 misdemeanors, and 30 juvenile charges against 136 alleged family perpetrators. Although

51% of the alleged perpetrators ended up in court, only 17% of the original 451 allegations were prosecuted for a felony. Only 36% of the victims had to appear in court. Of the 77 felony complaints filed, 48 (62%) ended in convictions and 43 of these individuals served time (average 6.8 years). Thus, relatively few of the perpetrators for whom probable cause of sexual abuse was demonstrated actually served time in prison for their crimes, and this rate itself would be even lower when we consider all of the perpetrators of sexual abuse who were never even taken to court.

The adjudication of child sexual abuse cases in criminal courts adds additional levels of complexity to the already complex process of adjudicating a child maltreatment case. Testifying in court can be a terrifying experience for the child victim (Somer & Szwarcberg, 2001), who may also be subjected to multiple and embarrassing interviews about the abusive events (Ghetti, Alexander, & Goodman, 2002). Medical testimony is important, but most sexually abused children have no specific or diagnostic medical evidence of sexual contact (DeJong, 1998). There are controversies over the role of expert witnesses (Allen & Miller, 1995) and the competency of children to testify in criminal cases (McGough, 1997). In addition, when the child is too young or too disturbed to testify, there are issues surrounding the acceptability of hearsay evidence (McGough, 1997; also see Golding, Alexander, & Stewart, 1999).

Many modifications have been recommended concerning the processing of criminal child maltreatment cases, with the specific goals of reducing the stress for the child and increasing the number of convictions of perpetrators. There has been much research, including mock trial research, on many of these innovations, which range from simple procedures such as familiarizing child victims with the courthouse prior to the beginning of the trial and "vertical prosecution" (i.e., having the same attorney work on a case throughout its prosecution) to procedures diverging more dramatically from traditional criminal practices—for example, videotaping of the children's testimony in a room away from the alleged perpetrator; allowing hearsay evidence concerning the abused children's report of their abuse rather than making them testify directly in court; using anatomically detailed dolls to allow children to show exactly what was done to them; and allowing the children to have a support person (e.g., a non-offending parent) with them in court. (e.g., Goodman et al., 1999; McGough, 1997).

What is known about the extent to which innovations are used, and their effects on the successful prosecution of cases? Evidence from a survey of prosecuting attorneys in 41 states revealed that the most common types of cases involving child witnesses were child sexual abuse/sexual assault cases (88%) (Goodman et al., 1999). The types of innovation most commonly used by the reporting prosecutors were vertical prosecution, the presence of a support person in the courtroom, and a tour of the courtroom for the children—all of which were relatively easy and inexpensive to implement. Much less frequently used were videotaped testimonies and interviewing child witnesses behind one-way mirrors—which were seen as too expensive, not necessary, and jeopardizing the case because they contribute to jurors' distrust of the child testimony. Similarly, the use of anatomically detailed dolls to elicit children's testimony was seen as harmful to the case and not necessary. Regarding outcomes, the increased use of vertical prosecution and faster trials were both associated with increases in guilty pleas and guilty verdicts. By contrast, using one-way mirrors in an interview room was negatively associated with both guilty pleas and convictions.

Box 12.1   The Processing of the Ann and Marie Case

When Marie first disclosed her father's sexual abuse to a counselor, she experienced a great deal of pressure from her sister Ann to keep quiet and then refused to repeat her claims during a CPS investigation. Consequently the case was closed as unfounded. Later, when Marie disclosed the abuse to a counselor, CPS was again called, and the father was arrested. The girls' mother was very ambivalent about pressing legal charges against the father for his incestuous behavior with his daughters. She was particularly anxious to avoid the public shame that would come with having her husband's sexual abuse as part of the public record. A probation officer interviewed both girls to determine both the extent of the harm they had experienced and the further trauma that might accompany the outcome of a trial. Ann, the 14-year-old daughter, said she would feel guilty about sending him to prison but also felt he should not get away with what he had done. Marie, the 15-year-old, initially felt much less ambivalence, asserting that her father should go to jail. She later backed down from this desire, saying he had shown improvements in therapy. It is not clear whether she had really changed her mind or was again experiencing excessive pressure from her family.

SOURCE: Cohen & Mannarino, 2000.

Consider the case of Ann and Marie in Box 12.1. Their incestuous experience at the hands of their father was first described in Chapter 5. At the time that the report describing their case was published, the case against the father was still pending. What do you think the court experience is likely to be like for each girl? What kind of accommodations might be made to make the experience less traumatic?

Special Issue: Interparental Violence as Child Maltreatment

As discussed in Chapter 6, there is increasing recognition that child abuse and intimate partner violence are tightly intertwined. Based on a review of the relevant literature, Appel and Holden (1998) concluded that the rate of co-occurrence of child abuse and interparental violence across the United States population is approximately 6%; however, they also concluded that within a sample of violent homes, the co-occurrence rate appears to be about 40%, even with a very conservative definition of child abuse. Indeed, within clinical samples of battered women or abused children, the overlap is sometimes as high as 100%, and there is a high correlation between a father hitting a mother and a mother hitting a child (Straus, 1990b). Even when children are not abused by the parent who is battering a spouse, or by the parent who is being battered, there is increasing concern that merely observing interparental violence can have deleterious effects on children (see Chapter 6). Consequently, in increasing numbers, children are being removed from the custody of mothers who are staying with or returning to their batterers. Some researchers argue that one aspect of this child removal policy seems to be an assumption among child welfare workers that mothers are the primary caregivers of children and that any trauma to the children is more the mother's fault than the father's, even if the trauma stems from the father beating the mother (Mills, 2000). Moreover, critics argue, many child welfare workers seem to believe there is only one acceptable way for abused women to respond to their abuser's violence, and that is to leave the abusive partner permanently. If battered women fail to respond to this mandate, some child welfare workers have argued that their children should be removed permanently from their custody (Magen, 1999, in Mills, 2000).

Not surprisingly, this trend has generated much controversy. Battered women view the policy as adding another layer of abuse to their lives—specifically, abuse at the hands of a criminal justice/CPS system that takes their children away from them instead of providing adequate assistance in the process of their trying to escape from their batterers. One worker in the field comments, "I worry that in our collective alarm about the effects of domestic violence on children we will move too quickly to new policies and laws extending the mandate of CPS to situations of domestic violence that would not otherwise be in the system. I think we do not yet know enough to do this wisely.

It also seems to me very difficult to ask child protective workers to serve simultaneously as women's advocates" (Fleck-Henderson, 2000, p. 334).

Box 12.2   Child Abuse and Domestic Violence

DSS became involved with the Smith family after Joe assaulted Sandra with a knife, inflicting several wounds. Joe was arrested and a criminal case against him was initiated. Sandra, who reported to the caseworker that she feared for her life, obtained a restraining order; however, she refused to testify because she did not want him to go to jail. Joe was given three years probation and mandated to attend a batterers' treatment program. Sandra's caseworker helped Sandra and her two small children enter a shelter, but Sandra, who was the only person of her racial/ethnic background in the shelter, was soon asked to leave. She was then unable to find a place that would take her and her children, so she returned home. Shortly after that, Joe rejoined the family.

SOURCE: Fleck-Henderson, 2000. From Fleck-Henderson, A., "Domestic violence in the child protection system: Seeing double," in Children and Youth Services Review, 22, pp. 333–354, with permission from Elsevier.

Consider the case described in Box 12.2. If you were a CPS worker assigned to this family, what would your reaction be? Do you think the children should be removed from the home? If you were a battered women's advocate, what would your response be? Would you want to remove the children from the home? Are there other possibilities? In this particular case, the DSS worker, the domestic violence specialist, and a trained consulting team ultimately agreed that the risk of danger to the children was too high and had them moved to foster care (Fleck-Henderson, 2000). Do you agree with their choice?

Special Issue: Maternal Substance Abuse During Pregnancy as Child Maltreatment

Another controversial practice that has increased markedly in recent years involves removal of infants from mothers who used illegal drugs during their pregnancies (see also Chapter 6). One national study revealed that the number of children in foster care who had been exposed to drugs prenatally had increased from 17% in 1986 to 55% in 1991 (U.S. General Accounting Office, 1994, cited in Sagatun-Edwards & Saylor, 2000). The late 1980s was a time of political conservatism in the United States. It was also a time when the government took the stance that individuals and families were responsible for their own problems, and it was not the government's responsibility to provide treatment programs for drug abusers or to address the societal conditions contributing to the use of illegal drugs (Paone & Alperen, 1998). Because the government tended to prosecute criminals rather than correct social ills, thousands of women who abused drugs while pregnant lost custody of their children, and, in some cases, were subjected to criminal prosecution.

By the mid-1990s, 70% of states had prosecuted women for taking illicit drugs while pregnant (Center for Reproductive Law and Policy, 1996, in Paone & Alperen, 1998). Because some experts believed that penal laws were being misused in these prosecutions and that the very filing of charges may have violated the women's constitutional rights, courts are now generally dropping the charges or ruling in favor of the defendants. However, several states, including South Carolina, Minnesota, Florida, Oklahoma, and Massachusetts, have officially added *fetal abuse through drug exposure* to the legally reportable forms of child abuse (Sagatun-Edwards & Saylor, 2000). What is your view on this issue? Should women who abuse drugs prenatally be prosecuted for child maltreatment? Are the women's rights being violated by such laws?

When women ingest illegal drugs during pregnancy, there is a good chance that the fetus will be injured (see e.g., Chapter 6 for the effects of cocaine and alcohol on the fetus). In many cases, the baby may be removed from the mother immediately following birth. Is taking the baby away immediately following birth in the best interests of the child? There are at least two main elements to the controversy over the removal of the child from a woman who abused drugs during pregnancy: (1) Although early medical studies seemed to indicate that exposure of unborn infants to illegal drugs carried significant risks for the fetus, later research has illustrated just how difficult it is to separate any effects due to the exposure of one particular drug from effects due to smoking, alcohol use, poor nutritional status, poor health, sexually transmitted diseases, and inadequate prenatal care (see Chapter 6)—all associated with poverty; (2) The women whose babies are taken from them are disproportionately single mothers from minority groups who are unemployed, undereducated, and poor (Sagatun-Edwards & Saylor, 2000). One set of investigators (Whiteford & Vitucci, 1997) has argued that the war on drugs has turned into a war on women, with low-income women of color being particularly victimized by the prosecutors of this war. They report that women who are Black, have limited resources, use illegal drugs while pregnant, and give birth in public hospitals are the group most likely to be targeted for prosecution. Indeed, approximately 70% of the women prosecuted for drug use while pregnant are women of color. Whiteford and Vitucci (1997) interpret these research findings as showing that laws jailing pregnant women for their addictions are concerned less with protecting the unborn than with punishing women for being poor, pregnant, and addicted. Do you agree?

In a California study of how drug-exposed infants were processed in the social service and juvenile justice systems following a positive toxicology screen (Sagatun-Edwards, Saylor, & Shifflett, 1995), there was an overrepresentation of Black and Latino cases and an underrepresentation of White and Asian cases as compared to their relative distributions in the urban county covered by the court. Only 22% of the women were married, and most were unemployed and had not finished high school. The baby's father was incarcerated or awaiting a hearing in criminal court in 25% of the cases. During the process of the court hearings, a total of 52 cases (~20%) were ultimately returned home—34 cases following reunification services and 18 at the initial dispositional hearing. A study addressing ongoing dispositions of drug-exposed infant cases in California indicated that although minority cases were disproportionately represented and rated as at a higher risk than White cases in the initial (dispositional) hearing, when controlling for other risk factors it was mothers' compliance with court orders and attendance at hearings that significantly predicted subsequent court decisions (e.g., whether the child could be reunified with the mother or was placed into a permanency planning program aimed at adoption) (Sagatun-Edwards & Saylor, 2000).

Box 12.3   Deirdre

At the age of 19, Deirdre was the mother of three and pregnant with her fourth child. Her first child was born positive for cocaine and placed with Deirdre's mother. Her second child was also born positive for cocaine, and a judge ordered Deirdre to enter drug treatment or lose that child as well. A local treatment center evaluated Deirdre and recommended a two-year residential program, but there were no openings in the program at that time. Deirdre waited six months for an opening to become available and then entered the program. Her child was placed in foster care and was shuttled back and forth through

(Continued)

(Continued)

several foster homes over the next few years. Deirdre left the drug treatment program after three months and became pregnant again. Fearful that her third baby would be taken away because she was still using drugs, she did not seek prenatal care. However, when she was six months pregnant she was arrested for shoplifting, and the judge ruled that to protect her unborn baby she had to finish her pregnancy in jail. Following the birth of her third child, she entered a day treatment program that provided neither transportation nor child care. Because she could not find anyone to take care of her child, she dropped out of the program. This was a violation of her parole, so the judge ordered her back to jail and had her son removed from her care. Pregnant again, Deirdre requested drug treatment, but because she had lost custody of her child she was no longer eligible for Medicaid and could not afford the program.

SOURCE: Whiteford & Vitucci, 1997.

Consider the case of Deirdre described in Box 12.3. From an ecological perspective, what factors contributed to the fetal abuse that ultimately resulted in the removal of her children? In your view, to what extent was removal of her children in the best interest of those children? What alternatives are there to the criminal justice approach that might better serve Deirdre and her children? As Whiteford and Vitucci (1997) point out, sometimes the only way that poor minority women can get desired treatment for their addictions is to get pregnant and/or arrested. Are there better solutions?

The Response to Wife Abuse

Both the battering of wives and wife rape are considered crimes in all 50 states (Bergen, 1996; Buzawa & Buzawa, 1996). When wife beating came to public attention in the 1970s, the criminal justice system was criticized for not responding appropriately—for example, for not handling "domestic disputes" in a way that prevented future violence and for not responding quickly to women's calls for help. One of the most notorious cases was that of Tracey Thurman of Torrington, Connecticut. Over a period of several years, Tracey was repeatedly beaten and threatened by her husband Charles, who worked in a local diner where he knew and served many of Torrington's police officers. After one encounter when Charles publicly screamed threats at Tracey and broke the car's windshield while she was inside the car, he was arrested for breach of peace, given a suspended sentence of six months and a two-year "conditional discharge," and ordered to stay completely away from Tracey. His threats and intimidation continued, and Tracey's constant appeals to the Torrington Police Department for protection were ignored. Finally, in 1983, when Charles went to the home where Tracey was staying, she again called the police for help. By the time a single officer arrived, Charles had stabbed Tracey numerous times in the chest, neck, and throat. In the presence of the officer, he kicked Tracey in the head twice. Even after three other officers arrived, he was left free to continue to threaten her, and kicked her again in the head while she was lying on a stretcher. Left partially paralyzed and scarred by his attack, Tracey successfully sued the Torrington Police Department for failure to protect her constitutional rights and was awarded $2.3 million in damages. In 1985, Connecticut passed a mandatory arrest law, and police departments around the country began developing better protocols for dealing with domestic assault cases (*Thurman v. City of Torrington*, 1984).

During the past 30 years, the criminal justice response to wife battering has changed considerably—all states have enacted legislation designed to modify police and court responses to wife abuse. For instance, most police precincts have adopted mandatory arrest policies in intimate violence cases, and prosecutors do not necessarily need a victim's consent to press charges against a batterer. Although it has been argued that these policies may actually deter victims from calling the police in incidents of battering (because they may not want their spouses arrested and prosecuted), the criminal justice system, and police in particular, appear to have become more concerned and helpful in intimate violence cases (Buzawa & Buzawa, 1996).

There are three main points at which abused women may come into contact with the criminal justice system: (1) when they are seeking protection orders (also called restraining orders) to keep their partners from battering them; (2) when an incident of abuse has led to an encounter with police that can lead to mandatory or discretionary reporting of the abuse; and (3) during and after filing criminal charges against the abuser.

Protection Orders

Abused women and men can seek protection orders to document that abuse is occurring and to notify abusers that if they violate the protection order they are subject to criminal prosecution (Malecha et al., 2003). Protection orders, which can be either civil or criminal, are intended to prevent violent or threatening acts against, harassment against, contact with, communication with, or physical proximity to, another person (Baker, 2002). The exact status of the individuals who can obtain these orders—for example, whether same-sex domestic partners are included—varies somewhat from state to state. Protection orders are a way of trying to end the abuse without having the perpetrator criminally charged or sentenced to jail for the maltreatment; thus, the goal is prevention rather than punishment. One provision of the 1994 Violence Against Women Act (VAWA; see Chapter 1) was to make protection orders accessible, affordable, and enforceable across state lines, and to establish penalties for individuals who crossed state lines in order to continue abusive behavior.

Thousands of women, and numerous men, apply for protection orders every year. More than 10,000 applications were filed in the city of Philadelphia alone in the year 1996 (Zoellner et al., 2000, cited in Malecha et al., 2003); however, fewer than half of the women who apply for protection orders receive them, primarily because they stop the process. Why might they do this? In one sample of 150 women who applied for protection orders (Malecha et al., 2003), 54% completed the process and 28% decided not to complete it. (The remaining women did not complete the process for reasons beyond their control.) The only significant differences found between the two groups were that women who dropped the protection order process were more likely to currently be in the relationship with their batterer and living with their abusers three months after their initial application for the protection order. Thus, whether they are still being battered or not, women continuing to live with their abusers following their efforts to get protection are the ones most likely to discontinue the effort.

Do protection orders actually give the protection they are designed to provide? The data are mixed. In a national longitudinal study of abused women who had taken out civil protection orders, the majority of respondents said they believed the protection orders protected them against repeated incidents of physical and psychological abuse (Keilitz et al., 1998). Initially, 72% of the women said their lives had improved. At follow-up, 85% reported life improvement, more than 90% reported feeling better about themselves, and 80% felt safer. More than 70% of participants in the initial interviews and 65% in the follow-up interviews reported no continuing problems. On the other hand, it was clear that protection orders were less effective against men with a history of violent offenses; women in these cases generally reported more violations of the protection order than women dealing with men without a history of violent offenses. The researchers suggested that criminal prosecution might be necessary with repeat offenders who were not deterred by the civil protection orders. Consider the case in Box 12.4. Do you think such a tragedy could occur again today? How can incidents like this be prevented?

Box 12.4   The Pinder Case

On the evening of March 10, 1989, Officer Johnson responded to a call reporting a domestic disturbance at the home of Carol Pinder. When he arrived at the scene, Johnson discovered that Pinder's former boyfriend, Don Pittman, had broken into her home. Pinder told Officer Johnson that when Pittman broke in, he pushed her, punched her, and threw various objects at her. Pittman was also screaming and threatening both Pinder and her children, saying he would murder them all. A neighbor, Darnell Taylor, managed to subdue Pittman and restrain him until the police arrived. Johnson then placed Pittman under arrest. After confining Pittman in the squad car, Johnson returned to the house to speak with Pinder again. Pinder explained to Officer Johnson that Pittman . . . had just been released from prison after being convicted of attempted arson at Pinder's residence some ten months earlier. She was naturally afraid for herself and her children and wanted to know whether it would be safe for her to return to work that evening. Officer Johnson assured her that Pittman would be locked up overnight. He further indicated that Pinder had to wait until the next day to swear out a warrant against Pittman because a county commissioner would not be available to hear the charges before morning. Based on these assurances, Pinder returned to work. That same evening, Johnson brought Pittman before Dorchester County Commissioner George Ames, Jr. for an initial appearance. Johnson only charged Pittman with trespassing and malicious destruction of property having a value of less than three hundred dollars . . . [and] simply released Pittman on his own recognizance and warned him to stay away from Pinder's home. Upon his release, [Pittman] returned to Pinder's house and set fire to it. Pinder was still at work, but her three children were home asleep and died of smoke inhalation.

SOURCE: *Pinder v. Johnson*, 1995, paragraphs 2–4.

Arrest

In the early days of the domestic violence movement, police typically took an arrest-avoidance position, which was seen by many feminists as denying battered women equal protection under the law (Walker, 1993). In the famous Minneapolis Domestic Violence Experiment, Sherman and Berk

(1984) found that subsequent offending, including domestic assault and property damage, was reduced by nearly half when the offenders were arrested and jailed. In response to this study, the U.S. attorney general's office recommended that the no-arrest policies of states be replaced, and currently all states except Arkansas and Washington, D.C., have codified mandatory arrest or pro-arrest policies. Many states have passed legislation allowing, and often requiring, police to make warrantless arrests in response to misdemeanor intimate violence not committed in their presence, and many police departments have adopted arrest-preferred policies in intimate violence cases (Buzawa & Buzawa, 1996). Even in jurisdictions with mandatory arrest policies, police generally have some discretion about whether to make an arrest on a domestic disturbance call.

Considerable research has been done on the "extralegal" factors that tend to influence arrest practices. For example, police are more likely to make an arrest on a domestic disturbance call when serious injuries have occurred (Connolly, Huzurbazar, & Routh-McGee, 2000); when the complainant is a White, affluent, older, or suburban woman (Avakame & Fyfe, 2001); or when the assailant is not married to the victim (Buzawa, Austin, & Buzawa, 1995; Fyfe, Klinger, & Flavin, 1997). In cases where each partner is reporting violence by the other, police are more likely to make arrests when the partners are married than when the relationship is nonspousal (Connolly et al., 2000). They are also more likely to make dual arrests (i.e., to arrest both parties) when the victim has been drinking and is confrontational with the police (Stewart & Maddren, 1997).

Does a policy of mandatory arrest operate to reduce intimate violence? Here again, the evidence is mixed. Pro-arrest policies have led to a substantial increase in dual arrests and a decrease in domestic disturbance calls in at least some states (e.g., Massachusetts, Connecticut, Rhode Island, Washington) (Martin, 1997). Some observers suggest that the increase in dual arrests shows that oftentimes violence is mutual between the partners; however, others argue that the frequency of dual arrests is one more way of blaming the victim and further victimizing women. Some suggest that the overall declines in calls have occurred because women have become afraid that if they call for help, their partners will make accusations that will lead to their own arrests as well as their abusers'; others argue that the decline is due to women's fear that their wishes that their partners not be arrested will be ignored because of mandatory arrest policies that do not consider the victim's wishes; and still others argue that the declines reflect actual decreases in the incidence of intimate violence. Thus, the results of these studies can be interpreted in many different ways and cannot be taken as evidence that mandatory arrest policies are having an ameliorating effect on the incidence of intimate violence.

The pioneer field experiment on the impact of mandatory arrest (Sherman & Berk, 1984) showed that, according to victim interviews, recidivism rates following mandatory arrest (19%) were substantially lower than the rates among men who had experienced merely physical separation (33%) or officer mediation (37%). However, similar studies in Omaha, Nebraska (Dunford, 1992), and Charlotte, North Carolina (Hirschel & Hutchinson, 1992), found that arrest did not deter recidivism more than separation or mediation did. In addition, studies in Milwaukee (Sherman et al., 1992), Dade County, Florida (Pate & Hamilton, 1992), and Colorado Springs (Berk et al., 1992) suggest that arrest may affect different groups of men differently and that lumping these groups together may mask those effects. For example, it appears that arrest reduces subsequent violence in employed men, but not unemployed men, and in White men, but not Black men. Further, the study in Milwaukee showed that violence *increased* over the long term in cases in which the perpetrator had been arrested. Buzawa and Buzawa (1993), based on their review of the research on the deterrent effects of arrest on perpetrators, concluded that although arrest for the purpose of deterring an offender from future violence does not always work, that does not necessarily mean that arrest per se fails to work. In their view, it may be important to bring the perpetrator into the criminal justice system, where there are at least possibilities of interventions that may reduce recidivism. They also note that some types of offenders may not be influenced by any sort of intervention except incarceration.

Box 12.5   Inappropriate Police Response: The Maria Macias Case

---

Maria Macias reported to the sheriff's department in April 1995 that her estranged husband, Avelino, physically and sexually abused her and their children. She received a temporary restraining order (TRO), ordering him to stay away from them. On January 21, 1996, Avelino forced his way into Maria's apartment. When the sheriff's deputy arrived at the scene, he told Maria she had to file a complaint to have Avelino arrested. In violation of state law, the deputy did not file a domestic violence report. On January 23, Avelino threatened Maria at a housecleaning job. A sheriff's deputy was dispatched but could not find the home. Later, while Maria visited a friend, Avelino parked his car behind Maria's, preventing her from leaving. Sheriff's deputies arrived and told him to move his car but did not arrest him. Twenty

(Continued)

(Continued)

minutes later, deputies were recalled to the scene because Avelino harassed and threatened to kill Maria. The officers did not file a report. On February 1, Maria filed a complaint that Avelino was hanging around her apartment. The deputy told Maria that Avelino was outside the TRO's zone of protection. On February 21, Maria reported that Avelino was threatening to kill her but was told that he could not be arrested unless officers caught him doing this. She was told to document any stalking behaviors. On February 23, Maria twice informed the sheriff's department that Avelino was stalking her. The deputy told the dispatcher that he could not keep filing reports every time she called, then telephoned Avelino, warning him to stay away from Maria. On February 28, Maria reported to a deputy that her husband was following her to her housecleaning jobs. The deputy confronted Avelino and told him to leave Maria alone. The abuse culminated on April 15, when Avelino shot and killed Maria and himself, and wounded Maria's mother. At the scene, deputies found two notebooks filled with Maria's documentation of Avelino's abuse.

SOURCE: Blackwell & Vaughn, 2003.

Clearly, arrest cannot do anything to deter perpetrators from future assaults if arrest is not used as a tactic when it could appropriately be used. Consider the case of Maria Macias in Box 12.5. What might be some reasons that laws concerning arrest in intimate violence cases did not help Maria? What do you think could be a solution to failures such as these?

As was the case with restraining orders, and in contrast to Maria's experience, there is evidence that many women find the police response to their intimate violence problems to be helpful. In a study of Boston intimate violence victims who had sought police help either by telephone or by coming to a station, nearly all victims who wanted help getting or enforcing a restraining order or wanted police to arrest the offender felt that the police provided the desired assistance (Apsler, Cummins, & Carl, 2003).

Processing Domestic Violence Cases

An example of the processing of domestic violence cases through criminal court comes from a useful case study of a random sample of 452 intimate violence cases in Sacramento County Criminal Court during a relatively recent one-year period (Kingsnorth et al., 2001). Criminal charges were filed in 70.5% of the cases with an uncooperative victim (i.e., who did not want to press charges) and 82.5% of the cases with cooperative victims. The high level of filing charges reflected the "no-drop" policy regarding domestic violence cases in that office. Of these, 260 were fully prosecuted (i.e., neither rejected nor dismissed); 257 perpetrators pled guilty, and three went to trial, of whom two were found guilty. Only 6% (15 cases) of the fully prosecuted and convicted cases received prison terms.

In their analysis of the factors that predicted proceeding with a case at each level of the criminal justice process, Kingsnorth et al. (2001) found that when there were dual arrests (i.e., when both parties of the violent incident were arrested at the time of the domestic dispute) prosecutors were more likely *not* to file a case for prosecution. Presumably when both parties are arrested, prosecutors view the likelihood of successfully assigning responsibility as remote. By contrast, the victim receiving hospital treatment and a greater number of witnesses of the abuse were positively associated with the prosecutor's willingness to file charges. As was true of the initial decision to file a case, defendants who were under the influence of alcohol or other drugs at the time of the offense were more likely to have their cases fully prosecuted, as were those who had inflicted more severe injuries on their victims. Hospital treatment of the victim, the availability of crime scene photos of the victim, and the severity of the attack all contributed significantly to the level of charges filed. Prior prison terms, prior domestic violence arrests, and perpetrator substance abuse were also positively associated with the level of severity of the charges filed (felony vs. misdemeanor).

A number of barriers to successful use of the criminal justice system by battered women have been identified, and these may influence the women's desire to prosecute the case: (1) fear of retaliation by their batterer (or someone else seeking vengeance on his behalf); (2) victim-blaming attitudes among criminal justice personnel; (3) resistance within the criminal justice system to treating domestic assaults as seriously as other assaults of equal impact; (4) poverty and all its associated factors (e.g., transportation problems, inability to take time off from work); (5) fear of abandonment by one's religious or ethnic community; and (6) general distrust of the criminal justice system (Hart, 1993). In addition, others have pointed out that the mandatory arrest and prosecution policies of most jurisdictions have the effect of making a battered woman feel ignored, voiceless, faceless, and powerless during the prosecution process (Mills, 2003).

Given such barriers, one might expect that battered women would not be particularly satisfied with their experiences in the criminal justice system. In a national study of victim satisfaction with criminal justice system responses to assaults by intimate partners versus assaults by others (Byrne et al., 1999), victims of intimate partner assault were less likely to report feeling satisfied with their experiences with the police officers (65.5% vs. 84.5%), the prosecutors (51.9% vs. 73.4%), the victim assistance staff (67.7% vs. 81.6%), the judges (53% vs. 71.6%), and the criminal justice system overall (37% vs. 52%). With regard to police officers, intimate partner assault victims were less likely to say the officers had demonstrated an interest in their feelings or tried to gather all the evidence necessary for the case. With regard to prosecutors, partner victims were: (1) significantly less likely than the non-partner victims to believe that their opinions had been taken into account by the prosecutor when decisions about the case were made (56.2% vs. 88.2%); and (2) much less likely to report that the prosecutor's office had encouraged them to attend the grand jury hearings (40.4% vs. 100.0%) and the sentencing hearing (47.2% vs. 83.2%). Moreover, when perpetrators entered a guilty plea, significantly more perpetrators of partner assault were allowed to plead guilty to a lesser crime than non-partner perpetrators.

Husband Abuse and the Criminal Justice System

Although we have demonstrated that men are subject to maltreatment by their intimate partners (see Chapter 8), there is little research on the experience of abused husbands or abusive wives in the criminal justice system. According to one source, "An arrest is not routinely made when the perpetrator is a woman. Why? Police officers occasionally suggest a range of excuses. These run the gamut from 'he would be embarrassed' to 'she didn't hurt him that badly' . . . There is no readily available law enforcement documentation of the numerous abuse calls where the male is the victim and no action is taken" (Gosselin, 2000, p. 6). Consider the case in Box 12.6 of a man who reportedly experienced physical abuse from his wife, never hit back, and one night called 911. Do you think such cases are typical for men who call 911 to report their wives' abuse? Why do you think the criminal justice system responded in that way? What are the possible ramifications of this type of response for the male victim and his wife?

Box 12.6   An Abused Man Calls 911

One evening in the summer of '95, she had been drinking some wine and decided we needed to have a "talk." This one was pretty bad. Typically she would start out with a calm sounding voice, which would escalate through several pitches and intensities until she reached the violent stage. This episode escalated very quickly, probably because of the wine. She was screaming and throwing kitchen implements at me and I headed for the balcony. She ran out the apartment door to head me off, and I went back in and locked the door behind her. Then I called 911. That was a very bad mistake. The police came and arrested ME. The officer that arrested me explained that they were required to make an arrest in every domestic violence call. He also explained that there were small children in the house who needed their mother, so I was going to jail. Never mind that it was ME who called them, or that it was ME who was scratched up and bloody. I was taken to jail and charged with two misdemeanor charges. I asked for a no-contact order so that I wouldn't have to be worried about her hunting me down as she had in the past. They obliged and placed a no-contact order on ME! Then they turned me loose. It was 2:00 A.M. I was 15 miles from home. No money or wallet. No car. They wouldn't even give me a ride somewhere. I walked into town. I couldn't get legal help either. They said that I would be able to talk to a public defender on the day of my trial . . . When it came time to go to court, I had a lawyer that I paid for myself, which I really couldn't afford. My lawyer, my wife, and the prosecutor went into a private meeting. When they came out, the prosecutor said they had enough evidence to convict and that they were going ahead with it. My lawyer said for me to sign a plea bargain to get rid of one of the charges . . . I left the courthouse convicted and on probation for a year.

SOURCE: Anonymous Case from the Domestic Abuse Helpline for Men, 2003, paragraphs 7–10.

Reports from men who attempt to use the criminal justice system to stop their wives' abuse vary as to the extent to which they find the system helpful. Some report that their wives have been arrested for abusing them, but others report that the abused husbands are arrested, prosecuted, and sentenced for being batterers. Sometimes male victims are required to enter batterers' treatment programs as part of their sentences, even if there is no evidence that they hit their wives but substantial evidence that they have been injured by their wives (Hines, Brown, & Dunning, 2004).

As mentioned previously, because of mandatory arrest policies in cases of intimate violence, dual arrests have increased dramatically in recent years, and thus many more women are being arrested for domestic violence. In response, several groups of researchers have investigated possible gender differences in male versus female arrestees in domestic violence cases, in part to ascertain whether female perpetrators should be treated differently than male perpetrators. For example, of 2,670 cases of domestic violence in a large Midwestern city (14% with female defendants) in 1997 (Melton & Belknap, 2003), males were significantly more likely than females to have had a prior arrest for domestic violence, whereas females were more likely than males to have been arrested as part of a dual-arrest case. Males were also more likely to have threatened their wives with either lethal or nonlethal harm; to have attempted to prevent the victim from calling the police; and to have shoved, pushed, grabbed, dragged, pulled the hair of, physically restrained, and strangled the victim. By contrast, females were more likely to have hit the victim with an object, thrown an object at the victim, struck the victim with a vehicle, bitten the victim, and used a weapon against the victim. In addition, there were no gender differences in the perpetrator's likeliness to have slapped, punched, hit, knifed, or stabbed the victim, and there were no gender differences in injury rates for cuts, abrasions, broken bones, or broken teeth. According to these researchers, the males' violence seemed to be more severe than the females' violence; however, the females' violence was also quite severe.

Similar conclusions were drawn from a study of 6,704 male and female domestic violence offenders (16.8% female) between December 1997 and March 2001 in Shelby County, Tennessee (Henning & Feder, 2004). Specifically, male offenders were more likely to have a history of domestic violence arrests, whereas female offenders were more likely to have been involved in a dual arrest. Male offenders were more likely to have violated a protective order, to have used substances prior to the incident, and to have a victim who feared him. They were also more likely to have used more serious assaults against their wives, such as choking and forcing sexual activity, and to have assaulted their wives more frequently. Male offenders were more likely to threaten homicide, to involve the children in domestic disputes, to have a history of both violent and nonviolent criminal behavior, and to own a gun. On the other hand, female offenders were more likely to have used a weapon during the incident, and there were no gender differences in injury rates, frequency or severity of psychological abuse, suicidal threats, stalking behaviors, or juvenile arrest rates.

Both of the above studies show that males' domestic violence behavior may be more severe than females' domestic violence behavior and that females make up only a minority of those arrested for domestic violence. In about one third of cases in which a female is arrested, the females are arrested as part of a dual-arrest scenario; thus, it is unknown whether these women were acting in self-defense, whether they were part of a mutual violence situation, or whether they were the primary perpetrators of violence. One study, however, investigated gender differences in 45 male and 45 female domestic violence offenders who were deemed the primary perpetrators of violence and mandated to attend treatment as part of their probation in northern California (Busch & Rosenberg, 2004). Although male offenders were more likely to have a history of domestic violence offending, among perpetrators with prior offenses there were no gender differences in the number of previous arrests. In addition, although men used more violent acts in the arrest incident, there were no gender differences in the likelihood of using a severely violent act. There were also no gender differences in the injury rates of the victims; however, when females injured their victims they tended to use a weapon or object, but when males injured their victims they tended to use their bodies alone. Also, although male perpetrators had a history of more nonviolent criminal offenses, there were no gender differences in histories of prior violence outside the home, and the majority of female perpetrators had criminal histories. In addition, there were no gender differences in substance abuse problems, the use of substances at the time of arrest, or the types of substances that the perpetrators abused. One significant gender difference that emerged was that more female offenders reported victimization from their spouses than male offenders did (24% vs. 7%).

Overall, mandatory arrest policies have resulted in the arrest and conviction of increasingly more female perpetrators of domestic violence. Studies on female arrestees show that they can be very violent against their male partners—they tend to use weapons to injure their partners, and the rates of injuries in male and female victims are similar. Although female perpetration occurs less frequently and may be less severe in some cases, these studies show the importance of studying the female perpetrator of domestic violence because the services she and her victim need may be somewhat different than the reverse situation.

Maltreated Elders in the Criminal Justice System

Elder maltreatment was originally conceptualized as more of a social and medical problem than as a criminal problem (Payne, 2002); however, during the early 1990s, elder abuse and neglect were increasingly criminalized. As part of this process, police were expected to detect elder abuse cases and judges were expected to make judgments concerning those cases, despite the fact that neither group received adequate guidance or training in the handling of elder abuse cases. Although criminal sentences for elder abusers increased on the basis of the assumption that this criminal justice response would have a deterrent effect, there has been little research supporting the deterrent rationale (Payne, 2002).

The first statutes designed to protect elders from maltreatment were modeled after the child abuse statutes. As is true of the legal response to child maltreatment, elder maltreatment can be addressed in either civil or criminal courts (Heisler & Quinn, 1995). In the criminal justice system, older victims of maltreatment, like victims of child maltreatment, are not asked to "press charges" against their abuser; that is the decision of the prosecuting attorney. Moreover, just as the judicial system may make decisions concerning the best placement of an abused child, many state elder abuse laws provide for the imposition of a guardianship over victims of elder abuse who are not deemed competent of protecting themselves from, for example, "undue influence" (such as when a family member or other caregiver is draining their financial resources). Moreover, in cases of both child and elder maltreatment, the criminal justice system cannot intervene unless criminal behavior has occurred or is occurring (Heisler & Quinn, 1995). Finally, as is the case with child maltreatment and intimate violence, not all cases of elder maltreatment are reported to adult protective services (APS). On the other hand, a huge disparity between efforts to deal with child versus elder maltreatment comes in the commitment of financial resources: Of the federal funding available for victims of abuse, 93.3% goes to child abuse, 6.7% to intimate violence, and only 0.08% to elder abuse (Hamilton, 2001).

There are no national data on the processing of elder maltreatment reports through the civil or criminal justice systems. Although a few localized studies provide some evidence concerning the nature of elder maltreatment cases, they typically do not give a breakdown of the processing of familial versus extrafamilial abuse cases. In one of the few studies with data on the role of the criminal justice system in processing cases, a survey of officers in four Alabama police precincts in 1991 and 1992 revealed that the percentage of officers who reported observing elder maltreatment ranged from 32.7% for exploitation to 43.2% for neglect. The main predictors of whether observed cases were reported included the race of the officers (Black officers were less likely to report), greater percentage of contacts with the elderly, the belief that maltreatment was not minor, and knowledge of the punishments for failure to report. In general, officers reported dissatisfaction with the Alabama system of APS; however, familiarity with the law, confidence in the ability to detect and report maltreatment, and knowledge of punishments for failure to report raised satisfaction slightly (Daniels, Baumhover, Formby, & Clark-Daniels, 1999).

Box 12.7   Elder Maltreatment Cases in the APS System

In Connecticut, a 95-year-old widow was admitted to the hospital in a semicomatose state, suffering from dehydration and malnutrition. A physician observed her son fondling his mother inappropriately, punching her, and verbally abusing her. An investigation disclosed serious neglect and possible sexual abuse by the son. Adult protective services (APS) secured a restraining order against the son, became temporary conservator of the woman, and placed her in a long-term care facility.

In Iowa, after an 83-year-old woman moved in with her son following surgery, he obtained a voluntary guardianship and power of attorney (POA) and appropriated more than $150,000 worth of her resources. APS was able to substantiate an elder abuse report on the son, finding that he used undue influence to get his mother, while confused, to sign the voluntary guardianship and POA. A civil suit against the son restored the woman's resources.

In Kansas, a mentally ill young woman physically and verbally abused her 90-year-old grandmother and 68-year-old mother. This woman was very volatile and threatening, sometimes hit her mother, brought strange men who stole things into the home, and forged checks on her mother's account. The police had been called several times, but each time the woman was able to convince them that nothing was going on, and then she would retaliate against the two women. An APS worker got her court-ordered into the state psychiatric hospital. When she was later released, APS obtained a restraining order to keep her away from the two older women, and set up a plan in which she could avoid criminal prosecution only if she had no contact with the two women and continued to participate in mental health treatment. The worker also found alternative housing for the two women so that the daughter would not know where they lived.

SOURCE: Hamilton, 2001.

Although there is little systematic data on the processing of domestic elder abuse cases through the justice system, numerous case studies provide examples of the process—some of which are presented in Box 12.7. Do the procedures in these cases appear appropriate to you? Do you think the criminal penalties should have been more or less severe in some cases? Are there alternatives to a criminal justice approach that may have served the victims better?

Summary

It was not until the second half of the 20th century that a variety of forms of maltreatment in families became criminalized. Criminalization has not put an end to violence and exploitation in families, but it has provided some legal and moral clout to victims and has legitimized their claims that family relationship status does not entitle people to victimize each other. There continues to be considerable dissatisfaction with what appears to be a bias against the poor, minorities, and married victims of intimate violence within the criminal justice system. There are also many workers within the social and protective service system who believe it is a mistake to allow the criminal justice system to play too large a role in society's response to family violence. Many of the concerns raised in the consideration of these issues are as much political, economic, and ethical as they are empirical, yet good social science research can make a sound contribution to decisions about how best to intervene in and prevent family violence. The biggest question may be the extent to which both government leaders and individual taxpayers have the will to do what is necessary to address the issues effectively.

References

Aber, J. L., Allen, J. P., Carlson, V., & Cicchetti, D. (1989). The effects of maltreatment on development during early childhood: Recent studies and their theoretical, clinical, and policy implications. In D. Cicchetti & V. Carlson (Eds.), *Child maltreatment: Theory and research on the causes and consequences of child abuse and neglect* (pp. 579–619). New York: Cambridge University Press.

Aborampah, O. M. (1989). Black male-female relationships: Some observations. *Journal of Black Studies, 19,* 320–342.

Abraham, S. (1997). *Revenge: A dish best served cold: A personal story.* Retrieved December 8, 2003, from http://www.menweb.org/scottrev.htm

Abrahams, J., & Hoey, H. (1994). Sibling incest in a clergy family: A case study. *Child Abuse and Neglect, 18,* 1029–1035.

Adams, C. (1993). I just raped my wife! What are you going to do about it, pastor? In E. Buchwald, P. Fletcher, & M. Roth (Eds.), *Transforming a rape culture* (pp. 57–86). Minneapolis: Milkweed.

Adler, N. A., & Schutz, J. (1995). Sibling incest offenders. *Child Abuse and Neglect, 19,* 811–819.

Agbayani-Siewert, P., & Flanagan, A. Y. (2001). Filipino American dating violence: Definitions, contextual justifications, and experiences of dating violence. *Journal of Human Behavior in the Social Environment, 3,* 115–133.

Agnew, R., & Huguley, S. (1989). Adolescent violence towards parents. *Journal of Marriage and the Family, 51,* 699–711.

Aguilar, R. J., & Nightingale, N. N. (1994). The impact of specific battering experiences on the self-esteem of abused women. *Journal of Family Violence, 9,* 35–45.

Aldarondo, E., & Straus, M. A. (1994). Screening for physical violence in couple therapy: Methodological, practical, and ethical considerations. *Family Process, 33,* 425–439.

Alessdandri, S. M. (1992). Mother-child interactional correlates of maltreated and nonmaltreated children's play behavior. *Development and Psychopathology, 4,* 257–270.

Alexander, P. C. (1992). Application of attachment theory to the study of sexual abuse. *Journal of Consulting and Clinical Psychology, 60,* 185–195.

317

Alexander, P. C., & Lupfer, S. L. (1987). Family characteristics and long-term consequences associated with sexual abuse. *Archives of Sexual Behavior, 16*, 235–245.

Alkhateeb, S. (n.d.). *Ending domestic violence in Muslim families.* Retrieved July 2003, from http://www.themodernreligion.com/index2.html

Allen, C., & Leventhal, B. (1999). History, culture, and identity: What makes GLBT battering different. In B. Leventhal & S. Lundy (Eds.), *Same-sex domestic violence: Strategies for change* (pp. 73–81). Thousand Oaks, CA: Sage.

Allen, D. M., & Tarnowski, K. G. (1989). Depressive characteristics of physically abused children. *Journal of Abnormal Child Psychology, 17*, 1–11.

Allen, R. J., & Miller, J. (1995). The expert as educator: Enhancing the rationality of verdicts in child abuse prosecutions. *Psychology, Public Policy and Law, 1*, 323–328.

Allport, G. W. & Ross, J. M. (1967). Personal religious orientation and prejudice. *Journal of Personality and Social Psychology, 5*, 432–443.

Alsdurf, P., & Alsdurf, J. M. (1988). Wife abuse and Scripture. In A. L. Horton & J. A. Williamson (Eds.), *Abuse and religion: When praying isn't enough.* (pp. 221–227). Lexington, MA: Lexington Books.

Altemeier, W., O'Connor, S., Vietze, P., Sandler, H., & Sherrod, K. (1982). Antecedents of child abuse. *Journal of Pediatrics, 100*, 823–829.

Alvy, K. T. (1987). *Black parenting.* New York: Irvington.

American Academy of Child and Adolescent Psychiatry (2000). *Policy statement: Juvenile death sentences.* Retrieved March 2004, from http://www.aacap.org/publications/policy/pS42.htm

American Academy of Family Physicians (2004). *Violence* (position paper). Retrieved July 26, 2003, from http://www.aafp.org/x7132.xml

American Association of University Affiliated Programs for Persons with Developmental Disabilities (2001). *Testimony for the Committee on Education and the Workforce Select Education Subcommittee United State House of Representatives Hearing on Child Abuse Prevention and Treatment Act.* Retrieved May 4, 2004, from http://www.aucd.org/legislative_affairs/testi mony_capta.htm

American Civil Liberties Union (1999). *ACLU fact sheet: Overview of lesbian and gay parenting, adoption and foster care.* Retrieved September 2003, from http://www.aclu.org/LesbianGayRights/LesbianGayRights.cfm?ID=9212&c=104

American Professional Society on the Abuse of Children (1995). *Guidelines for the psychosocial evaluation of suspected psychological maltreatment in children and adolescents.* Chicago: Author.

American Psychological Association (2003). *Elder abuse and neglect: In search of solutions.* Retrieved October 1, 2003, from http://www.apa.org/pi/aging/eldabuse.html

American Religion Data Archive (2002). *Custom analysis: Religion from General Social Survey, 2002.* Retrieved July 2003, from http://www.thearda.com

Ammerman, R. T. (1991). The role of the child in physical abuse: A reappraisal. *Violence and Victims, 6*, 87–101.

Ammerman, R. T., Lubetsky, M. J., & Stubenbort, K. F. (2000). Maltreatment of children with disabilities. In R. T. Ammerman & M. Hersen (Eds.), *Case studies in family violence* (2nd ed.) (pp. 231–258). Dordrecht, Netherlands: Kluwer.

Anderson, P. B. (1998). Women's motives for sexual initiation and aggression. In P. B. Anderson & C. Struckman-Johnson (Eds.), *Sexually aggressive women: Current perspectives and controversies* (pp. 79–93). New York: Guilford Press.

Anetzberger, G. J. (1987). *Etiology of elder abuse by adult offspring.* Springfield, IL: Charles C. Thomas.

Anetzberger, G. J. (1998). Psychological abuse and neglect: A cross-cultural concern to older Americans. In Archstone Foundation (Ed.), *Understanding and combating elder abuse in minority communities* (pp. 141–151). Long Beach, CA: Archstone Foundation.

Anetzberger, G. J., Korbin, J. E., & Tomita, S. K. (1996). Defining elder mistreatment in four ethnic groups across two generations. *Journal of Cross-Cultural Gerontology, 11,* 187–212.

Appel, A. E., & Holden, G. W. (1998). The occurrence of spouse and physical child abuse: A review and appraisal. *Journal of Family Psychology, 12,* 578–599.

Apsler, R., Cummins, M. R., & Carl, S. (2003). Perceptions of the police by female victims of domestic partner violence. *Violence Against Women, 9,* 1318–1335.

Arias, I. (1999). Women's responses to physical and psychological abuse. In X. B. Arriaga & S. Oskamp (Eds.), *Violence in intimate relationships* (pp. 139–162). Thousand Oaks, CA: Sage.

Arias, I., & Johnson, P. (1989). Evaluations of physical aggression among intimate dyads. *Journal of Interpersonal Violence, 4,* 298–307.

Arias, I., & Pope, K. T. (1999). Psychological abuse: Implications for adjustment and commitment to leave violent partners. Article for miniseries on psychological abuse. *Violence and Victims, 14,* 55–67.

Asherah, K. L. (2003). *The myth of mutual abuse.* Retrieved May 7, 2004, from http://www.nwnetwork.org/articles/4.html

Asser, S. M., & Swan, R. (1998). Child fatalities from religion-motivated medical neglect. *Pediatrics, 101,* 625–629.

Associated Press (1998). *No bail for father charged with murdering daughter.* Retrieved February 23, 2004, from http://www.cnn.com/US/9812/01/missing.girl.02/

Astor, R. A. (1994). Children's moral reasoning about family and peer violence: The role of provocation and retribution. *Child Development, 65,* 1054–1067.

Aubrey, M., & Ewing, C. P. (1989). Student and voter subjects: Differences in attitudes towards battered women. *Journal of Interpersonal Violence, 4,* 289–297.

Augoustinos, M. (1987). Developmental effects of child abuse: Recent findings. *Child Abuse and Neglect, 11,* 15–27.

Austin, J., & Dankwort, J. (1998). *A review of standards for batterer intervention programs.* Retrieved October 27, 2003, from http://www.vaw.umn.edu/final documents/vawnet/standard.htm

Avakame, E. F., & Fyfe, J. J. (2001). Differential police treatment of male-on-female spousal violence: Additional evidence on the leniency thesis. *Violence Against Women, 7,* 22–45.

Ayoub, C., & Jacewitz, M. M. (1982). Families at risk of poor parenting: A descriptive study of sixty at risk families in a model prevention program. *Child Abuse and Neglect, 6,* 413–422.

Ayoub, C., Willett, J. B., & Robinson, D. S. (1992). Families at risk of child maltreatment: Entry-level characteristics and growth in family functioning during treatment. *Child Abuse and Neglect, 16,* 495–511.

Azar, S. T., & Siegel, B. R. (1990). Behavioral treatment of child abuse: A developmental perspective. *Behavior Modification, 14,* 279–300.

Bachman, R. (1993). The double edged sword of violent victimization against the elderly: Patterns of family and stranger perpetration. *Journal of Elder Abuse and Neglect, 5,* 59–76.

Bachman, R. (2000). A comparison of annual incidence rates and contextual characteristics of intimate perpetrated violence against women from the National Crime Victimization Survey (NCVS) and the National Violence Against Women Survey (NVAW). *Violence Against Women, 6,* 839–867.

Back, S., & Lips, H. M. (1998). Child sexual abuse: Victim age, victim gender, and observer gender as factors contributing to attributions of responsibility. *Child Abuse and Neglect, 22,* 1239–1252.

Badgett, M. V. L. (1995). The wage effects of sexual orientation discrimination. *Industrial and Labor Relations Review, 48,* 726–739.

Baker, D. J. (2002). *Civil protection order: How is this piece of paper going to protect me?* School of Law Enforcement Supervision, Session XX, Criminal Justice Institute. Retrieved January 2004, from http://www.cji.net/CJI/CenterInfo/lemc/papers/BakerDelores.pdf

Baltes, P. B., Reese, H. W., & Nesselroade, J. R. (1977). *Life-span developmental psychology: Introduction to research methods.* Monterey, CA: Brooks/Cole.

Barak, G. (2003). *Violence and nonviolence: Pathways to understanding.* Thousand Oaks, CA: Sage.

Bard, L., Carter, D., Cerce, D., Knight, R., Rosenberg, R., & Schneider, B. (1983). *A descriptive study of rapists and child molesters: Developmental, clinical and criminal characteristics.* Bridgewater, MA: Mimeo.

Barling, J., O'Leary, K. D., Jouriles, E. N., Vivian, D., & MacEwen, K. E. (1987). Factor similarity of the Conflict Tactics Scales across samples, spouses, and sites: Issues and implications. *Journal of Family Violence, 2,* 37–54.

Barnett, D., Ganiban, J., & Cicchetti, D. (1999). Maltreatment, negative expressivity, and the development of type D attachments from 12 to 24 months of age. *Monographs of the Society for Research in Child Development, 64,* 97–118.

Barnett, O. W. (2000). Why battered women do not leave, part 1: External inhibiting factors within society. *Trauma, Violence & Abuse, 1,* 343–372.

Baron, S., & Welty, A. (1996). Elder abuse. *Journal of Gerontological Social Work, 25,* 33–57.

Bartholet, E. (1999). *Nobody's children: Abuse and neglect, foster drift and the adoption alternative.* Boston: Beacon.

Bartkowski, J. (1996). Beyond Biblical literalism and inerrancy: Conservative Protestants and the hermeneutic interpretation of Scripture. *Sociology of Religion, 57,* 259–272.

Bartkowski, J. P., & Ellison, C. G. (1995). Divergent models of childrearing in popular manuals: Conservative Protestants vs. the mainstream experts. *Sociology of Religion, 56,* 21–34.

Bartkowski, J. P., & Wilcox, W. B. (2000). Conservative Protestant child discipline: The case of parental yelling. *Social Forces, 79,* 265–290.

Basile, K. (1999). Rape by acquiescence: The ways in which women "give in" to unwanted sex with their husbands. *Violence Against Women, 5,* 1036–1058.

Basile, K. C. (2002). Attitudes toward wife rape: Effects of social background and victim status. *Violence and Victims, 17,* 341–354.

Basile, S. (2004). Comparison of abuse alleged by same- and opposite-gender litigants as cited in requests for abuse prevention orders. *Journal of Family Violence, 19,* 59–68.

Bath, H. I., & Haapala, D. A. (1993). Intensive family preservation services with abused and neglected children: An examination of group differences. *Child Abuse and Neglect, 17,* 213–225.

Baumrind, D. (1997). Necessary distinctions. *Psychological Inquiry, 8,* 176–182.

Beasley, R., & Stoltenberg, C. D. (1992). Personality characteristics of male spouse abusers. *Professional Psychology: Research and Practice, 23,* 310–317.

Becker, J., Kaplan, M., Cunningham-Rathner, B., & Kavoussi, R. (1986). Characteristics of adolescent incest sexual perpetrators. *Journal of Family Violence, 1,* 85–97.

Beckett, C. (2003). *Child protection: An introduction.* Thousand Oaks, CA: Sage.

Beeghly, M., & Cicchetti, D. (1994). Child maltreatment, attachment, and the self system: Emergence of an internal state lexicon in toddlers at high social risk. *Development and Psychopathology, 6,* 5–30.

Belsky, J. (1980). Child maltreatment: An ecological integration. *American Psychologist, 35,* 320–335.

Belsky, J. (1993). Etiology of child maltreatment: A developmental-ecological approach. *Psychological Bulletin, 114,* 413–434.

Belsky, J., & Vondra, J. (1989). Lessons from child abuse: The determinants of parenting. In D. Cicchetti & V. Carlson (Eds.), *Child maltreatment: Theory and research on the causes and consequences of child abuse and neglect* (pp. 153–202). New York: Cambridge University Press.

Benda, B. B. (1995). The effect of religion on adolescent delinquency revisited. *Journal of Research in Crime and Delinquency, 32,* 446–466.

Benda, B. B., & Corwyn, R. B. (2002). The effect of abuse in childhood and in adolescence on violence among adolescents. *Youth & Society, 33,* 339–365.

Benda, B. B., & Toombs, N. J. (2000). Religiosity and violence: Are they related after considering the strongest predictors? *Journal of Criminal Justice, 28,* 483–496.

Benjet, C., & Kazdin, A. E. (2003). Spanking children: The controversies, findings, and new directions. *Clinical Psychology Review, 23,* 197–224.

Bergen, R. K. (1996). *Wife rape: Understanding the response of survivors and service providers.* Thousand Oaks, CA: Sage.

Bergen, R. K. (1998). The reality of wife rape: Women's experiences of sexual violence in marriage. In R. K. Bergen (Ed.), *Issues in intimate violence* (pp. 237–250). Thousand Oaks, CA: Sage.

Berger, L. M. (2004). Income, family structure, and child maltreatment risk. *Children and Youth Services Review, 26,* 725–748.

Berger, R. D. (2001). *Successfully investigating acquaintance sexual assault: A national training manual for law enforcement.* Retrieved May 4, 2004, from www.vaw.umn.edu/documents/acquaintsa/acquaintsa.pdf

Bergeron, L. R. (2001). An elder abuse case study: Caregiver stress or domestic violence? You decide. *Journal of Gerontological Social Work, 34,* 47–63.

Berk, R. A., Campbell, A., Klap, R., & Western, B. (1992). The deterrent effect of arrest: A Bayesian analysis of four field experiments. *American Sociological Review, 57,* 698–708.

Berk, R. A., Newton, P. J., & Berk, S. F. (1986). What a difference a day makes: An empirical study of the impact of shelters for battered women. *Journal of Marriage and the Family, 48,* 481–490.

Berkeley Planning Associates (1997). *Disabled women rate caregiver abuse and domestic violence number one issue.* Berkeley, CA: Department of Education.

Berliner, L. (1988). Corporal punishment: Institutionalized assault or justifiable discipline? *Journal of Interpersonal Violence, 3,* 222–223.

Berliner, L., & Elliott, D. M. (2002). Sexual abuse of children. In J. E. Myers, L. Berliner, J. Briere, C. T. Hendrix, C. Jenny, & T. A. Reid (Eds.), *The APSAC handbook on child maltreatment* (2nd ed.) (pp. 55–78). Thousand Oaks, CA: Sage.

Berrick, J. D. (1988). Parental involvement in child abuse prevention training: What do they learn? *Child Abuse and Neglect, 12,* 543–553.

Berrick, J. D., & Gilbert, N. (1991). *With the best of intentions: The child sexual abuse prevention movement.* New York: Guilford.

Berrill, K. (1990). Anti-gay violence and victimization in the United States: An overview. *Journal of Interpersonal Violence, 5,* 274–294.

Besharov, D. J. (1993). Overreporting and underreporting are twin problems. In R. J. Gelles & D. R. Loseke (Eds.), *Current controversies in family violence* (pp. 257–272). Thousand Oaks, CA: Sage.

Besinger, B. A., Garland, A. F., Litrownik, A. J., & Landsverk, J. A. (1999). Caregiver substance abuse among maltreated children placed in out-of-home care. *Child Welfare, 78,* 221–239.

Bethke, T., & DeJoy, D. (1993). An experimental study of factors influencing the acceptability of dating violence. *Journal of Interpersonal Violence, 8,* 36–51.

Bickel, B. (2000). *The Kayla McLean murder.* Retrieved February 23, 2004, from http://crime.about.com/library/weekly/aa043000a.htm

Billingham, R. E., & Sack, A. R. (1986). Courtship violence and the interactive status of the relationship. *Journal of Adolescent Research, 1,* 315–325.

Binggeli, N. J., Hart, S. N., & Brassard, M. R. (2001). *Psychological maltreatment of children.* Thousand Oaks, CA: Sage.

Biringen, Z., & Robinson, J. (1991). Emotional availability in mother-child interactions: A reconceptualization for research. *American Journal of Orthopsychiatry, 61,* 258–271.

Bishop, S. J., & Leadbeater, B. J. (1999). Maternal social support patterns and child maltreatment: Comparison of maltreating and nonmaltreating mothers. *American Journal of Orthopsychiatry, 69,* 172–181.

Bishop, S. J., Murphy, J. M., Hicks, R., Quinn, D., Lewis, P. D., Grace, M. P., & Jellinek, M. S. (2001). The youngest victims of child maltreatment: What happens to infants in a court sample? *Child Maltreatment, 6,* 243–249.

Blackwell, B. S., & Vaughn, M. S. (2003). Police civil liability for inappropriate response to domestic assault victims. *Journal of Criminal Justice, 31,* 129–146.

Block, C. R., & Chrisakos, A. (1995). Intimate partner homicide in Chicago over 29 years. *Crime and Delinquency, 41,* 496.

Block, R. W. (2003). Child fatalities. In J. E. B. Myers, L. Berliner, J. Briere, C. T. Hendrix, C. Jenny, & T. A. Reid (Eds.), *The APSAC handbook on child maltreatment* (2nd ed.) (pp. 293–301). Thousand Oaks, CA: Sage.

Bograd, M. (1984). Family systems approaches to wife battering. *American Journal of Orthopsychiatry, 54,* 558–568.

Bolen, R. M. (1998). Predicting risk to be sexually abused: A comparison of logistic regression to event history analysis. *Child Maltreatment, 3,* 157–170.

Bolen, R. M. (2000). Extrafamilial child sexual abuse: A study of perpetrator characteristics and implications for prevention. *Violence Against Women, 6,* 1137–1169.

Boney-McCoy, S., & Finkelhor, D. (1995). Psychosocial sequelae of violent victimization in a national youth sample. *Journal of Consulting and Clinical Psychology, 63,* 726–736.

Bonner, B. L., Crow, S. M., & Logue, M. B. (1999). Fatal child neglect. In H. Dubowitz (Ed.), *Neglected children: Research, practice, and policy* (pp. 156–173). Thousand Oaks, CA: Sage.

Bookwala, J., Frieze, I. H., Smith, C., & Ryan, K. (1992). Predictors of dating violence: A multivariate analysis. *Violence and Victims, 7,* 297–311.

Booth, A., & Dabbs, J. M. (1993). Testosterone and men's marriages. *Social Forces, 72,* 463–477.

Bottoms, B. L., Shaver, P. R., Goodman, G. S., & Qin, J. (1995). In the name of God: A profile of religion-related child abuse. *The Society for the Psychological Study of Social Issues, 51,* 85–111.

Bousha, D. M., & Twentyman, C. T. (1984). Mother-child interactional style in abuse, neglect, and control groups: Naturalistic observation in the home. *Journal of Abnormal Psychology, 1,* 106–114.

Bower, M., & Knutson, J. F. (1996). Attitudes toward physical discipline as a function of disciplinary history and self-labeling as physically abused. *Child Abuse and Neglect, 20,* 689–699.

Bowker, L. H. (1983). Marital rape. A distinct syndrome? *Social Casework, June,* 347–352.

Bowker, L. H. (1988). Religious victims and their religious leaders: Services delivered to one thousand battered women by the clergy. In A. L. Horton & J. A. Williamson (Eds.), *Abuse and religion: When praying isn't enough* (pp. 229–234). Lexington, MA: Lexington Books.

Bowker, L. H., Arbitell, M., & McFerron, J. R. (1988). On the relationship between wife beating and child abuse. In K. Yllo & M. Bograd (Eds.), *Feminist perspectives on wife abuse* (pp. 158–174). Newbury Park, CA: Sage.

Bowlby, J. (1969/1982). *Attachment and loss: Attachment.* New York: Basic Books.

Bowlby, J. (1973). *Attachment and loss: Separation* (Vol. 2). London: The Hogarth Press.

Brachfield, S., Goldberg, S., & Sloman, J. (1980). Parent-infant interaction in free play at 8 and 12 months: Effects of prematurity and immaturity. *Infant Behavior in Development, 3,* 289–305.

Brand, P. A., & Kidd, A. H. (1986). Frequency of physical aggression in heterosexual and female homosexual dyads. *Psychological Reports, 59,* 1307–1313.

Brandl, B., & Cook-Daniels, L. (2002). *Domestic abuse in later life.* Retrieved December 11, 2003, from http://www.vaw.umn.edu/documents/vawnet/arlaterlife/arlaterlife.html

Brassard, M. R., Hart, S. N., & Hardy, D. B. (2000). Psychological and emotional abuse of children. In R. T. Ammerman & M. Hersen (Eds.), *Case studies in family violence* (2nd ed.) (pp. 293–319). New York: Kluwer.

Brennan, P., Mednick, S. A., & Kandel, E. (1993). Congenital determinants of violent and property offending. In D. J. Pepler & K. H. Rubin (Eds.), *The development and treatment of childhood aggression* (pp. 81–92). Hillsdale, NJ: Lawrence Erlbaum.

Brenner, R. A., Overpeck, M. D., Trumbel, A. C., DerSimonlan, R., & Berendes, H. (1999). Deaths attributable to injuries in infants, United States, 1983–1991. *Pediatrics, 103,* 968–974.

Brezina, T. (1999). Teenage violence toward parents as an adaptation to family strain: Evidence from a national survey of male adolescents. *Youth and Society, 30,* 416–444.

Briere, J. (1987). Predicting self-reported likelihood of battering: Attitudes and childhood experiences. *Journal of Research in Personality, 21,* 61–69.

Briere, J., & Gil, E. (1998). Self-mutilation in clinical and general population samples: Prevalence, correlates, and functions. *American Journal of Orthopsychiatry, 68,* 609–620.

Briere, J., & Runtz, M. (1987). Post-sexual abuse trauma: Data and implications for clinical practice. *Journal of Interpersonal Violence, 8,* 367–379.

Briere, J., & Runtz, M. (1988). Multivariate correlates of childhood psychological and physical maltreatment among university women. *Child Abuse and Neglect, 12,* 331–341.

Briere, J., & Runtz, M. (1989). University males' sexual interest in children: Predicting potential indices of pedophilia in a nonforensic sample. *Child Abuse and Neglect, 13,* 65–75.

Briere, J., & Runtz, M. (1990). Differential adult symptomology associated with three types of child abuse histories. *Child Abuse and Neglect, 14,* 357–364.

Briere, J., Woo, R., McRae, B., Foltz, J., & Sitzman, R. (1997). Lifetime victimization history, demographics, and clinical status in female psychiatric emergency room patients. *Journal of Nervous and Mental Disease, 185,* 95–101.

Bristowe, E., & Collins, J. (1989). Family mediated abuse of noninstitutionalized frail elderly men and women in British Columbia. *Journal of Elder Abuse and Neglect, 1,* 45–64.

Britner, P. A., & Mossler, D. G. (2002). Professionals' decision-making about out-of-home placements following instances of child abuse. *Child Abuse and Neglect, 26,* 317–332.

Bronfenbrenner, U. (1979). *The ecology of human development: Experiments by nature and design.* Cambridge. MA: Harvard University Press.

Brown v. Board of Education of Topeka, 347 U.S. 483–96 (1954). pp. 488–492.

Brown, E. J., & Kolko, D. J. (1999). Child victims' attributions about being physically abused: An examination of factors associated with symptom severity. *Journal of Abnormal Child Psychology, 27,* 311–322.

Brown, G. R., & Anderson, B. (1991). Psychiatric morbidity in adult inpatients with childhood histories of sexual and physical abuse. *American Journal of Psychiatry, 148,* 55–61.

Brown, J., Cohen, P., Johnson, J. G., & Salzinger, S. (1998). A longitudinal analysis of risk factors for child maltreatment: Findings of a 17-year prospective study of officially recorded and self-reported child abuse and neglect. *Child Abuse and Neglect, 22,* 1065–1078.

Brown, P. D., O'Leary, K. D., & Feldbau, S. R. (1997). Dropout in a treatment program for self-referring wife abusing men. *Journal of Family Violence, 12,* 365–387.

Brown, S. E. (1984). Social class, child maltreatment, and delinquent behavior. *Criminology, 22,* 259–278.

Browne, A. (1987). *When battered women kill.* New York: Macmillan Free Press.

Browne, A., & Finkelhor, D. (1986). Impact of child sexual abuse: A review of the research. *Psychological Bulletin, 18,* 66–77.

Browne, K., & Herbert, M. (1997). *Preventing family violence.* New York: John Wiley & Sons.

Browne, K. D., & Hamilton, C. (1998). Physical violence between young adults and their parents: Associated with a history of child maltreatment. *Journal of Family Violence, 13,* 59–79.

Brutz, J. L., & Allen, C. M. (1986). Religious commitment, peace activism, and marital violence in Quaker families. *Journal of Marriage and the Family, 48,* 491–502.

Brutz, J. L., & Ingoldsby, B. B. (1984). Conflict resolution in Quaker families. *Journal of Marriage and the Family, 46,* 21–26.

Bufkin, J., & Eschholz, S. (2000). Images of sex and rape: A content analysis of popular film. *Violence Against Women, 6,* 1317–1344.

Bugental, D. B., Mantyla, S. M., & Lewis, J. (1989). Parental attributions as moderators of affective communication to children at risk for physical abuse. In

D. Cicchetti & V. Carlson (Eds.), *Child maltreatment: Theory and research on the causes and consequences of child abuse and neglect*. New York: Cambridge University Press.

Buntain-Ricklefs, J. J., Kemper, K. J., Bell, M., & Babonis, T. (1994). Punishments: What predicts adult approval. *Child Abuse and Neglect, 18*, 945–955.

Burgess, R. L., & Conger, R. D. (1978). Family interactions in abusive, neglectful, and normal families. *Child Development, 49*, 1163–1173.

Burgess, R. L., Leone, J. M., & Kleinbaum, S. M. (2000). Social and ecological issues in violence toward children. In R. T. Ammerman & M. Hersen (Eds.), *Case studies in family violence* (2nd ed.) (pp. 15–38). New York: Plenum.

Burke, L. K., & Follingstad, D. R. (1999). Violence in lesbian and gay relationships: Theme, prevalence, and correlational factors. *Clinical Psychology Review, 19*, 487–512.

Burns, K., Chethik, L., Burns, W. J., & Clark, R. (1991). Dyadic disturbances in cocaine-abusing mothers and their infants. *Journal of Clinical Psychology, 47*, 316–319.

Burton, D. L., Miller, D. L., & Shill, C. T. (2002). A social learning theory comparison of the sexual victimization of adolescent sexual offenders and nonsexual offending male delinquents. *Child Abuse and Neglect, 26*, 893–907.

Busch, A. L., & Rosenberg, M. S. (2004). Comparing women and men arrested for domestic violence: A preliminary report. *Journal of Family Violence, 19*, 49–57.

Bushman, B. J., & Anderson, C. J. (2001). Media violence and the American public: Scientific facts versus media misinformation. *American Psychologist, 56*, 477–489.

Buzawa, E., Austin, T. L., & Buzawa, C. G. (1995). Responding to crimes of violence against women: Gender differences versus organizational imperatives. *Crime and Delinquency, 41*, 443–466.

Buzawa, E. S., & Buzawa, C. G. (1993). The scientific evidence is not conclusive: Arrest is no panacea. In R. J. Gelles & D. R. Loseke (Eds.), *Current controversies on family violence* (pp. 337–356). Newbury Park, CA: Sage.

Buzawa, E. S., & Buzawa, C. G. (1996). *Domestic violence: The criminal justice system response*. Thousand Oaks, CA: Sage.

Byrne, C. A., Kilpatrick, D. G., Howley, S. S., & Beatty, D. (1999). Female victims of partner versus nonpartner violence: Experiences with the criminal justice system. *Criminal Justice and Behavior, 26*, 275–292.

Cabral, A., & Coffey, D. (1999). Creating courtroom accessibility. In B. Leventhal & S. Lundy (Eds.), *Same-sex domestic violence: Strategies for change* (pp. 57–69). Thousand Oaks, CA: Sage.

Caesar, P. L. (1988). Exposure to violence in the families-of-origin among wife-abusers and maritally nonviolent men. *Violence and Victims, 3*, 49–63.

Caetano, R., & Cunradi, C. (2003). Intimate partner violence and depression among Whites, Blacks, and Hispanics. *Annals of Epidemiology, 13*, 661–665.

Cahill, L. T., Kaminer, R. K., & Johnson, P. G. (1999). Developmental, cognitive, and behavioral sequalae of child abuse. *Child and Adolescent Psychiatric Clinics of North America, 8*, 827–843.

Caliso, J. A., & Milner, J. S. (1992). Childhood history of abuse and child abuse screening. *Child Abuse and Neglect, 16,* 647–659.

Campbell, J. C., & Alford, P. (1989). The dark consequences of marital rape. *American Journal of Nursing, 89,* 946–949.

Campbell, R. (1989). *How to really love your child.* Wheaton, IL: Victor Books.

Campbell Reay, A. M., & Browne, K. D. (2002). The effectiveness of psychological interventions with individuals who physically abuse or neglect their elderly dependents. *Journal of Interpersonal Violence, 17,* 416–431.

Candib, L. M. (1999). Incest and other harms to daughters across cultures. *Women's Studies International Forum, 22,* 185–201.

Cantos, A. L., Neale, J. M., & O'Leary, K. D. (1997). Assessment of coping strategies of child abusing mothers. *Child Abuse and Neglect, 21,* 631–636.

Cappell, C., & Heiner, R. B. (1990). The intergenerational transmission of family aggression. *Journal of Family Violence, 5,* 135–151.

Carey, G., & Goldman, D. (1997). Genetics of antisocial behavior. In D. Stoff, J. Brelling, & J. Demiser (Eds), *Handbook of Antisocial Personality* (pp. 243–254). New York: Wiley.

Carey, T. A. (1994). Spare the rod and spoil the child. Is this a sensible justification for the use of punishment in child rearing? *Child Abuse and Neglect, 18,* 1005–1010.

Carter, P. (2001). *HHS reports new child abuse and neglect statistics.* HHS News Release. Retrieved March 2003, from http://www.hhs.gov/news/press/2001pres/20010402.html

Cascardi, M., Langhinrichsen, J., & Vivian, D. (1992). Marital aggression: Impact, injury, and health correlates for husbands and wives. *Archives of Internal Medicine, 152,* 1178–1184.

Cascardi, M., & O'Leary, D. (1992). Depressive symptomatology, self-esteem, and self-blame in battered women. *Journal of Family Violence, 7,* 249–259.

Cascardi, M., O'Leary, K. D., Lawrence, E. E., & Schlee, K. A. (1995). Characteristics of women physically abused by their spouses and who seek treatment regarding marital conflict. *Journal of Consulting and Clinical Psychology, 63,* 616–623.

Cate, R. M., Henton, J. M., Koval, J., Christopher, F. S., & Lloyd, S. (1982). Premarital abuse: A social psychological perspective. *Journal of Family Issues, 3,* 79–90.

Cazenave, N. A., & Straus, M. A. (1990). Race, class, network embeddedness, and family violence: A search for potent support systems. In M. A. Straus & R. J. Gelles (Eds.), *Physical violence in American families: Risk factors and adaptations to violence in 8, 145 families* (pp. 321–339). New Brunswick, NJ: Transaction.

Cazenave, N. A., & Zahn, M. A. (1992). Women, murder, and male domination: Police reports of domestic homicide in Chicago and Philadelphia. In E. C. Viano (Ed.), *Intimate violence: Interdisciplinary perspectives* (pp. 83–96). Washington, D.C.: Hemisphere Publishing.

Center for Domestic Violence Prevention (2003). Retrieved September 16, 2003, from http://www.cdvp.org/about/programs/index.html

Center for Research on Women with Disabilities (1999). *National Study of Women with Physical Disabilities*. Retrieved May 4, 2004, from http://www.bcm.edu/crowd/national_study/national_study.html

Centers for Disease Control and Prevention (1997). *Rates of homicide, suicide, and firearm-related death among children—26 industrialized countries*. Morbidity and Mortality Weekly Report. Retrieved May 20, 2004, from http://www.cdc.gov/mmwr/preview/mmwrhtml/00046149.htm

Centers for Disease Control and Prevention (2002). *Variations in homicide risk during infancy–United States, 1989–1998*. Morbidity and Mortality Weekly Report. Retrieved May 20, 2004, from http://www.cdc.gov/mmwr/preview/mmwrhtml/mm5109a3.htm

Chance, T. (2003, July). *Our children are dying: Understanding and improving national maltreatment fatality data*. Paper presented at the 8th International Family Violence Research Conference, Portsmouth, NH.

Chandy, J. M., Blum, R. W., & Resnick, M. D. (1996). Gender-specific outcomes for sexually abused adolescents. *Child Abuse and Neglect, 20,* 1219–1231.

Charles, A. V. (1986). Physically abused parents. *Journal of Family Violence, 1,* 343–355.

Chasnoff, I. J., & Lowder, L. A. (1999). Prenatal alcohol and drug use and risk for child maltreatment: A timely approach to intervention. In H. Dubowitz (Ed.), *Neglected children: Research, practice, and policy* (pp. 132–155). Thousand Oaks, CA: Sage.

Child Trends Data Bank (2002). *Infant homicide*. Retrieved August 16, 2004, from http://www.childtrendsdatabank.org/pdf/72_PDF.pdf

Child Welfare League of America (n.d.). *Promoting Safe and Stable Families (PSSF) Program*. Retrieved October 2003, from http://www.cwla.org/advocacy/pssf.htm

Child Welfare League of America (2002). *Juvenile offenders and the death penalty: Is justice served?* Retrieved January 2004, from http://www.cwla.org/programs/juvenilejustice/juveniledeathpenalty.pdf

Children's Defense Fund (2003). *Number of black children in extreme poverty hits record high*. Retrieved December 12, 2003, from http://www.childrensdefense.org/pressreleases/2003/030430.asp

Childs, H. W., Hayslip, B., Jr., Radika, L. M., & Reinberg, J. A. (2000). Young and middle-aged adults' perceptions of elder abuse. *Gerontologist, 40,* 75–85.

ChildStats (2004). *America's children in brief: Key national indicators of well-being, 2004*. Retrieved August 16, 2004, from http://childstats.gov/americaschildren/pdf/ac04brief.pdf

Chu, J. A., & Dill, D. L. (1990). Dissociative symptoms in relation to childhood physical and sexual abuse. *American Journal of Psychiatry, 147,* 887–892.

Claussen, A. H., & Crittenden, P. M. (1991). Physical and psychological maltreatment: Relations among the types of maltreatment. *Child Abuse and Neglect, 15,* 5–18.

Clements-Nolle, K., Marx, R., Guzman, R., & Katz, M. (2001). HIV prevalence, risk behaviors, health care use, and mental health status of transgender persons: Implications for public health intervention. *American Journal of Public Health, 91,* 915–921.

Cohen, D. (1996). Law, social policy, and violence: The impact of regional cultures. Law, social policy, and violence. *Journal of Personality and Social Psychology, 70*, 961–978.

Cohen, D. (2000). Caregiver stress increases risk of homicide-suicide. *Geriatric Times.* Retrieved December 5, 2003, from http://www.geriatrictimes.com/g001225.html

Cohen, J. A., Berliner, L., & Mannarino, A. P. (2000). Treating traumatized children: A research review. *Trauma, Violence, and Abuse, 1*, 29–46.

Cohen, D., Llorente, M., & Eisdorfer, C. (1998). Homicide-suicide in older persons. *American Journal of Psychiatry, 155*, 390–396.

Cohen, J. A., & Mannarino, A. P. (2000). Incest. In R.T. Ammerman & M. Hersen (Eds.), *Case studies in family violence* (2nd ed.) (pp. 209–230). New York: Kluwer.

Cohen, L. J., Gans, S. W., McGeoch, P. G., Poznnsky, O., Itskovich, Y., Murphy, S., Klein, E., Cullen, K., & Galynker, I. I. (2002). Impulsive personality traits in male pedophiles versus healthy controls: Is pedophilia an impulsive-aggressive disorder? *Comprehensive Psychiatry, 43*, 127–134.

Cohn, A. H., & Daro, D. (1987). Is treatment too late: What ten years of evaluative research tell us. *Child Abuse and Neglect, 11*, 433–442.

Cole, E. (1982). Sibling incest: The myth of benign sibling incest. *Women and Therapy, 1*, 79–89.

Collins, J. J., & Messerschmidt, P. M. (1993). Epidemiology of alcohol-related violence. *Alcohol Health and Research World, 17*, 93–101.

Comijs, H. C., Penninx, B. W., Knipscheer, K. P., & van Tilburg, W. (1999). Psychological distress in victims of elder mistreatment: The effects of social support and coping. *Journals of Gerontology: Series B: Psychological Sciences and Social Sciences, 54B*, 240–245.

Connolly, C., Huzurbazar, S., & Routh-McGee, T. (2000). Multiple parties in domestic violence situations and arrest. *Journal of Criminal Justice, 28*, 181–188.

Coohey, C. (1995). Neglectful mothers, their mothers, and partners: The significance of mutual aid. *Child Abuse and Neglect, 19*, 885–895.

Coohey, C. (2000). The role of friends, in-laws, and other kin in father-perpetrated child abuse. *Child Welfare, 79*, 373–402.

Cook, P.W. (1997). *Abused men: The hidden side of domestic violence.* Westport, CT: Praeger.

Corliss, H. L., Cochran, S. D., & Mays, V. M. (2002). Reports of parental maltreatment during childhood in a United States population based survey of homosexual, bisexual and heterosexual adults. *Child Abuse and Neglect, 26*, 1165–1178.

Cormier, B. M., Angliker, C. C. J., Gagne, P. W., & Markus, B. (1978). Adolescents who kill a member of the family. In J. M. Eekelaar & S. N. Katz (Eds.), *Family violence: An international and interdisciplinary study* (pp. 466–478). Toronto: Butterworth.

Cornell, C. P., & Gelles, R. J. (1982). Adolescent to parent violence. *Urban and Social Change Review, 15*, 8–14.

Coyne, A. C., Reichman, W. E., & Berbig, L. J. (1993). The relationship between dementia and elder abuse. *American Journal of Psychiatry, 150*, 643–646.

Craig Shea, M. E. (1998). When the tables are turned: Verbal sexual coercion among college women. In P. B. Anderson & C. Struckman-Johnson (Eds.), *Sexually aggressive women: Current perspectives and controversies* (pp. 94–104). New York: Guilford Press.

Crittenden, P. M. (1992). Children's strategies for coping with adverse home environments: An interpretation using attachment theory. *Child Abuse and Neglect, 16,* 329–343.

Crittenden, P. M. (1999). Child neglect: Causes and contributors. In H. Dubowitz (Ed.), *Neglected children: Research, practice, and policy* (pp. 47–68). Thousand Oaks, CA: Sage.

Crittenden, P. M., Claussen, A. H., & Sugarman, D. B. (1994). Physical and psychological maltreatment in middle childhood and adolescence. *Development and Psychopathology, 6,* 145–164.

Crittenden, P. M., & DiLalla, D. L. (1988). Compulsive compliance: The development of an inhibitory coping strategy in infancy. *Journal of Abnormal Child Psychology, 16,* 585–599.

Cross, T. P., Walsh, W. A., Simone, M., & Jones, L. M. (2003). Prosecution of child abuse: A meta-analysis of rates of criminal justice decisions. *Trauma, Violence, & Abuse, 4,* 323–340.

Crosse, S., Kaye, E., & Ratnofsky, A. (1993). *A report on the maltreatment of children with disabilities.* Rockville, MD: Westat, Inc.

Crossmaker, M. (1991). Behind locked doors: Institutional sexual abuse. *Sexuality and Disability, 9,* 201–218.

Crouch, J. L., & Behl, L. E. (2001). Relationships among parental beliefs in corporal punishment, reported stress, and physical child abuse potential. *Child Abuse and Neglect, 25,* 413–419.

Cullen, B. J., Smith, P. H., Funk, J. B., & Haaf, R. A. (2000). A matched cohort comparison of a criminal justice system's response to child sexual abuse: A profile of perpetrators. *Child Abuse and Neglect, 24,* 569–577.

Culp, R. E., Little, V., Letts, D., & Lawrence, H. (1991). Maltreated children's self-concept: Effects of a comprehensive treatment program. *American Journal of Orthopsychiatry, 61,* 114–121.

Cunradi, C. B., Caetano, R., Clark, C., & Schafer, J. (2000). Neighborhood poverty as a predictor of intimate partner violence among White, Black, and Hispanic couples in the United States. *Annals of Epidemiology, 10,* 297–308.

Curry, M. A., Hassouneh-Phillips, D., & Johnston-Silverberg, A. (2001). Abuse of women with disabilities: An ecological model and review. *Violence Against Women, 7,* 60–79.

Curtis, P. A., & McCullough, C. (1993). The impact of alcohol and other drugs on the child welfare system. *Child Welfare, 72,* 533–542.

Cuskey, W. R., & Wathey, B. (1982). *Female addiction.* Lexington, MA: Lexington Books.

Cwik, M. S. (1997). Peace in the home? The response of rabbis to wife abuse within American Jewish congregations. *Journal of Psychology and Judaism, 21,* 5–81.

Cyr, M., Wright, J., McDuff, P., & Perron, A. (2002). Intrafamilial sexual abuse: Brother-sister incest does not differ from father-daughter and stepfather-stepdaughter incest. *Child Abuse and Neglect, 26,* 957–973.

Dadds, M., Smith, M., Weber, Y., & Robinson, A. (1991). An exploration of family and individual profiles following father-daughter incest. *Child Abuse and Neglect, 15,* 575–586.

Dallam, S. J., Gleaves, D. H., Cepeda-Benito, A., Silberg, J. L., Kraemer, H. C., & Spiegel, D. (2001). The effects of child sexual abuse: Comment on Rind, Tromovitch, and Bauserman (1998). *Psychological Bulletin, 127,* 715–733.

Daly, M., Wilson, M., Salmon, C. A., Hiraiwa-Hasegawa, M., & Hasegawa, T. (2001). Siblicide and seniority. *Homicide Studies, 5,* 30–45.

Daniels, R. S., Baumhover, L. A., Formby, W. A., & Clark-Daniels, C. L. (1999). Police discretion and elder mistreatment: A nested model of observation, reporting, and satisfaction. *Journal of Criminal Justice, 27,* 209–225.

Daro, D. (1988). *Confronting child abuse.* New York: Free Press.

Daro, D. (1991). Prevention programs. In C. Hollin & K. Howells (Eds.), *Clinical approaches to sex offenders and their victims* (pp. 285–306). New York: John Wiley.

Daro, D. (2002). Public perception of child sexual abuse: Who is to blame? *Child Abuse and Neglect, 26,* 1131–1133.

Daro, D., & Cohn, A. (1988). Child maltreatment evaluation efforts: What have we learned? In G. Hotaling, D. Finkelhor, J. Kirkpatrick, & M. A. Straus (Eds.), *Coping with family violence: Research and policy perspectives* (pp. 275–287). Newbury Park, CA: Sage.

Daro, D., & Connelly, A. C. (2002). Child abuse prevention: Accomplishments and challenges. In J. E. B. Myers, L. Berliner, J. Briere, C. T. Hendrix, C. Jenny, & T. A. Reid (Eds.), *The APSAC handbook on child maltreatment* (2nd ed.) (pp. 431–448). Thousand Oaks, CA: Sage.

Daro, D., & Harding, K. (1999). Healthy Families America: Using research to enhance practice. *Future of Children, 9,* 152–176.

Dasgupta, S. D. (1999). Just like men? A critical view of violence by women. In M. F. Shepard & E. L. Pence (Eds.), *Coordinating community response to domestic violence: Lessons from Duluth and beyond* (pp. 195–222). Thousand Oaks, CA: Sage.

Dasgupta, S. D., & Warrier, S. (1996). In the footsteps of "Arundhati": Asian Indian women's experience of domestic violence in the United States. *Violence Against Women, 2,* 238–259.

D'Augelli, A. R. (1992). Lesbian and gay male undergraduates' experiences of harassment and fear on campus. *Journal of Interpersonal Violence, 7,* 383–395.

D'Augelli, A. R., & Grossman, A. H. (2001). Disclosure of sexual orientation, victimization, and mental health among lesbian, gay, and bisexual older adults. *Journal of Interpersonal Violence, 16,* 1008–1027.

Davidson, H. (1988). Failure to report child abuse: Legal penalties and emerging issues. In A. Maney & S. Wells (Eds.), *Professional responsibility in protecting children* (pp. 93–103). New York: Praeger.

Davidson, H. (1997). The legal aspects of corporal punishment in the home: When does physical discipline cross the line to become child abuse? *Children's Legal Rights Journal, 17,* 18–29.

Davis, R. C., & Medina-Ariza, J. (2001). *Results from an elder abuse prevention experiment in New York City.* U.S. Department of Justice, Office of Justice Programs, Research in Brief.

Day, N. L., & Richardson, G. A. (1994). Comparative teratogenicity of alcohol and other drugs. *Alcohol Health and Research World, 18,* 42–48.

DeBellis, M. D., Broussard, E. R., Herring, D. J., Wexler, S., Moritz, G., & Benitez, J.G. (2001). Psychiatric co-morbidity in caregivers and children involved in maltreatment: A pilot research study with policy implications. *Child Abuse and Neglect, 25,* 923–944.

DeJong, A. R. (1998). Impact of child sexual abuse medical examinations on the dependency and criminal systems. *Child Abuse and Neglect, 22,* 645–652.

DeKeseredy, W. S., & Schwartz, M. D. (2001). In C. M. Renzetti, J. L. Edleson, & R. K. Bergen (Eds.), *Sourcebook on violence against women* (pp. 23–34). Thousand Oaks, CA: Sage.

DeLozier, P. P. (1982). Attachment theory and child abuse. In M. Parkes & J. Stevenson-Hinde (Eds.), *The place of attachment in human behavior* (pp. 95–117). New York: Basic Books.

DePanifilis, D. (1999). Intervening with families when children are neglected. In H. Dubowitz (Ed.), *Neglected children: Research, practice, and policy* (pp. 211–236). Thousand Oaks, CA: Sage.

DeShaney v. Winnebago County Department of Social Services, 489 U.S. 189 (1989). Retrieved May 2004, from http://www.oyez.org/oyez/resource/case/634/

DiLalla, L. F., & Gottesman, I. I. (1991). Biological and genetic contributors to violence: Widom's untold tale. *Psychological Bulletin, 109,* 125–129.

DiPlacido, J. (1998). Minority stress among lesbians, gay men and bisexuals: A consequence of heterosexism, homophobia, and stigmatization. In G. Herek (Ed.), *Stigma and sexual orientation: Understanding prejudice against lesbians, gay men and bisexuals* (pp. 138–159). Thousand Oaks, CA: Sage.

Dobash, R. E., & Dobash, R. P. (1988). Research as social action: The struggle for battered women. In K. Yllo & M. Bograd (Eds.), *Feminist perspectives on wife abuse* (pp. 51–74). Newbury Park, CA: Sage.

Dobson, J. C. (1978). *Preparing for adolescence.* Ventura, CA: Gospel Light Publishing.

Dobson, J. C. (1987). *Parenting isn't for cowards.* Waco, TX: Word.

Dodge, K. A., Bates, J. E., & Pettit, G. S. (1990). Mechanisms in the cycle of violence. *Science, 250,* 1678–1683.

Dodge, K. A., Pettit, G. S., & Bates, J. E. (1994). Socialization mediators of the relation between socioeconomic status and child conduct problems. *Child Development, 65,* 649–665.

Doege, D. (2002a). *Battered, torn, but not broken: For many older women, love hurts.* Retrieved December 5, 2003, from http://www.jsonline.com/lifestyle/people/ju102/56906.asp

Doege, D. (2002b). *Women find solace, strength to leave in support group.* Retrieved December 5, 2003, from http://www.jsonline.com/lifestyle/people/ju102/56895.asp

Dolz, L., Cerezo, M. A., & Milner, J. S. (1997). Mother-child interactional patterns in high- and low-risk mothers. *Child Abuse and Neglect, 21,* 1149–1158.

Domestic Abuse Helpline for Men (2003). *No Name Story #1.* Retrieved May 4, 2004, from http://www.batteredmenshelpline.org/story_NoName1.html

Donahue, M. J., & Benson, P. L. (1995). Religion and the well-being of adolescents. *Journal of Social Issues, 51,* 145–160.

Donato, K. M., & Bowker, L. H. (1984). Understanding the help-seeking behavior of battered women: A comparison of traditional service agencies and women's groups. *International Journal of Women's Studies, 7,* 99–109.

Donnelly, D. A., Cook, K. J., & Wilson, L. (1999). Provision and exclusion: The dual face of services to battered women in three Deep South states. *Violence Against Women, 5,* 710–741.

Dopke, C. A., & Milner, J. S. (2000). Impact of child compliance on stress appraisals, attributions, and disciplinary choices in mothers at high and low risk for child physical abuse. *Child Abuse and Neglect, 24,* 493–504.

Douglas, H. (1991). Assessing violent couples. *Families in Society, 72,* 525–535.

Dowd, L. (2001). Female perpetrators of partner aggression: Relevant issues and treatment. *Journal of Aggression, Maltreatment, and Trauma, 5,* 73–104.

Downs, W. R., & Miller, B. A. (1998). Relationships between experiences of parental violence during childhood and women's self-esteem. *Violence and Victims, 13,* 63–77.

Drake, B., & Pandey, S. (1996). Understanding the relationship between neighborhood poverty and specific types of child maltreatment. *Child Abuse and Neglect, 20,* 1003–1018.

Drotar, D., Eckerle, D., Satola, J., Palotta, J., & Wyatt, B. (1990). Maternal interactional behavior with non-organic failure-to-thrive infants: A case comparison study. *Child Abuse and Neglect, 14,* 41–51.

Dubowitz, H. (Ed.). (1999). *Neglected children: Research, practice, and policy.* Thousand Oaks, CA: Sage.

Dubowitz, H., Klockner, A., Starr, R. H., & Black, M. M. (1998). Community and professional definitions of child neglect. *Child Maltreatment, 3,* 235–243.

Duminy, F. J., & Hudson, D. A. (1993). Assault inflicted by hot water. *Burns, 19,* 426–428.

Duncan, D. F. (1990). Prevalence of sexual assault victimization among heterosexual and gay/lesbian university students. *Psychological Reports, 66,* 65–66.

Duncan, J. W., & Duncan, G. M. (1971). Murder in the family. *American Journal of Psychiatry, 127,* 74–78.

Duncan, L. E., & Williams, L. M. (1998). Gender role socialization and male-on-male vs. female-on-male child sexual abuse. *Sex Roles, 39,* 765–785.

Dunford, F. W. (1992). The measurement of recidivism in cases of spouse assault. *Journal of Criminal Law and Criminology, 83,* 120–136.

Durst, D. (1991). Conjugal violence: Changing attitudes in two northern Native communities. *Community Mental Health Journal, 27,* 359–373.

Dutcher-Walls, P. (1999). Sociological directions in feminist biblical studies. *Social Compass, 46,* 441–453.

Dutton, D. G. (1985). An ecologically nested theory of male violence toward intimates. *International Journal of Women's Studies, 8,* 404–413.

Dutton, D. G. (1994). Behavioral and affective correlates of borderline personality organization in wife assaulters. *International Journal of Law and Psychiatry, 17,* 265–277.

Dutton, D. G. (1995). Male abusiveness in intimate relationships. *Clinical Psychology Review, 15,* 567–581.

Dutton, D. G. (1995). *The domestic assault of women: Psychological and criminal justice perspectives.* Vancouver: University of British Columbia Press.

Dutton, D. G., & Strachan, C. E. (1987). Motivational needs for power and spouse-specific assertiveness in assaultive and nonassaultive men. *Violence and Victims, 2,* 145–156.

Dutton, M. A. (1996). *Critique of the "battered woman syndrome" model.* Retrieved March 22, 2004, from http://www.vaw.umn.edu/documents/vawnet/bws/bws.html

Eckenrode, J., Laird, M., & Doris, J. (1993). School performance and disciplinary problems among abused and neglected children. *Developmental Psychology, 29,* 53–62.

Egeland, B., & Erickson, M. F. (1987). Psychologically unavailable caregiving. In M. Brassard, B. Germain, & S. Hart (Eds.), *Psychological maltreatment of children and youth* (pp. 110–120). Elmsford, NY: Pergamon.

Egeland, B., & Sroufe, L. A. (1981). Developmental sequelae of maltreatment in infancy. In B. Rizley & D. Cicchetti (Eds.), *New directions for child development: Developmental perspectives in child maltreatment* (pp. 77–92). San Francisco, CA: Jossey-Bass.

Egeland, B., Sroufe, L. A., & Erickson, M. F. (1983). Developmental consequences of different patterns of maltreatment. *Child Abuse and Neglect, 7,* 456–469.

Elliott, D. M. (1994). Impaired object relations in professional women molested as children. *Psychotherapy, 21,* 79–86.

Elliott, D. M., & Briere, J. (1992). Sexual abuse trauma among professional women: Validating the Trauma Symptom Checklist–40 (TSC–40). *Child Abuse and Neglect, 16,* 391–398.

Elliott, D. M., & Briere, J. (1994). Forensic sexual abuse evaluations of older children: Disclosures and symptomatology. *Behavioral Sciences and the Law, 12,* 261–277.

Elliott, M. (1993). *Female sexual abuse of children: The ultimate taboo.* New York: John Wiley & Sons.

Ellison, C. G., & Anderson, K. L. (2001). Religious involvement and domestic violence among U.S. couples. *Journal for the Scientific Study of Religion, 40,* 269–286.

Ellison, C. G., Bartkowski, J. P., & Anderson, K. L. (1999). Are there religious variations in domestic violence? *Journal of Family Issues, 20,* 87–113.

Ellison, C. G., Bartkowski, J. P., & Segal, M. L. (1996). Conservative Protestantism and the parental use of corporal punishment. *Social Forces, 74,* 1003–1028.

Ellison, C. G., Burr, J. A., & McCall, P. L. (2003). The enduring puzzle of Southern homicide: Is regional religious culture the missing piece? *Homicide Studies, 7,* 326–352.

Ellison, C. G., & Sherkat, D. E. (1993). Conservative Protestantism and support for corporal punishment. *American Sociological Review, 58,* 131–144.

Elwell, M. E., & Ephross, P. H. (1987). Initial reactions of sexually abused children. *Social Casework: The Journal of Contemporary Social Work, 68,* 109–116.

Emery, R. E. (1989). Family violence. *American Psychologist, 44,* 321–328.

Emery, R. E., & Laumann-Billings, L. (1998). An overview of the nature, causes and consequences of abusive family relationships: Toward differentiating maltreatment and violence. *American Psychologist, 53,* 121–135.

Englander, E. K. (1997). *Understanding violence.* Mahwah, NJ: Lawrence Erlbaum.

Erickson, M. F., & Egeland, B. (2002). Child neglect. In J. E. Myers, L. Berliner, J. Briere, C. T. Hendrix, C. Jenny, & T. A. Reid (Eds.), *The APSAC handbook on child maltreatment* (2nd ed.) (pp. 3–20). Thousand Oaks, CA: Sage.

Erickson, M. F., Egeland, B., & Pianta, R. C. (1989). The effects of maltreatment on the development of young children. In D. Cicchetti & V. Carlson (Eds.), *Child maltreatment: Theory and research on the causes and consequences of child abuse and neglect* (pp. 647–684). New York: Cambridge University Press.

Eron, L. D. (1987). The development of aggressive behavior from the perspective of a developing behaviorism. *American Psychologist, 42,* 425–442.

Eron, L. D. (1997). The development of antisocial behavior from a learning perspective. In D. M. Stoff, J. Breiling, & J. D. Maser (Eds.), *Handbook of antisocial behavior* (pp. 140–147). New York: John Wiley and Sons.

Eron, L. D., & Huesmann, L. R. (1980). Adolescent aggression and television. *Annals of the New York Academy of Sciences, 347,* 319–331.

Eron, L. D., & Huesmann, L. R. (1985). The role of television in the development of prosocial and antisocial behavior. In D. Olweus, M. Radke-Yarrow, & J. Block (Eds.), *Development of antisocial and prosocial behavior.* Orlando, FL: Academic Press.

European Network of Ombudsmen for Children (2001). *The European Network of Ombudsmen for Children (ENOC) seeks an end to all corporal punishment of children in Europe.* (Statement released at the European Seminar on Ending all Physical Punishment of Children, Barcelona, Sunday, October 19, 2001, and at the European Congress on Child Abuse and Neglect, Barcelona, October 19–22, 2001).

Evans, E. D., & Warren-Sohlberg, L. (1988). A pattern analysis of adolescent abusive behavior toward parents. *Journal of Adolescent Research, 3,* 201–216.

Ewoldt, C. A., Monson, C. M., & Langhinrichsen-Rohling, J. (2000). Attributions about rape in a continuum of dissolving marital relationships. *Journal of Interpersonal Violence, 15,* 1175–1182.

Ezzell, C. E., Swenson, C. C., & Brondino, M. J. (2000). The relationship of social support to physically abused children's adjustment. *Child Abuse and Neglect, 24,* 641–651.

Fagan, J., & Browne, A. (1994). Violence between spouses and intimates: Physical aggression between women and men in intimate relationships. In A. J. Reiss & J. A. Roth (Eds.), *Understanding and preventing violence* (Vol. 3) (pp. 115–292). Washington, D.C.: National Research Council, National Academy of Sciences.

Fajnzylber, P., Lederman, D., & Loayza, N. (2001). *Inequality and violent crime.* Retrieved August 21, 2001, from http://poverty.worldbank.org/library/topic/3362/12830/

Fals-Stewart, W. (2003). The occurrence of partner physical aggression on days of alcohol consumption: A longitudinal diary study. *Journal of Consulting and Clinical Psychology, 71,* 41–52.

Family Violence Prevention Fund (1993). *Men beating women: Ending domestic violence—A qualitative and quantitative study of public attitudes on violence against women.* San Francisco: Author.

Famularo, R., Fenton, T., & Kinscherff, R. T. (1992). Medical and developmental histories of maltreated children. *Clinical Pediatrics, 31,* 536–541.

Fantuzzo, J. W., delGaudio, W. A., Atkins, M., Meyers, R., & Noone, M. (1998). A contextually relevant assessment of the impact of child maltreatment on the social competencies of low-income urban children. *Journal of the American Academy of Child and Adolescent Psychiatry, 37,* 1201–1208.

Farrell, W. (1993). *The myth of male power.* New York: Berkley Books.

Farrington, K. (1986). The application of stress theory to the study of family violence: Principles, problems, and prospects. *Journal of Family Violence, 1,* 131–147.

FBI Uniform Crime Reports (2001). *Crime in the United States 2001.* Retrieved September 11, 2003, from http://www.fbi.gov/ucr/01cius.htm

Featherman, J. M. (1995). Jews and sexual child abuse. In L. A. Fontes (Ed.), *Sexual abuse in nine North American cultures: Treatment and prevention* (pp. 128–155). Thousand Oaks, CA: Sage.

Feldman, R. S., Salzinger, S., Rosario, M., Hammer, M., Alvarado, L., & Caraballo, L. (1989). *Parent and teacher ratings of abused and non-abused children's behavior.* New York: American Academy of Child Psychiatry.

Felson, R. B. (2002). *Violence and gender reexamined.* Washington, D.C.: American Psychological Association.

Felson, R. B., & Messner, S. F. (2000). The control motive in intimate partner violence. *Social Psychology Quarterly, 63,* 86–94.

Ferrara, F. F. (2002). *Childhood sexual abuse: Developmental effects across the lifespan.* Pacific Grove, CA: Brooks/Cole.

Ferrari, A. M. (2002). The impact of culture upon child rearing practices and definitions of maltreatment. *Child Abuse and Neglect, 26,* 793–813.

Ferraro, K. J. (2003). The words change, but the melody lingers: The persistence of the battered woman syndrome in criminal cases involving battered women. *Violence Against Women, 9,* 110–129.

Fiduccia, B. W., & Wolfe, L. R. (1999). *Violence against disabled women.* Retrieved May 4, 2004, from http://www.centerwomenpolicy.org

Fiebert, M. S., & Gonzalez, D. M. (1997). College women who initiate assaults on their male partners and the reasons offered for such behavior. *Psychological Reports, 80,* 583–590.

Finkelhor, D. (1979). *Sexually victimized children.* New York: Free Press.

Finkelhor, D. (1980). Sex among siblings: A survey of prevalence, variety, and effects. *Archives of Sexual Behavior, 9,* 171–193.

Finkelhor, D. (1984). *Child sexual abuse: New theory and research.* New York: Free Press.

Finkelhor, D. (1993). The main problem is still underreporting, not overreporting. In R. J. Gelles & D. R. Loseke (Eds.), *Current controversies on family violence* (pp. 273–287). Newbury Park, CA: Sage.

Finkelhor, D. (1994). Current information on the scope and nature of child sexual abuse. *Future of Children, 4,* 31–53.

Finkelhor, D., Hotaling, G. T., Lewis, J. A., & Smith, C. (1990). Sexual abuse in a national survey of adult men and women: Prevalence, characteristics, and risk factors. *Child Abuse and Neglect, 14,* 19–28.

Finkelhor, D., & Redfield, D. (1984). How the public defines sexual abuse. In D. Finkelhor (Ed.), *Child sexual abuse: New theory and research.* New York: Free Press.

Finkelhor, D., & Strapko, N. (1992). Sexual abuse prevention education: A review of evaluation studies. In D.J. Willis, E. Holden, & M. Rosenberg (Eds.), *Prevention of child maltreatment: Developmental and ecological perspectives* (pp. 150–167). New York: John Wiley.

Finkelhor, D., & Yllo, K. (1983). Rape in marriage: A sociological view. In D. Finkelhor, R. J. Gelles, G. T. Hotaling, & M.A. Straus (Eds.), *The dark side of families* (pp. 119–130). Beverly Hills, CA: Sage.

Finkelhor, D., & Yllo, K. (1985). *License to rape: Sexual abuse of wives.* New York: Holt, Rinehart, & Winston.

Finn, J. (1986). The relationship between sex role attitudes and attitudes supporting marital violence. *Sex Roles, 14,* 235–244.

Finn, M. A., & Stalans, L. J. (2002). Police handling of the mentally ill in domestic violence situations. *Criminal Justice & Behavior, 29,* 278–307.

Flanzer, J. P. (1993). Alcohol and other drugs are key causal agents of violence. In R. J. Gelles & D. R. Loseke (Eds.), *Current controversies on family violence* (pp. 171–181). Newbury Park, CA: Sage.

Fleck-Henderson, A. (2000). Domestic violence in the child protection system: Seeing double. *Children and Youth Services Review, 22,* 333–354.

Fleming, J., Mullen, P., & Bammer, G. (1997). A study of potential risk factors for sexual abuse in childhood. *Child Abuse and Neglect, 21,* 49–58.

Flisher, A. J., Kramer, R.A., Hoven, C. W., Greenwald, S., Alegria, M., Bird, H. R., Canino, G., Connell, R., & Moore, R. E. (1997). Psychosocial characteristics of physically abused children and adolescents. *Journal of the American Academy of Child and Adolescent Psychiatry, 36,* 123–131.

Flowers, R. B. (1984). Withholding medical care for religious reasons. *Journal of Religion and Health, 23,* 268–282.

Flynn, C. P. (1994). Regional differences in attitudes toward corporal punishment. *Journal of Marriage and the Family, 56,* 314–324.

Flynn, C. P. (1998). To spank or not to spank: The effect of situation and age of child on support for corporal punishment. *Journal of Family Violence, 13,* 21–37.

Follingstad, D. R., Brennan, A. F., Hause, E. S., Polek, D. S., & Rutledge, L. L. (1991). Factors moderating physical and psychological symptoms of battered women. *Journal of Family Violence, 6,* 81–95.

Follingstad, D. R., Rutledge, L. L., Berg, B. J., Hause, E. S., & Polek, D. S. (1990). The role of emotional abuse in physically abusive relationships. *Journal of Family Violence, 5,* 107–120.

Follingstad, D. R., Wright, S., Lloyd, S., & Sebastian, J. A. (1991). Sex differences in motivations and effects in dating violence. *Family Relations, 40,* 51–57.

Fontes, L. A., Cruz, M., & Tabachnick, J. (2001). Views of child sexual abuse in two cultural communities: An exploratory study among African Americans and Latinos. *Child Maltreatment, 6,* 103–117.

Ford, H. H., Schindler, C. B., & Medway, F. J. (2001). School professionals' attributions of blame for child sexual abuse. *Journal of School Psychology, 39,* 25–44.

Fox, J. A., & Zawitz, M. W. (2002). *Homicide trends in the United States.* U.S. Department of Justice. Office of Justice Programs. Bureau of Justice Statistics. Retrieved August 16, 2004, from http://www.ojp.usdoj.gov/bjs/homicide/homtrnd.htm

Frank, D. A., Augustyn, M., Knight, W. G., Pell, T., & Zuckerman, B. (2001). Growth, development, and behavior in early childhood following prenatal cocaine exposure: A systematic review. *Journal of the American Medical Association, 285,* 1613–1625.

Frank, D. A., Zuckerman, B. S., Amaro, H., Aboagye, K., Bauchner, H., Cabral, H., Fried, L., Hingson, R., Kayne, G., Levenson, S. M., Parker, S., Reece, H., & Vinci, R. (1988). Cocaine use during pregnancy: Prevalence and correlates. *Pediatrics, 82,* 888–895.

Fraser, B. (1986). A glance at the past, a gaze at the present, a glimpse at the future: A critical analysis of the development of child abuse reporting statutes. *Journal of Juvenile Law, 10,* 641–686.

Fray-Witzer, E. (1999). Twice abused: Same-sex domestic violence and the law. In B. Leventhal & S. Lundy (Eds.), *Same-sex domestic violence: Strategies for change* (pp. 19–42). Thousand Oaks, CA: Sage.

Freeman, A. C., Strong, M. F., Barker, L. T., & Haight-Liotta, S. (1996). *Priorities for future research: Results of BPA's Delphi survey of disabled women.* Berkeley: Berkeley Planning Associates. Retrieved May 4, 2004, from http://www.ncddr.org/rr/women/priorities.html

Freyd, J. J. (2002). Memory and dimensions of trauma: Terror may be "all-too-well remembered" and betrayal buried. In J. R. Conte (Ed.), *Critical issues in child sexual abuse: Historical, legal, and psychological perspectives* (pp. 139–173). Thousand Oaks, CA: Sage.

Fried, S. (2001, April). *Cognitive, economic, and ethnic diversity among maltreated children with disabilities abusing children.* Presented at the 13th National Conference on Child Abuse and Neglect. Retrieved December 2003, from http://nccanch.acf.hhs.gov/profess/conferences/cbconference/resourcebook/140.cfm

Friedrich, J. (1988). A model program for training religious leaders to work with abuse. In A. L. Horton & J. A. Williamson (Eds.), *Abuse and religion: When praying isn't enough* (pp. 181–187). Lexington, MA: Lexington Books.

Friedrich, W. N., Dittner, C. A., Action, R., Berliner, L., Butler, J., Damon, L., Davies, W. H., Gray, A., & Wright, J. (2001). Child Sexual Behavior Inventory: Normative, psychiatric, and sexual abuse comparisons. *Child Maltreatment, 6,* 37–49.

Frieze, I. H. (1983). Investigating the causes and consequences of marital rape. *Signs, 8,* 532–553.

Frodi, A. M., & Lamb, M. E. (1980). Child abusers' responses to infants' smiles and cries. *Child Development, 51,* 238–241.

Fulmer, T., & O'Malley, T. (1987). *Inadequate care of the elderly: A health care perspective on abuse and neglect.* New York: Springer.

Fyfe, J. J., Klinger, D. A., & Flavin, J. (1997). Differential police treatment of male-on-female spousal violence. *Criminology, 35,* 455–473.

Gabler, M., Stern, S., & Miserandino, M. (1998). Latin American, Asian, and American cultural differences in perception of spousal abuse. *Psychological Reports, 83,* 587–592.

Gambrill, E. (1983). Behavioral interventions with child abuse and neglect. *Progress in Behavior Modification, 15,* 1–56.

Ganzarain, R. (1992). Narcissistic and borderline personality disorders in cases of incest. *Group Analysis, 25,* 491–494.

Garbarino, J., & Collins, C. C. (1999). Child neglect: The family with a hole in the middle. In H. Dubowitz (Ed.), *Neglected children: Research, practice, and policy* (pp. 1–23). Thousand Oaks, CA: Sage.

Garbarino, J., Guttman, E., & Seeley, J. (1986). *The psychologically battered child: Strategies for identification, assessment, and intervention.* San Francisco: Jossey-Bass.

Garbarino, J., & Kostelny, K. (1992). Child maltreatment as a community problem. *Child Abuse and Neglect, 16,* 455–464.

Garbarino, J., & Sherman, D. (1980). High-risk neighborhoods and high-risk families: The human ecology of child maltreatment. *Child Development, 51,* 188–198.

Gaudin, J. M. (1993). Effective interventions with neglectful families. *Criminal Justice and Behavior, 20,* 66–89.

Gaudin, J. M. (1999). Child neglect: Short-term and long-term outcomes. In H. Dubowitz (Ed.), *Neglected children: Research, practice, and policy* (pp. 89–108). Thousand Oaks, CA: Sage.

Gaudin, J. M., Polansky, N. A., Kilpatrick, A. C., & Shilton, P. (1996). Family functioning in neglectful families. *Child Abuse and Neglect, 20,* 363–377.

Gebo, E. (2002). A contextual exploration of siblicide. *Violence and Victims, 17,* 157–168.

Gelles, R. J. (1974). *The violent home: A study of physical aggression between husbands and wives.* Beverly Hills, CA: Sage.

Gelles, R. J. (1990). Violence and pregnancy: Are pregnant women at greater risk of abuse? In M. A. Straus & R. J. Gelles (Eds.), *Physical violence in American families: Risk factors and adaptations to violence in 8,145 families* (pp. 279–286). New Brunswick, NJ: Transaction.

Gelles, R. J. (1993). Alcohol and other drugs are associated with violence—they are not its cause. In R. J. Gelles & D. R. Loseke (Eds.), *Current controversies on family violence* (pp. 182–196). Newbury Park, CA: Sage.

Gelles, R. J. (1999). Male offenders: Our understanding from the data. In M. Harway & J. M. O'Neil (Eds.), *What causes men's violence against women?* (pp. 36–48). Thousand Oaks, CA: Sage.

Gelles, R. J. (1999). Policy issues in child neglect. In H. Dubowitz (Ed.), *Neglected children: Research, practice, and policy* (pp. 278–298). Thousand Oaks, CA: Sage.

Gelles, R. J., & Cornell, C. P. (1990). *Intimate violence in families.* Newbury Park, CA: Sage.

Gelles, R. J., & Harrop, J. W. (1989). Violence, battering, and psychological distress among women. *Journal of Interpersonal Violence, 4,* 400–420.

Gelles, R. J., & Straus, M. A. (1990). The medical and psychological costs of family violence. In M. A. Straus & R. J. Gelles (Eds.), *Physical violence in American families: Risk factors and adaptations to violence in 8,145 families* (pp. 425–430). New Brunswick, NJ: Transaction.

George, C. (1996). A representational perspective of child abuse and prevention: Internal working models of attachment and caregiving. *Child Abuse and Neglect, 20,* 411–424.

Gershoff, E. T. (2002). Corporal punishment by parents and associated child behaviors and experiences: A meta-analytic and theoretical review. *Psychological Bulletin, 128,* 539–579.

Gershoff, E. T., Miller, P. C., & Holden, G. W. (1999). Parenting influences from the pulpit: Religious affiliation as a determinant of parental corporal punishment. *Journal of Family Psychology, 13,* 307–320.

Ghetti, S., Alexander, K. W., & Goodman, G. S. (2002). Legal involvement in child sexual abuse cases: On sequences and interventions. *International Journal of Law and Psychiatry, 25,* 235–251.

Giardino, A. P., Hudson, K. M., & Marsh, J. (2003). Providing medical evaluations for possible child maltreatment to children with special health care needs. *Child Abuse and Neglect, 27,* 1179–1186.

Gil, V. E. (1988). In thy father's house: Self-report findings of sexually abused daughters from Conservative Christian homes. *Journal of Psychology and Theology, 16,* 144–152.

Giles-Sims, J. (1983). *Wife battering: A systems theory approach.* New York: Guilford Press.

Giles-Sims, J. (1998). The aftermath of partner violence. In J. L. Jasinski & L. M. Williams (Eds.), *Partner violence: A comprehensive review of 20 years of research* (pp. 44–72). Thousand Oaks, CA: Sage.

Giles-Sims, J., Straus, M. A., & Sugarman, D. B. (1995). Child, maternal and family characteristics associated with spanking. *Family Relations, 44,* 170–176.

Gillham, B., Tanner, G., Cheyne, B., Freeman, I., Rooney, M., & Lambie, A. (1998). Unemployment rates, single parent density, and indices of child poverty: Their relationship in different categories of child abuse and neglect. *Child Abuse and Neglect, 22,* 79–90.

Gioglio, G., & Blakemore, P. (1982). *Elder abuse in New Jersey: The knowledge and experience of abuse among elder New Jerseyans.* Trenton, NJ: New Jersey Division on Aging.

Giordano, P. C., Millhollin, T. J., Cernkovich, S. A., Pugh, M. D., & Rudolph, J. L. (1999). Delinquency, identity, and women's involvement in relationship violence. *Criminology, 37,* 17–37.

Giovannoni, J., & Billingsley, A. (1970). Child neglect among the poor: A study of parental adequacy in families of three ethnic groups. *Child Welfare, 49,* 196–204.

Giovannoni, J. M., & Becerra, R. M. (1979). *Defining child abuse.* New York: Free Press.

Girshick, L. B. (2002). No sugar, no spice: Reflections on research on woman-to-woman sexual violence. *Violence Against Women, 8,* 1500–1520.

Gladue, B. A., Boechler, M., & McCaul, K. D. (1989). Hormonal response to competition in human males. *Aggressive Behavior, 15,* 409–422.

Gleason, W. J. (1993). Mental disorders in battered women: An empirical study. *Violence and Victims, 8,* 53–68.

Globus-Goldberg, N. (2001). *An emotional abuse survivor's story.* Retrieved December 8, 2003, from http://www.womanabuseprevention.com/html/ nancy_s_story.html

Glock, C. Y., & Stark, R. (1965). *Religion and society in tension.* Chicago: Rand McNally.

Goddard, A. B., & Hardy, T. (1999). Assessing the lesbian victim. In B. Leventhal & S. Lundy (Eds.), *Same-sex domestic violence: Strategies for change* (pp. 193–200). Thousand Oaks, CA: Sage.

Godkin, M. A., Wolf, R. S., & Pillemer, K. A. (1989). A case-comparison analysis of elder abuse and neglect. *International Journal of Aging and Human Development, 28,* 207–225.

Gold, E. R. (1986). Long-term effects of sexual victimization in childhood: An attributional approach. *Journal of Consulting and Clinical Psychology, 54,* 471–475.

Goldberg, R. T., Pachas, W. N., & Keith, D. (1999). Relationship between traumatic events in childhood and chronic pain. *Disability and Rehabilitation: An International Multidisciplinary Journal, 21,* 23–30.

Golding, J. M., Alexander, M. C., & Stewart, T. L. (1999). On the social psychology of hearsay evidence. *Psychology, Public Policy, and Law, 5,* 473–484.

Gomby, D., Culross, P., & Behrman, R. (1999). Home visiting: Recent program evaluations: Analysis and recommendations. *Future of Children, 9,* 4–26.

Gondolf, E. W. (1988). The effect of batterer counseling on shelter outcome. *Journal of Interpersonal Violence, 3,* 275–289.

Gondolf, E. W. (1990). An exploratory survey of court-mandated batterer programs. *Response to the Victimization of Women and Children, 13,* 7–11.

Gondolf, E. W. (1995). *Discharge criteria for batterer programs.* Retrieved November 13, 2003, from http://www.mincava.umn.edu/papers/gondolf/discharg.htm

Gondolf, E. W. (1999). Characteristics of court-mandated batterers in four cities. *Violence Against Women, 5,* 1277–1293.

Gondolf, E. W., & Fisher, E. R. (1988). *Battered women as survivors: An alternative to treating learned helplessness.* Lexington, MA: Lexington Books.

Gondolf, E. W., & Foster, R. A. (1991). Preprogram attrition in batterer programs. *Journal of Family Violence, 6,* 337–349.

Goodman, G. S., Bottoms, B. L., Redlich, A., Shaver, P. R., & Diviak, K. (1998). Correlates of multiple forms of victimization in religion-related child abuse cases. *Journal of Aggression, Maltreatment, and Trauma, 2,* 273–295.

Goodman, G. S., Quas, J. A., Bulkley, J., & Shapiro, C. (1999). Innovations for child witnesses: A national survey. *Psychology, Public Policy, and the Law, 5,* 255–281.

Gordon, M. (1989). The family environment of sexual abuse: A comparison of natal and stepfather abuse. *Child Abuse and Neglect, 13,* 121–130.

Gosselin, D. K. (2000). *Heavy hands: An introduction to the crimes of domestic violence.* Upper Saddle River, NJ: Prentice-Hall.

Gottesman, I. I., Goldsmith, H. H., & Carey, G. (1997). A developmental and genetic perspective on aggression. In N. L. Segal, G. E. Weisfeld, & C. C. Weisfeld (Eds.), *Uniting psychology and biology: Integrative perspectives on human development* (pp. 107–130). Washington, D.C.: American Psychological Association.

Gottlich, V. (1994). Beyond granny bashing: Elder abuse in the 1990s. *Clearinghouse Review, 28,* 371–381.

Gowan, J. (1993). *Effects of neglect on the early development of children: Final report.* Washington, D.C.: National Clearinghouse on Child Abuse and Neglect, National Center on Child Abuse and Neglect, Administration for Children and Families.

Grasmick, H. G., Bursik, R. J., Jr., & Kimpel, M. (1991). Protestant fundamentalism and attitudes toward corporal punishment of children. *Violence and Victims, 6,* 283–298.

Graziano, A. M., & Namaste, K. A. (1990). Parental use of physical force in child discipline: A survey of 679 college students. *Journal of Interpersonal Violence, 5,* 449–463.

Green, A. H. (1984). Child abuse by siblings. *Child Abuse and Neglect, 8,* 311–317.

Green, A. H. (2000). Child neglect. In R. T. Ammerman & M. Hersen (Eds.), *Case studies in family violence* (2nd ed.) (pp. 157–176). New York: Kluwer.

Greenberg, J. R., McKibben, M., & Raymond, J. A. (1990). Dependent adult children and elder abuse. *Journal of Elder Abuse and Neglect, 2,* 73–86.

Greenblat, C. S. (1985). "Don't hit your wife . . . unless . . .": Preliminary findings on normative support for the use of physical force by husbands. *Victimology, 10,* 221–241.

Greenfeld, L. A. (1996). *Child victimizers: Violent offenders and their victims.* Washington, D.C.: U.S. Department of Justice.

Greenfeld, L. A., Rand, M. R., Craven, D., Klaus, P. A., Perkins, C. A., Ringel, C., et al. (1998). *Violence by intimates: Analysis of data on crimes by current or former spouses, boyfriends, and girlfriends.* U.S. Department of Justice, Bureau of Justice Statistics. Retrieved August 16, 2004, from http://www.ojp.usdoj.gov/bjs/pub/pdf/vi.pdf

Greenwald, J. (1994). Romantic and sexual self-esteem, a new construct: Its relationship to current and early childhood sexual experiences (Doctoral dissertation, University of Nebraska, 1993). *Dissertation Abstracts International, 54,* 3852.

Greenwood, G. L., Relf, M. V., Huang, B., Pollack, L. M., Canchola, J. A., & Catania, J. A. (2002). Battering victimization among a probability-based sample of men who have sex with men. *American Journal of Public Health, 92,* 1964–1969.

Griffin, L. W. (1999a). Elder mistreatment in the African American community: You just don't hit your mama!!! In T. Tatara (Ed.), *Understanding elder abuse in minority communities* (pp. 13–26). Philadelphia: Brunner/Mazel.

Griffin, L. W. (1999b). Understanding elder abuse. In R. L. Hampton (Ed.), *Family violence: Prevention and treatment* (pp. 260–287). Thousand Oaks, CA: Sage.

Gross, A. B., & Keller, H. R. (1992). Long-term consequences of childhood physical and psychological maltreatment. *Aggressive Behavior, 18,* 171–185.

Groth, A. N., Hobson, W., & Gary, R. (1982). Child molester: Clinical observations. In J. Conte & D. Shore (Eds.), *Social work and child sexual abuse* (pp. 129–144). New York: Hayworth Press.

Gruskin, E. P., Hart, S., Gordon, N., & Ackerson, L. (2001). Patterns of cigarette smoking and alcohol use among lesbians and bisexual women enrolled in a large health maintenance organization. *American Journal of Public Health, 91,* 976–979.

Gül, S. S., & Gül, H. (2000). The question of women in Islamic revivalism in Turkey: A review of the Islamic press. *Current Sociology, 4,* 1–26.

Hagan, J. (1993). Beyond the classics: Reform and renewal in the study of crime and inequality. *Journal of Research in Crime and Delinquency, 30,* 485–490.

Hamberger, L. K., & Arnold, J. (1990). The impact of mandatory arrest on domestic violence perpetrator counseling services. *Family Violence and Sexual Assault Bulletin, 6,* 11–12.

Hamberger, L. K., & Hastings, J. E. (1991). Personality correlates of men who batter and nonviolent men: Some continuities and discontinuities. *Journal of Family Violence, 6,* 131–147.

Hamburg, D. (1992). *Today's children: Creating a future for a generation in crisis.* New York: Times Books.

Hamby, S. L. (1998). Partner violence: Prevention and intervention. In J. L. Jasinski & L. M. Williams (Eds.), *Partner violence: A comprehensive review of 20 years of research* (pp. 210–258). Thousand Oaks, CA: Sage.

Hamel, M., Gold, D. P., Andres, D., Reis, M., Dastoor, D., Grauer, H., & Bergman, H. (1990). Predictors and consequences of aggressive behavior by community-based dementia patients. *Gerontologist, 30*, 206–211.

Hamilton, A. R. (2001, June 14). *Testimony provided for the National Association of Adult Protective Services Administrators.* United States Senate Hearings on Saving Our Seniors: Preventing Elder Abuse, Neglect and Exploitation.

Hammer, R. (2003). Militarism and family terrorism: A critical feminist perspective. *The Review of Education, Pedagogy, and Cultural Studies, 25*, 231–256.

Haney, C. (1997). Violence and the capital jury: Mechanisms of moral disengagement and the impulse to condemn to death. *Stanford Law Review, 49*, 1447.

Hanson, R. K., & Slater, S. (1988). Sexual victimization in the history of sexual abusers: A review. *Annals of Sex Research, 1*, 485–499.

Harbin, H., & Madden, D. (1979). Battered parents: A new syndrome. *American Journal of Psychiatry, 136*, 1288–1291.

Harden, B. J. (1998). Building bridges for children: Addressing the consequences of exposure to drugs and to the child welfare system. In R. L. Hampton, V. Senatore, & T. P. Gullotta (Eds.), *Substance abuse, family violence, and child welfare* (pp. 18–61). Thousand Oaks, CA: Sage.

Harper, J. (1991). Children's play: The differential effects of intrafamilial physical and sexual abuse. *Child Abuse and Neglect, 15*, 89–98.

Harrell, A. (1991). *Evaluation of court-ordered treatment for domestic violence offenders.* Raleigh: North Carolina Supreme Court Library.

Harries, K. (1995). The last walk: A geography of execution in the United States, 1786–1985. *Political Geography, 14*, 473–495

Harrington, D., & Dubowitz, H. (1999). Preventing child maltreatment. In R. L. Hampton (Ed.), *Family violence: Prevention and treatment* (2nd ed.) (pp. 122–147). Thousand Oaks, CA: Sage.

Harris, L. M., Gergen, K. J., & Lannamann, J. W. (1986). Aggression rituals. *Communications Monographs, 53*, 252–265.

Harris, S. B. (1996). For better or for worse: Spouse abuse grown old. *Journal of Elder Abuse and Neglect, 8*, 1–33.

Harry, J. (1989). Parental physical abuse and sexual orientation in males. *Archives of Sexual Behavior, 18*, 251–261.

Hart, B. (1986). Lesbian battering: An examination. In K. Lobel (Ed.), *Naming the violence: Speaking out about lesbian battering* (pp. 173–189). Seattle: Seal Press.

Hart, B. (1993). Battered women and the criminal justice system. *American Behavioral Scientist, 36*, 624–628.

Haskett, M. E., Scott, S. S., Grant, R., Ward, C. S., & Robinson, C. (2003). Child-related cognitions and affective functioning of physically abusive and comparison parents. *Child Abuse and Neglect, 27*, 663–686.

Hassouneh-Phillips, D. (2001a). American Muslim women's experiences of leaving abusive relationships. *Health Care for Women International, 22*, 415–432.

Hassouneh-Phillips, D. (2001b). Polygamy and wife abuse: A qualitative study of Muslim women in America. *Health Care for Women International, 22*, 735–748.

Hassouneh-Phillips, D. (2001c). "Marriage is half of faith and the rest is fear of Allah": Marriage and spousal abuse among American Muslims. *Violence Against Women, 7,* 927–946.

Hassouneh-Phillips, D. (2003). Strength and vulnerability: Spirituality in abused American Muslim women's lives. *Issues in Mental Health Nursing, 24,* 681–694.

Hastings, B. M. (2000). Social information processing and the verbal and physical abuse of women. *Journal of Interpersonal Violence, 15,* 651–664.

Healey, K., Smith, C., & O'Sullivan, C. (1998). *Batterer intervention: Program approaches and criminal justice strategies.* U.S. Department of Justice/National Institute of Justice. Retrieved May 7, 2004, from http://www.ncjrs.org/pdf files/168638.pdf

Hearn, J. (1988). Commentary. Child abuse: Violences and sexualities towards young people. *Sociology, 22,* 531–544.

Heffer, R. W., & Kelly, M. L. (1987). Mothers' acceptance of behavioral interventions for children: The influence of parental race and income. *Behavior Therapy, 2,* 153–163.

Heide, K. M. (1995). *Why kids kill parents: Child abuse and adolescent homicide.* Thousand Oaks, CA: Sage.

Heinlein, G., & Beaupre, B. (2002). Shelters must detail help for men. *The Detroit News.* Retrieved September 29, 2002, from http://www.noexcuse4abuse.org/ news/detroit_shelter.html

Heisler, C. J., & Quinn, M. J. (1995). A legal perspective. *Journal of Elder Abuse and Neglect, 7,* 23–40.

Hemenway, D., Shinoda-Tagawa, T., & Miller, M. (2002). Firearm availability and female homicide victimization rates among 25 populous high-income countries. *Journal of the American Medical Women's Association, 57,* 100–104.

Hendricks, S. S. (1998). *Authority and the abuse of power in Muslim marriages.* Women's Conference of the 2nd International Islamic Unity Conference, Washington, DC. Retrieved July 2003, from http://www.themodernreligion .com/women/abuse-marriage.html

Henman, M. (1996). Domestic violence: Do men under report? *Forensic Update, 47,* 3–8.

Henning, K., & Feder, L. (2004). A comparison of men and women arrested for domestic violence: Who presents the greater threat? *Journal of Family Violence, 19,* 69–80.

Henton, J., Cate, R., Koval, J., Lloyd, S., & Christopher, S. (1983). Romance and violence in dating relationships. *Journal of Family Issues, 4,* 467–482.

Herbert, M. (1987). *Conduct disorders of childhood and adolescence.* Chichester, UK: Wiley.

Herek, G. M. (1989). Hate crimes against lesbians and gay men: Issues for research and policy. *American Psychologist, 44,* 948–955.

Herek, G. M., Gillis, J. R., Cogan, J. C., & Glunt, E. K. (1997). Hate crime victimization among lesbian, gay, and bisexual adults: Prevalence, psychological correlates, and methodological issues. *Journal of Interpersonal Violence, 12,* 195–215.

Herkov, M. J., Gynther, M. D., Thomas, S., & Myers, W. C. (1996). MMPI differences among adolescent inpatients, rapists, sodomists, and sexual abusers. *Journal of Personality Assessment, 66,* 81–90.

Herman-Giddens, M., Brown, G., Verbiest, S., Carlson, P., Hooten, E., Howell, E., & Butts, J. (1999). Underascertainment of child abuse mortality in the United States. *Journal of the American Medical Association, 282,* 463–467.

Herman-Giddens, M. E., Smith, J. B., Mittal, M., Carlson, M., & Butts, J. D. (2003). Newborns killed or left to die by a parent: A population-based study. *Journal of the American Medical Association, 289,* 1425–1429.

Hermin, J. (1981). *Father-daughter incest.* Cambridge, MA: Harvard University Press.

Herrenkohl, E. C., Herrenkohl, R. C., Egolf, B. P., & Russo, M. J. (1998). The relationship between early maltreatment and teenage parenthood. *Journal of Adolescence, 21,* 291–303.

Herrenkohl, E. C., Herrenkohl, R. C., & Toedter, L. J. (1983). Perspectives on the intergenerational transmission of abuse. In D. Finkelhor, R. J. Gelles, G. T. Hotaling, & M. A. Straus (Eds.), *The dark side of families: Current family violence research* (pp. 305–316). Newbury Park, CA: Sage.

Herrenkohl, R. C., Egolf, B. P., & Herrenkohl, E. C. (1997). Preschool age antecedents of adolescent assaultive behavior: Results from a longitudinal study. *American Journal of Orthopsychiatry, 67,* 422–432.

Herrenkohl, R. C., Herrenkohl, E. C., Egolf, B. P., & Wu, P. (1991). The developmental consequences of child abuse: The Lehigh Longitudinal Study. In R. H. Starr & D. A. Wolfe (Eds.), *The effects of child abuse and neglect* (pp. 57–81). New York: Guilford.

Hertel, B. R., & Hughes, M. (1987). Religious affiliation, attendance, and support for "pro-family" issues in the United States. *Social Forces, 65,* 858–882.

Herzberger, S. D., & Rueckert, Q. H. (1997). Attitudes as explanations for aggression against family members. In G. Kaufman Kantor & J. L. Jasinski (Eds.), *Out of darkness: Contemporary perspectives on domestic violence* (pp. 151–160). Thousand Oaks, CA: Sage.

Herzberger, S. D., & Tennen, H. (1985). The effect of self-relevance on judgments of moderate and severe disciplinary encounters. *Journal of Marriage and the Family, 47,* 311–318.

Hetherton, J. (1999). The idealization of women: Its role in the minimization of child sexual abuse by females. *Child Abuse and Neglect, 23,* 161–174.

Hines, D. A. (2001, August). *Effects of emotional abuse against men in intimate relationships.* Presented at the 109th Annual Convention of the American Psychological Association, San Francisco, CA.

Hines, D. A., Brown, J., & Dunning, E. (2004). Characteristics of callers to the Domestic Abuse Helpline for Men. *Journal of Family Violence.*

Hines, D. A., & Malley-Morrison, K. (2001). Psychological effects of partner abuse against men: A neglected research area. *Psychology of Men and Masculinity, 2,* 75–85.

Hines, D. A., & Malley-Morrison, K. (2003, July). *Abusive childhood experiences: Effects in late adolescent boys.* Presented at the 8th International Family Violence Research Conference, Portsmouth, NH.

Hines, D. A., & Saudino, K. J. (2002). Intergenerational transmission of intimate partner violence: A behavioral genetic perspective. *Trauma, Violence, & Abuse, 3,* 210–225.

Hines, D. A., & Saudino, K. J. (2003). Gender differences in psychological, physical, and sexual aggression among college students using the Revised Conflict Tactics Scales. *Violence and Victims, 18,* 197–218.

Hines, D. A., & Saudino, K. J. (2004a). Genetic and environmental influences on intimate partner aggression. *Violence and Victims,* in press.

Hines, D. A., & Saudino, K. J. (2004b). Personality and interpersonal relationship behaviors: The Big Five as predictors of intimate partner aggression. Manuscript submitted for publication.

Hirschel, J. D., & Hutchinson, I. W. (1992). Female spousal abuse and the police response: The Charlotte, North Carolina, experiment. *Journal of Criminal Law and Criminology, 83,* 73–119.

Hoagwood, K., & Stewart, J. M. (1989). Sexually abused children's perceptions of family functioning. *Child and Adolescent Social Work, 6,* 139–149.

Hoffman-Mason, C., & Bingham, R. P. (1988). Developing a sensitivity for culture and ethnicity in family violence. In A. L. Horton & J. A. Williamson (Eds.), *Abuse and religion: When praying isn't enough* (pp. 138–143). Lexington, MA: Lexington Books.

Holden, E. W., & Nabors, L. (1999). The prevention of child neglect. In H. Dubowitz (Ed.), *Neglected children: Research, practice, and policy* (pp. 174–190). Thousand Oaks, CA: Sage.

Holmes, W. C., & Slap, G. B. (1998). Sexual abuse of boys: Definition, prevalence, correlates, sequelae, and management. *Journal of the American Medical Association, 280,* 1855–1862.

Holtzworth-Munroe, A. (1992). Social skill deficits in maritally violent men: Interpreting the data using a social information processing model. *Clinical Psychology Review, 12,* 605–617.

Holtzworth-Munroe, A., Markman, H., O'Leary, K. D., Neidig, P., Leber, D., Heyman, R. E., Hulbert, D., & Smutzler, N. (1995). The need for marital violence prevention efforts: A behavioral-cognitive secondary prevention program for engaged and newly married couples. *Applied and Preventive Psychology: Current Scientific Perspectives, 4,* 77–88.

Holtzworth-Munroe, A., & Stewart, G. L. (1994). Typologies of male batterers: Three subtypes and the differences among them. *Psychological Bulletin, 116,* 476–497.

Homer, A. C., & Gilleard, C. (1990). Abuse of elderly people by their carers. *British Medical Journal, 301,* 1359–1362.

Hong, G. K., & Hong, L. K. (1991). Comparative perspectives in child abuse and neglect: Chinese versus Hispanics and Whites. *Child Welfare, 70,* 463–475.

Horner-Johnson, W., Drum, C. E., & Pobutsky, A. (2002). *Maltreatment and disability: Results from the Oregon community assessment survey.* Oregon Office on Disability & Health Policy Brief 2002 (1). Oregon Health & Sciences University, Portland, Oregon.

Horsburgh, B. (1995). Lifting the veil of secrecy: Domestic violence in the Jewish community. *Harvard Women's Law Journal, 18,* 171–209.

Horton, A. L., Wilkins, M. M., & Wright, W. (1988). Women who ended abuse: What religious leaders and religion did for these victims. In A. L. Horton, & J. A. Williamson (Eds.), *Abuse and religion: When praying isn't enough* (pp. 235–246). Lexington, MA: Lexington Books.

Hotaling, G. T., Straus, M. A., & Lincoln, A. J. (1990). Intrafamily violence and crime and violence outside the family. In M. A. Straus & R. J. Gelles (Eds.), *Physical violence in American families: Risk factors and adaptations to violence in 8,145 families* (pp. 431–472). New Brunswick, NJ: Transaction.

Hotaling, G. T., & Sugarman, D. B. (1986). An analysis of risk markers in husband to wife violence: The current state of knowledge. *Violence and Victims, 1,* 101–124.

Hotte, J. P., & Rafman, S. (1992). The specific effects of incest on prepubertal girls from dysfunctional families. *Child Abuse and Neglect, 16,* 273–283.

House Select Committee on Aging, U.S. Congress. (1991). *Elder abuse: What can be done?* Washington, D.C.: U.S. Government Printing Office.

Howell, M. J., & Pugliesi, K. L. (1988). Husbands who harm: Predicting spousal violence by men. *Journal of Family Violence, 3,* 15–27.

Howes, P. W., Cicchetti, D., Toth, S. L., & Rogosch, F.A. (2000). Affective, structural, and relational characteristics of maltreating families: A systems perspective. *Journal of Family Psychology, 14,* 95–110.

Hubbard, G. B. (1989). Mothers' perceptions of incest: Sustained disruption and turmoil. *Archives of Psychiatric Nursing, 3,* 34–40.

Hucker, S., Langevin, R., Wortzman, G., Bain, J., Handy, L., Chambers, J., & Wright, S. (1986). Neuropsychological impairment in pedophiles. *Canadian Journal of Behavioral Science, 18,* 440–448.

Hucker, S., Langevin, R., Wortzman, G., Dickey, R., Bain, J., Handy, L., Chambers, J., & Wright, S. (1988). Cerebral damage and dysfunction in sexually aggressive men. *Annals of Sex Research, 1,* 33–47.

Hudson, M. F., Armachain, W. D., Beasley, C. M., & Carlson, J. R. (1998). Elder abuse: Two Native American views. *The Gerontologist, 38,* 538–548.

Hudson, M. F., & Carlson, J. R. (1999). Elder abuse: Its meaning to Caucasians, African Americans, and Native Americans. In T. Tatara (Ed.), *Understanding elder abuse in minority populations* (pp. 187–204). Philadelphia: Brunner/Mazel.

Huesmann, L. R., Moise-Titus, J., Podolski, C. L., & Eron, L. D. (2003). Longitudinal relations between children's exposure to TV violence and their aggressive and violent behavior in young adulthood: 1977–1992. *Developmental Psychology, 39,* 201–221.

Humphries Lynch, S. (1997). Elder abuse: What to look for, how to intervene. *American Journal of Nursing, 97,* 27–32.

Hutchings, B., & Mednick, S. A. (1977). Criminality in adoptees and their adoptive and biological parents: A pilot study. In S. A. Mednick & K. O. Christiansen (Eds.), *Biosocial bases of criminal behavior* (pp. 127–141). New York: Gardner Press.

Hwalek, M. A., Neale, A. V., Goodrich, C. S., & Quinn, K. (1996). The association of elder abuse and substance abuse in the Illinois Elder Abuse System. *Gerontologist, 36,* 694–700.

In re Gault, 387 U.S. 1 (1967).

Ingraham v. Wright (430 U.S. 651, 51 L.Ed.2d 71) (1977).

Iovanni, L., & Miller, S. L. (2001). Criminal justice system responses to domestic violence: Law enforcement and the courts. In C. M. Renzetti, R. K. Bergen, & J. L. Edleson (Eds.), *Sourcebook on violence against women* (pp. 303–328). Thousand Oaks, CA: Sage.

Islamic Society of North America (2004). *The Islamic response to domestic violence.* Retrieved July 2004 from http://www.isna.net/dv/islamicresponse.asp

Island, D., & Letellier, P. (1991). *Men who beat the men who love them.* New York: Harrington Park Press.

Jaccoby, J. (2002). Traditional Judaism roars back. *The American Enterprise, 13,* 28–29.

Jackson, S., Thompson, R. A., Christiansen, E. H., Colman, R. A., Wyatt, J., & Buckendahl, C. W. (1999). Predicting abuse-prone parental attitudes and discipline practices in a nationally representative sample. *Child Abuse and Neglect, 23,* 15–29.

Jacobson, J. L., Jacobson, S. W., Sokol, R. J., & Ager, J. W. (1998). Relation of maternal age and pattern of pregnancy drinking to functionally significant cognitive deficits in infancy. *Alcoholism: Clinical and Experimental Research, 22,* 345–351.

Jacobson, N. S., Gottman, J. M., Gortner, E., Berns, S., & Shortt, J. W. (1996). Psychological factors in the longitudinal course of battering: When do the couples split up? When does the abuse decrease? *Violence and Victims, 11,* 371–392.

Jaudes, P. K., Ekwo, E., & Voorhis, J. V. (1995). Association of drug abuse and child abuse. *Child Abuse and Neglect, 19,* 1065–1075.

Jehu, D. (1988). *Beyond sexual abuse: Therapy with women who were childhood victims.* Chichester, UK: Wiley.

Jenness, V., & Broad, K. (1994). Antiviolence activism and the (in)visibility of gender in the gay/lesbian and women's movements. *Gender & Society, 8,* 402–423.

Jennings, J. L., & Murphy, C. M. (2000). Male-male dimensions of male-female battering: A new look at domestic violence. *Psychology of Men and Masculinity, 1,* 21–29.

Jennings, K. T. (1993). Female child molesters: A review of the literature. In M. Elliott (Ed.), *Female sexual abuse of children: The ultimate taboo* (pp. 241–257). New York: John Wiley & Sons.

Johnson, C. F. (2002). Physical abuse: Accidental versus intentional trauma in children. In J. E. B. Myers, L. Berliner, J. Briere, C. T. Hendrix, C. Jenny, & T. A. Reid (Eds.), *The APSAC handbook on child maltreatment* (2nd ed.) (pp. 249–268). Thousand Oaks, CA: Sage.

Johnson, C. F., & Showers, J. (1985). Injury variables in child abuse. *Child Abuse and Neglect, 9,* 207–215.

Johnson, J. G., Cohen, P., Brown, J., Smailes, E. M., & Bernstein, D. P. (1999). Childhood maltreatment increases risk for personality disorders during early adulthood. *Archives of General Psychiatry, 56,* 600–606.

Johnson, J. G., Smailes, E. M., Cohen, P., Brown, J., & Bernstein, D. P. (2000). Associations between four types of childhood neglect and personality disorder symptoms during adolescence and early adulthood: Findings of a community-based longitudinal study. *Journal of Personality Disorders, 14,* 171–187.

Johnson, M. P. (1995). Patriarchal terrorism and common couple violence: Two forms of violence against women. *Journal of Marriage and the Family, 57,* 283–294.

Johnson, T. C. (1988). Child perpetrators: Children who molest other children: Preliminary findings. *Child Abuse and Neglect, 12,* 219–229.

Jones, D. P. H. (2002). Editorial. Is sexual abuse perpetrated by a brother different from that committed by a parent? *Child Abuse and Neglect, 26,* 955–956.

Jones, J. S. (1994). Elder abuse and neglect: Responding to a national problem. *Annals of Emergency Medicine, 23,* 845–848.

Joseph, R. (1999). The neurology of traumatic dissociative amnesia: Commentary and literature review. *Child Abuse and Neglect, 23,* 715–727.

Jurich, A. P. (1990). Families who physically abuse adolescents. In S. M. Stith & M. B. Williams (Eds), *Violence hits home: Comprehensive treatment approaches to domestic violence: Springer series on social work, Vol. 19.* (pp. 126–150). New York: Springer.

Kahn, F. I., Welch, T. L., & Zillmer, E. A. (1993). MMPI-2 profiles of battered women in transition. *Journal of Personality Assessment, 60,* 100–111.

Kalichman, S. C., Benotsch, E., Rompa, D., Gore-Felton, C., Austin, J., Luke, W., DiFonzo, K., Buckles, J., Kyomugisha, F., & Simpson, D. (2001). Unwanted sexual experiences and sexual risks in gay and bisexual men: Associations among revictimization, substance use, and psychiatric symptoms. *Journal of Sex Research, 38,* 1–9.

Kalmuss, D. (1984). The intergenerational transmission of marital aggression. *Journal of Marriage and the Family, 46,* 11–19.

Kandel, E., Brennan, P., & Mednick, S. A. (1990). *Minor physical anomalies and parental modeling of physical aggression predict adult offending.* Unpublished manuscript, University of Southern California.

Kanuga, M., & Rosenfeld, W. D. (2004). Adolescent sexuality and the Internet: The good, the bad, and the URL. *Journal of Pediatric and Adolescent Gynecology, 17,* 117–124.

Kaplan, S., Pelcovitz, D., Salzinger, S., Mandel, F. S., & Weiner, M. (1998). Adolescent physical abuse: Risk for adolescent psychiatric disorders. *American Journal of Psychiatry, 155,* 949–959.

Kaplan, S. J., Pelcovitz, D., & Labruna, V. (1999). Child and adolescent abuse and neglect research: A review of the past 10 years: Part I. Physical and emotional abuse and neglect. *Journal of the American Academy of Child and Adolescent Psychiatry, 38,* 1214–1222.

Kashani, J. H., & Allan, W. D. (1998). *The impact of family violence on children and adolescents.* Thousand Oaks, CA: Sage.

Kashani, J. H., Daniel, A. E., Dandoy, A. C., & Holcomb, W. R. (1992). Family violence: Impact on children. *Journal of the American Academy of Child and Adolescent Psychiatry, 31,* 181–189.

Kasian, M., & Painter, S. L. (1992). Frequency and severity of psychological abuse in a dating population. *Journal of Interpersonal Violence, 7,* 350–364.

Kaufman, J., & Cicchetti, D. (1989). Effects of maltreatment on school-age children's socioemotional development: Assessments in day-camp setting. *Developmental Psychology, 25,* 516–524.

Kaufman, J., & Zigler, E. (1987). Do abused children become abusive parents? *American Journal of Orthopsychiatry, 57,* 186–192.

Kaufman, J. G., & Widom, C. S. (1999). Childhood victimization, running away, and delinquency. *Journal of Research in Crime and Delinquency, 36,* 347–370.

Kaufman Kantor, G., & Asdigian, N. L. (1996). When women are under the influence: Does drinking or drug use by women provoke beatings by men? In M. Galanter (Ed.), *Recent developments in alcoholism.* New York: Plenum.

Kaufman Kantor, G., & Straus, M. A. (1990). The "drunken bum" theory of wife beating. In M.A. Straus & R. J. Gelles (Eds.), *Physical violence in American families: Risk factors and adaptations to violence in 8,145 families* (pp. 203–226). New Brunswick, NJ: Transaction.

Kazak, A. E. (1989). Families of chronically ill children: A systems and social-ecological model of adaptation and challenge. *Journal of Consulting and Clinical Psychology, 57,* 25–30.

Kean, R. B., & Dukes, R. L. (1991). Effects of witness characteristics on the perception and reportage of child abuse. *Child Abuse and Neglect, 15,* 423–435.

Keiley, M. K., Howe, T. R., Dodge, K. A., Bates, J. E., & Pettit, G. E. (2001). The timing of child physical maltreatment: A cross-domain growth analysis of impact on adolescent externalizing and internalizing problems. *Development and Psychopathology, 13,* 891–912.

Keilitz, S. L., Davis, D., Efkeman, H. S., Flango, C., & Hannaford, P. L. (1998). *Civil protection orders: Victims' views on effectiveness.* National Institute of Justice research preview. Washington, D.C.: U.S. Department of Justice. Retrieved May 2004 from http://www.ncjrs.org/pdffiles/fs000191.pdf

Kelder, L. R., McNamara, J. R., Carlson, B. W., & Lynn, S. J. (1991). Perceptions of physical punishment: The relation to childhood and adolescent experiences. *Journal of Interpersonal Violence, 6,* 432–445.

Kelley, S. J. (1992). Parenting stress and child maltreatment in drug-exposed children. *Child Abuse and Neglect, 16,* 317–328.

Kelley, S. J. (2002). Child maltreatment in the context of substance abuse. In J. E. B. Myers, L. Berliner, J. Briere, C. T. Hendrix, C. Jenny, & T. A. Reid (Eds.), *The APSAC handbook on child maltreatment* (2nd ed.) (pp. 105–117). Thousand Oaks, CA: Sage.

Kelly, R. J., Wood, J. J., Gonzalez, L. S., MacDonald, V., & Waterman, J. (2002). Effects of mother-son incest and positive perceptions of sexual abuse experiences on the psychosocial adjustment of clinic-referred men. *Child Abuse and Neglect, 26,* 425–441.

Kemp, A., Rawlings, E. I., & Green, B. L. (1991). Post-traumatic stress disorder in battered women: A shelter sample. *Journal of Traumatic Stress, 4,* 137–149.

Kemp, N. T., & Malkinrodt, B. (1996). Impact of professional training on case conceptualization of clients with a disability. *Professional Psychology: Research and Practice, 27,* 378–385.

Kempe, C. H., Silverman, F. N., Steele, B. F., Droegemueller, W., & Silver, H. K. (1962). The battered-child syndrome. *Journal of the American Medical Association, 181,* 17–24.

Kendall-Tackett, K. A. (2003). *Treating the lifetime health effects of childhood abuse.* New York: Civic Research Institute.

Kendall-Tackett, K. A., Williams, L. M., & Finkelhor, D. (1993). The effects of sexual abuse on children: A review and synthesis of recent empirical studies. *Psychological Bulletin, 113,* 164–180.

Kilpatrick, D. G., Best, C. L., Saunders, B. E., & Veronen, L. J. (1988). Rape in marriage and in dating relationships: How bad is it for mental health? In R. A. Prentky & V. L. Quinsey (Eds.), *Human sexual aggression: Current perspectives* (pp. 335–344). New York: New York Academy of Sciences.

Kingsnorth, R. F., Macintosh, R. C., Berdahl, T., Blades, C., & Rossi, S. (2001). Domestic violence: The role of interracial/ethnic dyads in criminal court processing. *Journal of Contemporary Criminal Justice, 17,* 123–141.

Klaus, P. A. (2000). *Crimes against persons age 65 or older, 1992–97.* U.S. Department of Justice, Office of Justice Programs, Bureau of Justice Statistics, NCJ 176352. Retrieved January 2004, from http://www.ojp.usdoj.gov/bjs/pub/pdf/cpa6597.pdf

Klimes-Dougan, B., & Kistner, J. (1990). Physically abused preschoolers' response to peers' distress. *Developmental Psychology, 26,* 599–602.

Klitzman, R. L., Greenberg, J. D., Pollack, L. M., & Dolezal, C. (2002). MDMA ("ecstasy") use, and its association with high risk behaviors, mental health, and other factors among gay/bisexual men in New York City. *Drug and Alcohol Dependence, 66,* 115–125.

Knepper, K. (1995). Withholding medical treatment from infants: When is it child neglect? *University of Louisville Journal of Family Law, 33,* 1–54.

Knudsen, D. D. (1988). Child sexual abuse and pornography: Is there a relationship? *Journal of Family Violence, 3,* 253–267.

Kocher, M. S., & Kasser, J. R. (2000). Orthopaedic aspects of child abuse. *Journal of the American Academy of Orthopedic Surgeons, 8,* 10–20.

Koenig, A. L., Cicchetti, D., & Rogosch, F. A. (2000). Child compliance/noncompliance and maternal contributors to internalization in maltreating and non-maltreating dyads. *Child Development, 71,* 1018–1032.

Kolko, D. J. (1996). Individual cognitive behavioral treatment and family therapy for physically abused children and their offending parents: A comparison of clinical outcomes. *Child Maltreatment, 1,* 322–342.

Kolko, D. J. (2002). Child physical abuse. In J. E. B. Myers, L. Berliner, J. Briere, C. T. Hendrix, C. Jenny, & T. A. Reid (Eds.), *The APSAC handbook on child maltreatment* (2nd ed.) (pp. 21–54). Thousand Oaks, CA: Sage.

Kolko, D. J., Seleyo, J., & Brown, E. J. (1999). The treatment histories and service involvement of physically and sexually abusive families: Description, correspondence, and clinical correlates. *Child Abuse and Neglect, 23,* 459–476.

Korbin, J. E. (1980). The cultural context of child abuse and neglect. *Child Abuse and Neglect, 4,* 3–13.

Korbin, J. E., Coulton, C. J., Chard, S., Platt-Houston, C., & Su, M. (1998). Impoverishment and child maltreatment in African American and European American neighborhoods. *Development and Psychopathology, 10,* 215–233.

Korbin, J. E., Coulton, C. J., Lindstrom-Ufuti, H., & Spilsbury, J. (2000). Neighborhood views on the definition and etiology of child maltreatment. *Child Abuse and Neglect, 24,* 1509–1527.

Koski, P. R., & Mangold, W. D. (1988). Gender effects in attitudes about family violence. *Journal of Family Violence, 3,* 225–237.

Kowal, L. W., Kottmeier, C. P., Ayoub, C. C., Komives, J. A., Robinson, D. S., & Allen, J. P. (1989). Characteristics of families at risk of problems in parenting: Findings from a home-based secondary prevention program. *Child Welfare, 68,* 529–538.

Koyano, W. (1989). Japanese attitudes toward the elderly: A review of research findings. *Journal of Cross-Cultural Psychology, 4,* 335–345.

Kratcoski, P. C. (1985). Youth violence directed toward significant others. *Journal of Adolescence, 8,* 145–157.

Krob, M. J., Johnson, A., & Jordan, M. H. (1986). Burned and battered adults. *Journal of Burn Care and Rehabilitation, 7,* 529–531.

Krug, R. S. (1989). Adult male reports of childhood sexual abuse by mothers: Case descriptions, motivations, and long-term consequences. *Child Abuse and Neglect, 13,* 111–119.

Kuehnle, K., & Sullivan, A. (2001). Patterns of anti-gay violence: An analysis of incident characteristics and victim reporting. *Journal of Interpersonal Violence, 16,* 928–943.

Kumagai, F., & Straus, M. A. (1983). Conflict resolution tactics in Japan, India, and the USA. *Journal of Comparative Family Studies, 14,* 377–392.

Kurz, D. (1993). Physical assaults by husbands: A major social problem. In R. J. Gelles & D. R. Loseke (Eds.), *Current controversies in family violence* (pp. 257–272). Thousand Oaks, CA: Sage.

Lachs, M. S. (2003). *Elder justice: Medical forensic issues concerning abuse and neglect (draft report): Medical forensic roundtable discussion. Detection and diagnosis: What are the forensic markers for identifying physical and psychological signs of elder abuse and neglect?* Retrieved March 22, 2002, from http://www.ojp.usdoj.gov/nij/elderjust/elder_05.html

Lachs, M. S., & Pillemer, K. (1995). Current concepts: Abuse and neglect of elderly persons. *The New England Journal of Medicine, 332,* 437–443.

Lachs, M. S., Williams, C., O'Brien, S., Hurst, L., & Horwitz, R. (1996). Older adults: An 11-year longitudinal study of Adult Protective Service use. *Archives of Internal Medicine, 156,* 449–453.

Lachs, M. S., Williams, C., O'Brien, S., Hurst, L., & Horwitz, R. (1997). Risk factors for reported elder abuse and neglect: A nine-year observational cohort study. *Gerontologist, 37,* 469–474.

Lachs, M. S., Williams, C. S., O'Brien, S., Pillemer, K. A., & Charlson, M. E. (1998). The mortality of elder mistreatment. *Journal of the American Medical Association, 280,* 428–432.

Lahey, B., Conger, R., Atkenson, B., & Treiber, F. (1984). Parenting behavior and emotional status of physically abusive mothers. *Journal of Consulting and Clinical Psychology, 52,* 1062–1071.

Landis, J. (1956). Experiences of 500 children with adult sexual deviation. *Psychiatric Quarterly Supplement, 30,* 91–109.

Landolt, M. A., & Dutton, D. G. (1997). Power and personality: An analysis of gay male intimate abuse. *Sex Roles, 37,* 335–359.

Lang, R. A., Langevin, R., Van Santen, V., Billingsley, D., et al. (1990). Marital relations in incest offenders. *Journal of Sex and Marital Therapy, 16,* 214–229.

Langeland, W., & Hartgers, C. (1998). Child sexual and physical abuse and alcoholism: A review. *Journal of Studies on Alcohol, 59,* 336–348.

Langevin, R., Handy, L., Hook, H., Day, D., & Russon, A. (1983). Are incestuous fathers pedophilic and aggressive? In R. Langevin (Ed.), *Erotic preference, gender identity, and aggression.* New York: Erlbaum Associates.

Langevin, R., Wortzman, G., Dickey, R., Wright, P., & Handy, L. (1988). Neuropsychological impairment in incest offenders. *Annals of Sex Research, 1,* 401–415.

Langhinrichsen-Rohling, J., & Monson, C. M. (1998). Marital rape: Is the crime taken seriously without co-occurring physical abuse? *Journal of Family Violence, 13,* 433–443.

Langhinrichsen-Rohling, J., & Neidig, P. (1995). Violent backgrounds of economically disadvantaged youth: Risk factors for perpetrating violence? *Journal of Family Violence, 10,* 27–36.

Lanktree, C., Briere, J., & Zaidi, L. (1991). Incidence and impact of sexual abuse in a child outpatient sample: The role of direct inquiry. *Child Abuse and Neglect, 15,* 447–453.

LaRose, L., & Wolfe, D. (1987). Psychological characteristics of parents who abuse or neglect their children. In B. Lahey & A. E. Kazdin (Eds.), *Advances in clinical child psychology* (Vol. 10) (pp. 55–97). New York: Plenum.

Larzelere, R. (1986). Moderate spanking: Model or deterrent of children's aggression in the family? *Journal of Family Violence, 1,* 27–36.

Larzelere, R. E. (1994). Should the use of corporal punishment by parents be considered child abuse?: No. In M. A. Mason & E. Gambrill (Eds.), *Debating children's lives: Current controversies on children and adolescents* (pp. 204–209). Thousand Oaks, CA: Sage.

Laviola, M. (1992). Effects of older brother-younger sister incest: A study of the dynamics in 17 cases. *Child Abuse and Neglect, 16,* 409–421.

Lefkowitz, M., Eron, L., Walder, L., & Huesmann, L. (1977). *Growing up to be violent: A longitudinal study of the development of aggression.* New York: Pergamon.

Lehmann, M., & Santilli, N. R. (1996). Sex differences in perception of spousal abuse. *Journal of Social Behavior & Personality, 11,* 229–239.

Lemberg, J. (2002, July 21). Spouse abuse in South Asian marriages may be high. *Women's News.* Retrieved March 12, 2004, from http://www.womense news.org/article.cfm/dyn/aid/979/context/archive

Lesserman, J., Li, Z., Drossman, D., Toomey, T. C., Nachman, G., & Glogau, L. (1997). Impact of sexual and physical abuse dimensions on health status: Development of an abuse severity measure. *Psychosomatic Medicine, 59,* 152–160.

Letellier, P. (1999). Rape. In B. Leventhal & S. E. Lundy (Eds.), *Same-sex domestic violence.* Thousand Oaks, CA: Sage.

Levesque, R. J. (2001). *Culture and family violence: Fostering change through human rights law.* Washington, D.C.: American Psychological Association.

Levinson, D. (1989). *Family violence in a cross-cultural perspective.* Newbury Park, CA: Sage.

Lewis, D. O., Lovely, R., Yeager, C., & Femina, D. (1989). Toward a theory of the genesis of violence: A follow-up study of delinquents. *Journal of the American Academy of Child and Adolescent Psychiatry, 28,* 431–436.

Lewis, D. O., Shanok, S. S., & Balla, D. A. (1979). Parental criminality and medical histories of delinquent children. *American Journal of Psychiatry, 136,* 288–292.

Lewis, M. L., & Schaffer, S. (1981). Peer behavior and mother-infant interaction. In M. L. Lewis & L. A. Rosenblum (Eds.), *The uncommon child.* New York: Plenum.

Lie, G. Y., & Gentlewarrier, S. (1991). Intimate violence in lesbian relationships: Discussion of survey findings and practice implications. *Journal of Social Service Research, 15,* 41–59.

Lie, G. Y., Schilit, R., Bush, J., Montagne, M., & Reyes, L. (1991). Lesbians in currently aggressive relationships: How frequently do they report aggressive past relationships? *Violence and Victims, 6,* 121–135.

Lindner Gunnoe, M., Hetherington, E. M., & Reiss, D. (1999). Parental religiosity, parenting style, and adolescent social responsibility. *Journal of Early Adolescence, 19,* 199–225.

Linz, D., Donnerstein, E., & Penrod, S. (1988). Effects of long-term exposure to violent and sexually degrading depictions of women. *Journal of Personality and Social Psychology, 55,* 758–768.

Lipschitz, D. S., Winegar, R. K., Nicolaou, A. L., Hartnick, E., Wolfson, M., & Southwick, S. M. (1999). Perceived abuse and neglect as risk factors for suicidal behavior in adolescent inpatients. *The Journal of Nervous and Mental Disease, 187,* 32–39.

Livingston, L. R. (1986). Children's violence to single mothers. *Journal of Sociology and Social Welfare, 13,* 920–933.

Locke, L. M., & Richman, C. L. (1999). Attitudes toward domestic violence: Race and gender issues. *Sex Roles, 40,* 227–247.

Lockhart, L. L., White, B. W., Causby, V., & Issac, A. (1994). Letting out the secret: Violence in lesbian relationships. *Journal of Interpersonal Violence, 9,* 469–492.

Loeber, R., & Strouthamer-Loeber, M. (1986). Family factors as correlates and predictors of juvenile conduct problems and delinquency. In M. Tonry & N. Morris (Eds.), *Crime and justice: An annual review of the research* (Vol. 7). Chicago: University of Chicago Press.

Loftus, E., & Ketcham, K. (1994). *The myth of repressed memory: False memories and allegations of sexual abuse.* New York: St. Martin's Griffin.

Lohrmann-O'Rourke, S., & Zirkel, P. A. (1998). The case law on aversive interventions for students with disabilities. *Exceptional Children, 65,* 101–123.

Lombardi, E. (2001). Enhancing transgender health care. *American Journal of Public Health, 91,* 869–872.

Lombardi, E. L., & van Servellen, G. (2000). Building culturally sensitive substance use prevention and treatment programs for transgendered populations. *Journal of Substance Abuse Treatment, 19,* 291–296.

Longdon, C. (1993). A survivor's and therapist's viewpoint. In M. Elliott (Ed.), *Female sexual abuse of children: The ultimate taboo* (pp. 50–60). New York: John Wiley & Sons.

Lung, C. T., & Daro, D. (1996). *Current trends in child abuse reporting and fatalities: The results of the 1995 annual fifty-state survey.* Chicago: National Committee to Prevent Child Abuse.

Luntz, B., & Widom, C. S. (1994). Antisocial personality disorder in abused and neglected children grown up. *Journal of Psychiatry, 151,* 670–674.

Lutzker, J. R., Bigelow, K. M., Doctor, R. M., Gershater, R. M., & Greene, B. F. (1998). An ecobehavioral model for the prevention and treatment of child abuse and neglect. In J. R. Lutzker (Ed.), *Handbook on child abuse research and treatment* (pp. 239–266). New York: Plenum.

Lynch, M., & Cicchetti, D. (1998). An ecological-transactional analysis of children and contexts: The longitudinal interplay among child maltreatment, community violence, and children's symptomatology. *Development and Psychopathology, 20,* 235–257.

Macolini, R. M. (1995). Elder abuse policy: Consideration in research and legislation. *Behavioral Sciences and the Law, 13,* 349–363.

Madonna, P. G., VanScoyk, S., & Jones, D. P. (1991). Family interactions with incest and nonincest families. *American Journal of Psychiatry, 148,* 46–49.

Magen, R. (1999). In the best interest of battered women: Reconceptualizing allegations of failure to protect. *Child Maltreatment, 4,* 127–135.

Mahoney, A., & Donnelly, W. O. (2000). Adolescent-to-parent physical aggression in clinic-referred families: Prevalence and co-occurrence with parent-to-adolescent physical aggression. In *Victimization of Children and Youth: An*

International Research Conference. Family Research Laboratory, University of New Hampshire, Durham, NH.

Mahoney, A., Donnelly, W. O., Boxer, P., & Lewis, T. (2003). Marital and severe parent-to-adolescent physical aggression in clinic-referred families: Mother and adolescent reports on co-occurence and links to child behavior problems. *Journal of Family Psychology, 17,* 3–19.

Mahoney, A., Pargament, K. I., Jewell, T., Swank, A. B., Scott, E., Emery, E., & Rye, M. (1999). Marriage and the spiritual realm: The role of proximal and distal religious constructs in marital functioning. *Journal of Family Psychology, 13,* 1–18.

Main, M., & George, C. (1985). Responses of abused and disadvantaged toddlers to distress in agemates: A study in the day care setting. *Developmental Psychology, 21,* 407–412.

Makepeace, J. (1997). Courtship violence as a process: A developmental theory. In A. P. Cardarelli (Ed.), *Violence between intimate partners: Patterns, causes, and effects* (pp. 29–47). Boston: Allyn & Bacon.

Makepeace, J. M. (1986). Gender differences in courtship violence victimization. *Family Relations, 35,* 383–388.

Makepeace, J. M. (1981). Courtship violence among college students. *Family Relations, 30,* 97–100.

Malecha, A., McFarlane, J., Gist, J., Watson, K., Batten, E., Hall, I., & Smith, S. (2003). Applying for and dropping a protection order: A study with 150 women. *Criminal Justice Policy Review, 14,* 486–504.

Malinosky-Rummell, R., Ellis, J. T., Warner, J. E., Ujcich, K., Carr, R. E., & Hansen, D. J. (1991, November). *Individualized behavioral intervention for physically abusive and neglectful families: An evaluation of the family interaction skills project.* Paper presented at the 25th Annual Convention of the Association for the Advancement of Behavior Therapy, New York, NY.

Malinoksy-Rummell, R., & Hansen, D. J. (1993). Long-term consequences of childhood physical abuse. *Psychological Bulletin, 114,* 68–79.

Malley-Morrison, K., & Hines, D. A. (2003, July). *Religion, corporal punishment, and attitudes toward aggression.* Paper presented at the 8th International Family Violence Research Conference, Portsmouth, NH.

Malley-Morrison, K., & Hines, D. A. (2004). *Family violence in a cultural perspective: Defining, understanding, and combating abuse.* Thousand Oaks, CA: Sage.

Malley-Morrison, K., You, H. S., & Mills, R. B. (2000). Young adult attachment styles and perceptions of elder abuse: A cross-cultural study. *Journal of Cross-Cultural Gerontology, 15,* 163–184.

Mallon, G. P. (1998). *We don't exactly get the welcome wagon: The experiences of gay and lesbian adolescents in child welfare systems.* New York: Columbia University Press.

Malone, J., Tyree, A., & O'Leary, K. D. (1989). Generalization and containment: Different effects of past aggression for wives and husbands. *Journal of Marriage and the Family, 51,* 687–697.

Manly, J. T., Kim, J. E., Rogosch, F. A., & Cicchetti, D. (2001). Dimensions of child maltreatment and children's adjustment: Contributions of developmental timing and subtype. *Development and Psychopathology, 13,* 759–782.

Mann, C. R. (1996). *When women kill.* New York: State University of New York Press.

Mannarino, A. P., & Cohen, J. A. (1996). A follow-up study on factors that mediate the development of psychological symptomatology in sexually abused girls. *Child Maltreatment, 1,* 246–260.

Mannarino, A. P., Cohen, J. A., & Berman, S. P. (1994). The relationship between pre-abuse factors and psychological symptomatology in sexually abused girls. *Child Abuse and Neglect, 18,* 63–71.

Marcus, R. F., & Swett, B. (2003). Violence in close relationships: The role of emotion. *Aggression and Violent Behavior, 8,* 313–327.

Margolies, L., Becker, M., & Jackson-Brewer, K. (1987). Internalized homophobia: Identifying and treating the oppressor within. In Boston Lesbian Collective (Eds.), *Lesbian psychologies: Explorations & challenges* (pp. 229–241). Urbana, IL: University of Illinois Press.

Margolies, L., & Leeder, E. (1995). Violence at the door: Treatment of lesbian batterers. *Violence Against Women, 1,* 139–157.

Marrujo, B. & Kreger, M. (1996). Definition of roles in abusive lesbian relationships. In C. M. Renzetti & C. H. Miley (Eds.), *Violence in gay and lesbian domestic partnerships* (pp. 23–33). New York: Harrington Park Press/ Haworth Press.

Marshall, L. L. (1996). Psychological abuse of women: Six distinct clusters. *Journal of Family Violence, 11,* 379–409.

Marshall, L. L., & Rose, P. (1990). Premarital violence: The impact of origin violence, stress, and reciprocity. *Violence and Victims, 5,* 51–64.

Marshall, P. D., & Norgard, K. E. (1983). *Child abuse and neglect: Sharing responsibility.* New York: Wiley.

Martin, M. E. (1997). Police promise: Community policing and domestic violence victim satisfaction. *Policing, 20,* 519–529.

Martin, S. E. (1989). Research note: The response of the clergy to spouse abuse in a suburban county. *Violence and Victims, 4,* 217–225.

Martone, M., Jaudes, P. K., & Cavins, M. K. (1996). Criminal prosecution of child sexual abuse cases. *Child Abuse and Neglect, 20,* 457–464.

Matthews, J. K. (1993). Working with female sexual abusers. In M. Elliott (Ed.), *Female sexual abuse of children: The ultimate taboo* (pp. 61–78). New York: John Wiley & Sons.

May, P. A., & Gossage, P. (2001). Estimating the prevalence of fetal alcohol syndrome: A summary. *Alcohol Research & Health, 25,* 159–167.

Mayer, E., Kosmin, B. A., & Keysar, A. (2001). *American Religious Identification Survey 2001.* Retrieved September 2003, from http://www.gc.cuny.edu/studies/aris_index.htm

Maynard, C., & Wiederman, M. (1997). Undergraduate students' perceptions of child sexual abuse: Effects of age, sex, and gender-role attitudes. *Child Abuse and Neglect, 21,* 833–844.

McAlpin, J. P. (2003). N.J. draws fire over starvation case: Caseworkers fired; horrific abuse brings 2nd probe of agency. *Boston Globe*. Retrieved November 17, 2003, from http://www.boston.com/news/nation/articles/2003/10/28/

McClain, P. W., Sacks, J. J., Froehlke, R. G., & Ewigman, B. G. (1993). Estimates of fatal child abuse and neglect, United States, 1979 through 1988. *Pediatrics, 91*, 338–343.

McClennen, J. C., Summers, A. B., & Daley, J. G. (2002). The lesbian partner abuse scale. *Research on Social Work Practice, 12*, 277–292.

McCloskey, L. A. (1996). Socioeconomic and coercive power within the family. *Gender and Society, 10*, 449–463.

McCord, J. (1983). A forty-year perspective on effects of child abuse and neglect. *Child Abuse and Neglect, 7*, 265–270.

McCrone, E. R., Egeland, B., Kalkoske, M., & Carlson, E. A. (1994). Relations between early maltreatment and mental representations of relationships assessed with projective storytelling in middle childhood. *Development and Psychopathology, 6*, 99–120.

McFayden, R. G., & Kitson, W. J. H. (1996). Language comprehension and expression among adolescents who have experienced childhood physical abuse. *Journal of Child Psychology and Psychiatry and Allied Disciplines, 37*, 551–562.

McGough, L. S. (1997). Stretching the blanket: Legal reforms affecting child witnesses. *Learning and Individual Differences, 9*, 317–340.

McGregor, B. A., Carver, C. S., Antoni, M. H., Weiss, S., Yount, S., & Ironson, G. (2001). Internalized homophobia and distress among lesbian women who have been treated for breast cancer. *Psychology of Women Quarterly, 25*, 1–9.

McLaughlin, T. L., Heath, A. C., Bucholz, K. K., Madden, P. A., Bierut, L. J., Slutske, W. S., Dinwiddie, S., Statham, D. J., Dunne, M. P., & Martin, N. G. (2000). Childhood sexual abuse and pathogenic parenting in the childhood recollections of adult twin pairs. *Psychological Medicine, 30*, 1293–1302.

McNeely, R. L., Cook, P. W., & Torres, J. B. (2001). Is domestic violence a gender issue, or a human issue? *Journal of Human Behavior in the Social Environment, 4*, 227–251.

Mehta, M. D. (2001). Pornography in Usenet: A study of 9,800 randomly selected images. *CyberPsychology and Behavior, 4*, 695–703.

Meiselman, K. C. (1978). *Incest: A psychological study of causes and effects with treatment recommendations.* San Francisco: Jossey-Bass.

Mellott, R. N., Wagner, W. G., & Broussard, S. D. (1997). India vs. United States undergraduates' attitudes concerning child sexual abuse: The impact of survivor sex, survivor age, survivor response, respondent sex, and country of origin. *International Journal of Intercultural Relations, 21*, 305–318.

Melossi, D. (2001). The cultural embeddedness of social control: Reflections on the comparison of Italian and North-American cultures concerning punishment. *Theoretical Criminology, 5*, 403–424.

Melton, H. C., & Belknap, J. (2003). He hits, she hits: Assessing gender differences and similarities in officially reported intimate partner violence. *Criminal Justice and Behavior, 30*, 328–348.

Memon, K. (n.d.). *Wife abuse in the Muslim community*. Islamic Society of North America. Retrieved July 2003, from http://www.zawaj.com/articles/abuse_memon.html

Mercy, J. A., Krug, E. G., Dahlberg, L. L., & Zwi, A. B. (2003). Violence and health: The United States in a global perspective. *American Journal of Public Health, 93*, 256–261.

Meredith, W. H., Abbott, D. A., & Adams, S. L. (1986). Family violence: Its relation to marital and parental satisfaction and family strengths. *Journal of Family Violence, 1*, 299–305.

Merrill, G. S. (1998). Understanding domestic violence among gay and bisexual men. In R. K. Bergen (Ed.), *Issues in Intimate Violence* (pp. 129–141). Thousand Oaks, CA: Sage.

Merrill, G. S., & Wolfe, V. A. (2000). Battered gay men: An exploration of abuse, help seeking, and why they stay. *Journal of Homosexuality, 39*, 1–30.

Meyer, S. L., Vivian, D., & O'Leary, K. D. (1998). Men's sexual aggression in marriage. *Violence Against Women, 4*, 415–435.

Mignon, S. I. (1998). Husband battering: A review of the debate over a controversial social phenomenon. In N. A. Jackson & G. C. Oates (Eds.), *Violence in intimate relationships: Examining sociological and psychological issues* (pp. 137–160). Boston: Butterworth-Heinemann.

Mihalic, S. W., & Elliott, D. (1997). A social learning theory model of marital violence. *Journal of Family Violence, 12*, 21–47.

Milburn, M. A., & Conrad, S. D. (1996). *The politics of denial*. Cambridge, MA: MIT Press.

Miles-Doan, R. (1998). Violence between spouses and intimates: Does neighborhood context matter? *Social Forces, 77*, 623–645.

Miller, D. H., Greene, K., Causby, V., White, B. W., & Lockhart, L. L. (2001). Domestic violence in lesbian relationships. *Women and Therapy, 23*, 107–127.

Mills, L. G. (2000). Woman abuse and child protection: A tumultuous marriage (part I). *Children and Youth Services Review, 22*, 199–205.

Mills, L. G. (2003). *Insult to injury: Rethinking our responses to intimate abuse.* Princeton, NJ: Princeton University Press.

Mills, R. B., Vermette, V., & Malley-Morrison, K. (1998). Judgments about elder abuse and college students' relationship with grandparents. *Gerontology and Geriatrics Education, 19*, 17–30.

Milner, J. S. (1993). Social information processing and physical child abuse. *Clinical Psychology Review, 13*, 275–294.

Milner, J. S. (1994). Assessing physical child abuse risk: The child abuse potential inventory. *Clinical Psychology Review, 14*, 547–583.

Milner, J. S. (1998). Individual and family characteristics associated with intrafamilial child physical and sexual abuse. In P. K. Trickett & C. J. Schellenbach (Eds.), *Violence against children in the family and community* (pp. 141–170). Washington, D.C.: American Psychological Association.

Moeller, T. P., Bachman, G. A., & Moeller, J. R. (1993). The combined effects of physical, sexual, and emotional abuse during childhood: Long-term health consequences for women. *Child Abuse and Neglect, 17,* 623–640.

Moffitt, T. E., Caspi, A., Rutter, M., & Silva, P. A. (2001). *Sex differences in antisocial behaviour: Conduct disorder, delinquency and violence in the Dunedin longitudinal study.* Cambridge, UK: Cambridge University Press.

Molidor, C. E. (1995). Gender differences of psychological abuse in high school dating relationships. *Child and Adolescent Social Work Journal, 12,* 119–134.

Mollerstrom, W. W., Patchner, M. M., & Milner, J. S. (1992). Family functioning and child abuse potential. *Journal of Clinical Psychology, 48,* 445–454.

Molnar, B. E., Buka, S. L., Brennan, R. T., Holton, J. K., & Earls, F. (2003). A multilevel study of neighborhoods and parent-to-child physical aggression: Results from the Project on Human Development in Chicago neighborhoods. *Child Maltreatment, 8,* 84–97.

Moncher, F. J. (1996). The relationship of maternal adult attachment style and risk of physical child abuse. *Journal of Interpersonal Violence, 11,* 335–350.

Montes, M. P., de Paul, J., & Milner, J. S. (2001). Evaluations, attributions, affect, and disciplinary choices in mothers at high and low risk for child physical abuse. *Child Abuse and Neglect, 25,* 1015–1036.

Moon, A., & Benton, D. (2000). Tolerance of elder abuse and attitudes toward third-party intervention among African American, Korean American, and White elderly. *Journal of Multicultural Social Work, 6,* 283–303.

Moon, A., & Williams, O. (1993). Perceptions of elder abuse and help-seeking patterns among African American, Caucasian American, and Korean American elderly women. *The Gerontologist, 33,* 386–395.

Moon, A., Tomita, S. K., & Jung-Kamei, S. (2001). Elder mistreatment among four Asian American groups: An exploratory study on tolerance, victim blaming and attitudes toward third-party intervention. *Journal of Gerontological Social Work, 36,* 53–169.

Morse, B. J. (1995). Beyond the conflict tactics scale: Assessing gender differences in partner violence. *Violence and Victims, 10,* 251–272.

Morton, E., Runyan, C., Moracco, K. E., & Butts, J. D. (1998). Partner homicide-suicide involving female homicide victims: A population based study in North Carolina, 1988–1992. *Violence and Victims, 13,* 91–106.

Moskowitz, H., Griffith, J. L., DiScala, C., & Sege, R. D. (2001). Serious injuries and deaths of adolescent girls resulting from interpersonal violence: Characteristics and trends from the United States, 1989–1998. *Archives of Pediatrics & Adolescent Medicine, 155,* 903–908.

Mouton, C., Rovi, S., Furniss, K., & Lasser, N. (1999). The associations between health and domestic violence in older women: Results of a pilot study. *Journal of Women's Health and Gender-Based Medicine, 1,* 1173–1179.

Mulder, R. T., Beautrais, A. L., Joyce, P. R., & Fergusson, D. M. (1998). Relationship between dissociation, childhood sexual abuse, childhood physical abuse, and mental illness in a general population sample. *American Journal of Psychiatry, 155*, 806–811.

Mullen, P. E., Martin, J. L., Anderson, J. C., Romans, S. E., et al. (1993). Childhood sexual abuse and mental health in adult life. *British Journal of Psychiatry, 163*, 721–732.

Mullen, P. E., Martin, J. L., Anderson, J. C., Romans, S. E., Herbison, G. P. (1994). The effect of child sexual abuse on social, interpersonal and sexual function in adult life. *British Journal of Psychiatry, 165*, 35–47.

Murphy, C. M., & Cascardi, M. (1999). Psychological abuse in marriage and dating relationships. In R. L. Hampton (Ed.), *Family violence prevention and treatment* (2nd ed.) (pp. 198–226). Thousand Oaks, CA: Sage.

Murphy, C. M., & Hoover, S. A. (2001). Measuring emotional abuse in dating relationships as a multifactorial construct. In K. D. O'Leary & R. D. Maiuro (Eds.), *Psychological abuse in violent domestic relations* (pp. 29–46). New York: Springer.

Murphy, C. M., O'Farrell, T. J., Fals-Stewart, W., & Feehan, M. (2001). Correlates of intimate partner violence among male alcoholic patients. *Journal of Consulting and Clinical Psychology, 69*, 528–540.

Murphy, C. M., & O'Leary, K. D. (1989). Psychological aggression predicts physical aggression in early marriage. *Journal of Consulting and Clinical Psychology, 57*, 579–582.

Myers, J. E. (1998). *Legal issues in child abuse and neglect practice* (2nd ed.). Thousand Oaks, CA: Sage.

Myers, J. E. B. (2002). The legal system and child protection. In J. E. B. Myers, L. Berliner, J. Briere, C. T. Hendrix, C. Jenny, & T. A. Reid (Eds.), *The APSAC handbook on child maltreatment* (2nd ed.) (pp. 305–328). Thousand Oaks, CA: Sage.

Nahmiash, D., & Reis, M. (2000). Most successful intervention strategies for abused older adults. *Journal of Elder Abuse and Neglect, 12*, 53–70.

National Center for Victims of Crime (2003). *Spousal rape laws: 20 years later.* Victim Policy Pipeline. Retrieved January 4, 2004, from http://www.ncvc.org/ncvc/main.aspx?dbName=DocumentViewer&DocumentID=3270

National Center on Elder Abuse (1998). *National Elder Abuse Incidence Study of 1996.* Retrieved October 1, 2002, from http://www.aoa.gov/naic/publicaitionlist.html

National Center on Elder Abuse (1999). Violence against disabled women brief released. *National Center on Elder Abuse Newsletter, 2 (1)*, 6. Retrieved May 4, 2004, from http://www.elderabusecenter.org/pdf/newsletter/news21.pdf

National Coalition Against Domestic Violence (1996). *Open minds, open doors: Working with women with disabilities resource manual.* Harrisburg, PA: National Coalition Against Domestic Violence.

National Coalition Against Domestic Violence (2003). *About NCADV.* Retrieved September 16, 2003, from http://www.ncadv.org/about.htm

National Coalition of Anti-Violence Programs (2000). *Lesbian, gay, transgender and bisexual (LGTB) domestic violence in 1999: A report of the National Coalition of Anti-Violence Programs.* Retrieved May 7, 2004, from http://www.avp.org

National Coalition of Anti-Violence Programs (2001). *Lesbian, gay, bisexual and transgender domestic violence in 2000: A report of the National Coalition of Anti-Violence Programs.* Retrieved May 7, 2004, from http://www.avp.org

National Coalition of Anti-Violence Programs (2003). *Lesbian, gay, bisexual and transgender domestic violence in 2002: A report of the National Coalition of Anti-Violence Programs.* Retrieved May 7, 2004, from http://www.avp.org

National Coalition to Abolish Corporal Punishment in the Schools (2001). *Facts about corporal punishment.* Retrieved February 4, 2003, from http://www.stophitting.com/disatschool/facts.php

National Domestic Violence Hotline (2003). Retrieved September 11, 2003, from http://www.ndvh.org

National Gay and Lesbian Task Force (2002). *Domestic violence laws in the U.S.* Retrieved December 2003, from http://www.thetaskforce.org/downloads/domesticviolencelawsmap.pdf

National Institute on Drug Abuse (1994). *National pregnancy and health survey.* Rockville, MD: U.S. Department of Health and Human Services.

Nelsen, H. M., & Kroliczak, A. (1984). Parental use of the threat "God will punish": Replication and extension. *Journal for Scientific Study of Religion, 23,* 267–277.

Nelson, B. (1984). *Making an issue of child abuse.* Chicago: University of Chicago Press.

Nelson, E. C., Heath, A. C., Madden, P. A., Cooper, L., Dinwiddie, S. H., Bucholz, K. K., Glowinski, A., McLaughlin, T., Dunne, M. P., Statham, M. P., Statham, D. J., & Martin, N. G. (2002). Association between self-reported childhood sexual abuse and adverse psychosocial outcomes: Results from a twin study. *Archives of General Psychiatry, 59,* 139–145.

New, M. J., Stevenson, J., & Skuse, D. (1999). Characteristics of mothers of boys who sexually abuse. *Child Maltreatment, 4,* 21–31.

Newberger, E. H. (1999). *The men they will become: The nature and nurture of male character.* Cambridge: Perseus Publishing.

Newberger, E. H., & Bourne, R. (1978). The medicalization and legalization of child abuse. *American Journal of Orthopsychiatry, 48,* 593–607.

Newberger, E. H., Hampton, R. L., Marx, T. J., & White, K. M. (1986). Child abuse and pediatric social illness: An epidemiological analysis and ecological reformulation. *American Journal of Orthopsychiatry, 56,* 589–601.

Newby, J. H., Ursano, R. J., McCarroll, J. E., Martin, L. T., Norwood, A. E., & Fullerton, C. S. (2003). Spousal aggression by U.S. Army female soldiers toward employed and unemployed civilian husbands. *American Journal of Orthopsychiatry, 73,* 288–293.

Ney, P. G., Fung, T., & Wickett, A. R. (1994). The worst combinations of child abuse and neglect. *Child Abuse and Neglect, 18,* 705–714.

Norwood, R. (1988). *Letters from women who love too much: A closer look at relationship addiction and recovery.* New York: Pocket Books.

Nosek, M. A. (1996). Wellness among women with physical disabilities. *Sexuality and Disability, 14,* 165–181.

Nosek, M. A., & Howland, C. A. (1998). *Abuse and women with disabilities.* Violence Against Women online resources. Retrieved May 4, 2004, from http://www.vaw.umn.edu/documents/vawnet/disab/disab.html

O'Brien, M. J. (1991). Taking sibling incest seriously. In M. Patton (Ed.), *Family sexual abuse: Frontline research and evaluation* (pp. 75–92). Newbury Park, CA: Sage.

O'Farrell, T. J., Fals-Stewart, W., Murphy, M., & Murphy, C. M. (2003). Partner violence before and after individually based alcoholism treatment for male alcoholic patients. *Journal of Consulting and Clinical Psychology, 71,* 92–102.

O'Hearn, R. E., & Davis, K. E. (1997). Women's experience of giving and receiving emotional abuse: An attachment perspective. *Journal of Interpersonal Violence, 12,* 375–391.

O'Leary, K. D. (1988). Physical aggression between spouses: A social learning theory perspective. In V. B. Van Hasselt, R. L. Morrison, A. S. Bellack, & M. Hersen (Eds.), *Handbook of family violence* (pp. 31–56). New York: Plenum.

O'Leary, K. D., Barling, J., Arias, I., Rosenbaum, A., Malone, J., & Tyree, A. (1989). Prevalence and stability of physical aggression. *Journal of Consulting and Clinical Psychology, 57,* 263–268.

O'Leary, K. D., Malone, J., & Tyree, A. (1994). Physical aggression in early marriage: Prerelationship and relationship effects. *Journal of Consulting and Clinical Psychology, 62,* 594–602.

Oates, R. K., & Bross, D. C. (1995). What have we learned about treating child physical abuse? A literature review of the last decade. *Child Abuse and Neglect, 19,* 463–473.

Oates, R. K., Ryan, M. G., & Booth, S. M. (2000). Child physical abuse. In R. T. Ammerman & M. Hersen (Eds.), *Case studies in family violence* (2nd ed.) (pp. 133–156). New York: Kluwer.

Olds, D. L., Eckenrode, J., Henderson, C. R., Kitzman, H., Powers, J., Cole, R., Sidora, K., Morris, P., Pettitt, L., & Luckey, D. (1997). Long-term effects of home visitation on maternal life course and child abuse and neglect. *Journal of the American Medical Association, 278,* 637–642.

Olds, D. L., Henderson, C. R., Chamberlin, R., & Tatelbaum, R. (1986). Preventing child abuse and neglect: A randomized trial of home nurse visitation. *Pediatrics, 78,* 65–78.

Ondersma, S. J., Chaffin, M., Berliner, L., Cordon, I., Goodman, G. S., & Barnett, D. (2001). Sex with children is abuse: Comment on Rind, Tromovitch, and Bauserman (1998). *Psychological Bulletin, 127,* 707–714.

Ontario Consultants on Religious Tolerance (2003). *Child corporal punishment: Spanking, the results of studies.* Retrieved January 30, 2004, from http://www.religioustolerance.org/spankin5.htm

Orelove, F. P., Hollahan, D. J., & Myles, K. T. (2000). Maltreatment of children with disabilities: Training needs for a collaborative response. *Child Abuse and Neglect, 24,* 185–194.

Pagelow, M. D. (1984). *Family violence.* New York: Praeger.

Painter, K., & Farrington, D. P. (1998). Marital violence in Great Britain and its relationship to marital and non-marital rape. *International Review of Victimology, 5,* 257–276.

Paone, D., & Alperen, J. (1998). Pregnancy policing: Policy of harm. *International Journal of Drug Policy, 9,* 101–108.

Parker, H., & Parker, S. (1986). Father-daughter sexual abuse: An emerging perspective. *American Journal of Orthopsychiatry, 56,* 531–549.

Pate, A. M., & Hamilton, E. E. (1992). Formal and informal deterrents to domestic violence: The Dade County spouse assault experiment. *American Sociological Review, 57,* 691–707.

Patterson, C. J. (2003). *Lesbian and gay parenting.* Retrieved August 24, 2004, from http://www.apa.org/pi/parent.html

Paul, J. P., Catania, J., Pollack, L., & Stall, R. (2001). Understanding childhood sexual abuse as a predictor of sexual risk-taking among men who have sex with men: The Urban Men's Health Study. *Child Abuse and Neglect, 25,* 557–584.

Paulozzi, L. J., Saltzman, L. A., Thompson, M. J., & Holmgreen, P. (2001). Surveillance for homicide among intimate partners: United States, 1981–1998. *CDC Surveillance Summaries 50(SS-3),* 1–16.

Paveza, G. J., Cohen, D., Eisdorfer, C., Freels, S., Semla, T., et al. (1992). Severe family violence and Alzheimer's disease: Prevalence and risk factors. *Gerontologist, 32,* 493–497.

Payne, B. K. (2002). An integrated understanding of elder abuse and neglect. *Journal of Criminal Justice, 30,* 535–547.

Peacock, P. L. (1995). Marital rape. In V. R. Wiehe & A. L. Richards (Eds.), *Intimate betrayal: Understanding and responding to the trauma of acquaintance rape.* Thousand Oaks, CA: Sage.

Peek, C. W., Fischer, J. L., & Kidwell, J. S. (1985). Teenage violence toward parents: A neglected dimension of family violence. *Journal of Marriage and the Family, 47,* 1051–1058.

Pelletier, D., & Coutu, S. (1992). Substance abuse and family violence in adolescents. *Canada's Mental Health, 40,* 6–12.

Pelton, L. H. (1994). The role of material factors in child abuse and neglect. In G. B. Melton & F. D. Barry (Eds.), *Protecting children from abuse and neglect: Foundations for a new national strategy* (pp. 131–181). New York: Guilford.

Pennsylvania v. Ritchie (1987). Pennsylvania v. Ritchie, 480, U.S. 39 Retrieved January 2004 from http://laws.findlaw.com/us/480/39.html

Perez, C. M., & Widom, C. S. (1994). Childhood victimization and long term intellectual and academic outcomes. *Child Abuse and Neglect, 18,* 617–633.

Perkins, D. F., & Luster, T. (1999). The relationship between sexual abuse and purging: Findings from community-wide surveys of female adolescents. *Child Abuse and Neglect, 23,* 371–382.

Peterman, L. M., & Dixon, C. G. (2003). Domestic violence between same-sex partners: Implications for counseling. *Journal of Counseling and Development, 81,* 40–47.

Phillips, L. R. (1983). Abuse and neglect of the frail elderly at home: An exploration of theoretical relationships. *Journal of Advanced Nursing, 8,* 379–392.

Phillips, L. R., Torres de Ardon, E., & Briones, G. S. (2000). Abuse of female caregivers by care recipients: Another form of elder abuse. *Journal of Elder Abuse and Neglect, 12,* 123–143.

Pianta, R., Egeland, B., & Erickson, M. F. (1989). The antecedents of maltreatment: Results of the Mother-Child Interaction Research Project. In D. Cicchetti & V. Carlson (Eds.), *Child maltreatment: Theory and research on the causes and consequences of child abuse and neglect* (pp. 203–253). New York: Cambridge University Press.

Pierce, L. H., & Pierce, R. L. (1987). Incestuous victimization by juvenile sex offenders. *Journal of Family Violence, 2,* 351–364.

Pillemer, K. (1985). The dangers of dependency: New findings on domestic violence against elderly. *Social Problems, 33,* 146–158.

Pillemer, K. (1986). Risk factors in elder abuse: Results from a case-control study. In K. Pillemer & R.S. Wolf (Eds.), *Elder abuse: Conflict in the family* (pp. 236–263). Dover, MA: Auburn House.

Pillemer, K. (1993). The abused offspring are dependent: Abuse is caused by the deviance and dependence of abusive caregivers. In R. J. Gelles & D. R. Loseke (Eds.), *Current controversies on family violence* (pp. 237–249). Newbury Park, CA: Sage.

Pillemer, K., & Finkelhor, D. (1988). The prevalence of elder abuse: A random sample survey. *Gerontologist, 28,* 51–57.

Pillemer, K., & Finkelhor, D. (1989). Causes of elder abuse: Caregiver stress versus problem relatives. *American Journal of Orthopsychiatry, 59,* 179–187.

Pillemer, K., & Suitor, J. J. (1992). Violence and violent feelings: What causes them among family caregivers. *Journal of Gerontology, 47,* S165–S172.

Pillemer, K., & Wolf, R. (Eds.). (1986). *Elder abuse: Conflict in the family.* Dover, MA: Auburn House.

Pinder v. Johnson, 54 F.3d 1169 (CA 4 1995). Retrieved August 2003, from http://www.soc.umn.edu/~samaha/courses/4162_04sp/pindervjohnson_ed.htm

Piran, N., Lerner, P., Garfinkel, P. E., Kennedy, S. H., Brouillette, C. (1988). Personality disorders in anorexic patients. *International Journal of Eating Disorders, 7,* 589–599.

Pleck, E. (1989). Criminal approaches to family violence, 1640–1980. In L. Ohlin & M. Tonry (Eds.), *Family violence* (pp. 19–57). Chicago: University of Chicago Press.

Pleck, E., Pleck, J. H., Grossman, M., & Bart, P. B. (1977–78). The battered data syndrome: A comment on Steinmetz's article. *Victimology: An International Journal, 2,* 680–683.

Plomin, R., Chipuer, H. M., & Neiderhiser, J. M. (1994). Behavioral genetic evidence for the importance of nonshared environment. In E. M. Hetherington, D. Reiss, & R. Plomin (Eds.), *Separate social worlds of siblings: Importance of nonshared environment on development.* Hillsdale, NJ: Lawrence Erlbaum.

Plummer, C. A. (1993). Prevention is appropriate, prevention is successful. In R. J. Gelles & D. R. Loseke (Eds.), *Current controversies on family violence* (pp. 288–305). Newbury Park, CA: Sage.

Polansky, N. A. (1979). Help for the help-less. *Smith College Studies in Social Work, 49,* 169–191.

Polansky, N. A., Chalmers, M. A., Butenwieser, E., & Williams, D. P. (1981). *Damaged parents: An anatomy of child neglect.* Chicago: University of Chicago Press.

Polansky, N. A., Gaudin, J. M., Ammons, P. W., & Davis, K. B. (1985). The psychological ecology of the neglectful mother. *Child Abuse and Neglect, 9,* 265–275.

Polisi, C. E. (2003). Universal rights and cultural relativism: Hinduism and Islam deconstructed. *BC Journal of International Affairs* [online version]. Retrieved December 2003, from http://www.jhubc.it/bcjournal/articles/polisi.cfm

Pollack, S. D., Cicchetti, D., Hornung, K., & Reed, A. (2000). Recognizing emotion in faces: Developmental effects of child abuse and neglect. *Developmental Psychology, 36,* 679–688.

Pope, H. G., Jr., Hudson, J. I., Bodkin, A., & Oliva, P. (1998). Questionable validity of dissociative amnesia in trauma victims. *British Journal of Psychiatry, 172,* 210–215.

Portwood, S. G. (1998). The impact of individuals' characteristics and experiences on their definitions of child maltreatment. *Child Abuse and Neglect, 22,* 437–452.

Price, J. H., Islam, R., Gruhler, J., Dove, L., Knowles, J., & Stults, G. (2001). Public perceptions of child abuse and neglect in a midwestern urban community. *Journal of Community Health, 26,* 271–284.

Prince, J. E., & Arias, I. (1994). The role of perceived control and the desirability of control among abusive and nonabusive husbands. *American Journal of Family Therapy, 22,* 126–134.

Ptacek, J. (1998). Why do men batter their wives? In R. K. Bergen (Ed.), *Issues in intimate violence* (pp. 181–195). Thousand Oaks, CA: Sage.

Radhakrishnan, S., & Moore, C. A. (1989). *A source book in Indian philosophy.* Princeton, NJ: Princeton University Press.

Rahim, H. (2000). Virtue, gender and the family: Reflections on religious texts in Islam and Hinduism. *Journal of Social Distress and the Homeless, 9,* 187–199.

Raine, A. (1993). *The psychopathology of crime: Criminal behavior as a clinical disorder.* San Diego: Academic Press.

Ramsey-Klawsnik, H. (1995). Investigating suspected elder maltreatment. *Journal of Elder Abuse and Neglect, 7,* 41–67.

Rand, M. R. (1997). *Violence-related injuries treated in hospital emergency departments.* U.S. Department of Justice, Bureau of Justice Statistics. Retrieved November 2003, from http://www.ojp.usdoj.gov/bjs/pub/ascii/vrithed.txt

Rank, M. R. (2001). The effect of poverty on America's families: Assessing our research knowledge. *Journal of Family Issues, 22,* 882–903.

Rausch, K., & Knutson, J. F. (1991). The self-report of personal punitive childhood experiences and those of siblings. *Child Abuse and Neglect, 15,* 29–36.

Ray, K. C., Jackson, J. L., & Townsley, R. M. (1991). Family environments of victims of intrafamilial and extrafamilial child sexual abuse. *Journal of Family Violence, 6,* 365–374.

Raymond, J. (2002). Building a statewide network of services for older abused women. *Nexus, 8,* 10–11.

Reid, J., Macchetto, P., & Foster, S. (1999). *No safe haven: Children of substance-abusing parents.* New York: National Center on Addiction and Substance Abuse at Columbia University.

Reid, J. B., Kavenaugh, K., & Baldwin, D. V. (1987). Abusive parents' perceptions of child problem behavior: An example of parental bias. *Journal of Abnormal Psychology, 15,* 457–466.

Reis, M., & Nahmiash, D. (1997). Abuse of seniors: Personality, stress, and other indicators. *Journal of Mental Health and Aging, 3,* 337–356.

Rennison, C. M. (2001). *Intimate partner violence and age of victim, 1993–99.* U.S. Department of Justice, Office of Justice Programs, Bureau of Justice Statistics, Special Report, NCJ 187635. Retrieved December 2002, from http://www.ojp. usdoj.gov/bjs/abstract/ipva99.htm

Rennison, C. M. (2002). *Rape and sexual assault: Reporting to police and medical attention, 1992–2000.* Retrieved February 12, 2003, from http://www .ojp.usdoj.gov/bjs/abstract/rsarp00.htm

Rennison, C. M. (2003). *Intimate partner violence, 1993–2001.* U.S. Department of Justice, Office of Justice Programs, Bureau of Justice Statistics Special Report, NCJ 197838. Retrieved December 2003, from http://www.ojp.usdoj.gov/bjs/ abstract/ipv01.htm

Rennison, C. M., & Welchans, S. (2000). *Intimate partner violence.* U.S. Department of Justice, Office of Justice Programs, Bureau of Justice Statistics Special Report. NCJ 178247. Retrieved February 2002, from http://www.ojp. usdoj.gov/bjs/abstract/ipv.htm

Renzetti, C. M. (1989). Building a second closet: Third party responses to victims of lesbian partner abuse. *Family Relations, 38,* 157–163.

Renzetti, C. M. (1992). *Violent betrayal: Partner abuse in lesbian relationships.* Newbury Park, CA: Sage.

Renzetti, C. M. (1998). Violence and abuse in lesbian relationships: Theoretical and empirical issues. In R. K. Berger (Ed.), *Issues in intimate violence* (pp. 117–127). Thousand Oaks, CA: Sage.

Reppucci, N. D., & Haugaard, J. J. (1993). Problems with child sexual abuse prevention programs. In R. J. Gelles & D. R. Loseke (Eds.), *Current controversies on family violence* (pp. 306–322). Newbury Park, CA: Sage.

Resnick, M. D., Bearman, P. S., Blum, R. W., Bauman, K. E., Harris, K. M., Jones, J., et al. (1997). Protecting adolescents from harm: Findings from the National Longitudinal Study on Adolescent Health. *Journal of the American Medical Association, 278,* 823–832.

Riak, J. (2002). *Where are the Christians who oppose corporal punishment? An invitation from PTAVE, December 2002.* Retrieved December 30, 2003, from http://www.nospank.net/cnp.htm

Richardson, G. A., Day, N. L., & McGauhey, P. J. (1993). The impact of prenatal marijuana and cocaine use on the infant and child. *Clinical Obstetrics and Gynecology, 36,* 302–318.

Rind, B., & Tromovitch, P. (1994). A meta-analytic review of findings from national samples on psychological correlates of child sexual abuse. *Journal of Sex Research, 34,* 237–255.

Rind, B., Tromovitch, P., & Bauserman, R. (1998). A meta-analytic examination of assumed properties of child sexual abuse using college samples. *Psychological Bulletin, 124,* 22–53.

Robins, R. W., Caspi, A., & Moffitt, T. E. (2002). It's not just who you're with, it's who you are: Personality and relationship experiences across multiple relationships. *Journal of Personality, 70,* 925–964.

Rogers, C. (1999). Personal story. In B. Leventhal & S. Lundy (Eds.), *Same-sex domestic violence: Strategies for change* (pp. 11–15). Thousand Oaks, CA: Sage.

Rohde, M. (2002). *Transgenderism—Frequently asked questions.* Retrieved October 2003, from http://prism.truman.edu/transgender.htm

Rollins, B. C., & Oheneba-Sakyi, Y. (1990). Physical violence in Utah households. *Journal of Family Violence, 5,* 301–309.

Rooney, J., & Hanson, R. K. (2001). Predicting attrition from treatment programs for abusive men. *Journal of Family Violence, 16,* 131–149.

Rorty, M., Yager, J., & Rossotto, E. (1994). Childhood sexual, physical, and psychological abuse in bulimia nervosa. *American Journal of Psychiatry, 151,* 1122–1126.

Rose, S. J. (1999). Reaching consensus on child neglect: African American mothers and child welfare workers. *Children and Youth Services Review, 21,* 463–479.

Rose, S., & Meezan, W. (1995). Child neglect: A study of the perceptions of mother and child welfare workers. *Children and Youth Services Review, 17,* 471–486.

Rosen, K. H., Matheson, J. L., & Smith, S. (2003). Negotiated time-out: A de-escalation tool for couples. *Journal of Marital and Family Therapy, 29,* 291–298.

Rosenbaum, A. (1991). The neuropsychology of marital aggression. In J. S. Milner (Ed.), *Neuropsychology of aggression* (pp. 167–180). Boston: Kluwer.

Rosenbaum, A., & Hodge, S. K. (1989). Head injury and marital aggression. *American Journal of Psychiatry, 146,* 1048–1051.

Rosenbaum, A., & O'Leary, K. D. (1981). Marital violence: Characteristics of abusive couples. *Journal of Consulting and Clinical Psychology, 49,* 63–71.

Rosenberg, M. S., Giberson, R. S., Rossman, B. B., & Acker, M. (2000). The child witness of family violence. In R. T. Ammerman & M. Hersen (Eds.), *Case studies in family violence* (2nd ed.) (pp. 259–292). New York: Kluwer.

Rosenthal, A. J. (1999). *Domestic violence statistics.* Retrieved December 5, 2003, from http://www.domesticabuseaware.org

Ross, L. F., & Aspinwall, T. J. (1997). Religious exemptions to the immunization statutes: Balancing public health and religious freedom. *Journal of Law, Medicine & Ethics, 25,* 202–209.

Rouse, L. P. (1990). The dominance motive in abusive partners: Identifying couples at risk. *Journal of College Student Development, 31,* 330–335.

Rudd, J. M., & Herzberger, S. D. (1999). Brother-sister incest/father-daughter incest: A comparison of characteristics and consequences. *Child Abuse and Neglect, 23*, 915–928.

Ruggiero, K. J., McLeer, S. V., & Dixon, J. F. (2000). Sexual abuse characteristics associated with survivor psychopathology. *Child Abuse and Neglect, 24*, 951–964.

Russ, B. (2003). *First report on the governor's commission on child protection: Strengthening child protection services in New Hampshire.* State of New Hampshire.

Russell, D. E. (1982, March). *Rape, child sexual abuse, sexual harassment in the workplace: An analysis of the prevalence, causes and recommended solutions.* Final report for the California Commission on Crime Control and Violence Prevention.

Russell, D. E. (1986). *The secret trauma: Incest in the lives of girls and women.* New York: Basic Books.

Russell, D. E. (1990). *Rape in marriage.* Indianapolis: Indiana University Press.

Russell, D. E. (1995). The prevalence, trauma, and sociocultural causes of incestuous abuse of females: A human rights issue. In R. J. Kleber & C. R. Figley (Eds.), *Beyond trauma: Cultural and societal dynamics. Plenum series on stress and coping* (pp. 171–186). New York: Plenum.

Russell, D. H. (1984). A study of juvenile murderers of family members. *International Journal of Offender Therapy and Comparative Criminology, 28*, 177–192.

Russo, A. (1999). Lesbians organizing lesbians against battering. In B. Leventhal & S. Lundy (Eds.), *Same-sex domestic violence: Strategies for change* (pp. 83–96). Thousand Oaks, CA: Sage.

Rutter, M. (1985). Resilience in the face of adversity: Protective factors and resistance to psychiatric disorder. *British Journal of Psychiatry, 147*, 598–611.

Sackett, L. A., & Saunders, D. G. (2001). The impact of different forms of psychological abuse on battered women. In K. D. O'Leary & R. D. Maiuro (Eds.), *Psychological abuse in violent domestic relations* (pp. 197–212). New York: Springer.

Sagatun-Edwards, I., & Saylor, C. (2000). Drug-exposed infant cases in juvenile court: Risk factors and court outcomes. *Child Abuse and Neglect, 24*, 925–937.

Sagatun-Edwards, I., Saylor, C., & Shifflett, B. (1995). Drug-exposed infants in the social welfare system and the juvenile court. *Child Abuse and Neglect, 19*, 83–91.

Salzinger, S., Feldman, R. S., & Hammer, M. (1993). The effects of physical abuse on children's social relationships. *Child Development, 64*, 169–187.

Sarbo, A. (1985). Childhood molestations: Adult personality correlates and predictors of trauma (Doctoral dissertation, Georgia State University, 1984). *Dissertation Abstracts International, 45*, 3960.

Saunders, B. E., Kilpatrick, D. G., Hanson, R. F., Resnick, H. S., & Walker, M. E. (1999). Prevalence, case characteristics, and long-term psychological correlates of child rape among women: A national survey. *Child Maltreatment, 4*, 187–200.

Saunders, D. G. (1994). Post-traumatic stress symptom profiles of battered women: A comparison of survivors in two settings. *Violence and Victims, 9,* 31–44.

Saunders, D. G. (1986). When battered women use violence: Husband-abuse or self-defense? *Victims and Violence, 1,* 47–60.

Saunders, D. G. (1992). A typology of men who batter: Three types derived from cluster analysis. *American Journal of Orthopsychiatry, 62,* 264–275.

Saunders, D. G., & Browne, A. (2000). Intimate partner homicide. In R. T. Ammerman & M. Hersen (Eds.), *Case studies in family violence* (2nd ed.) (pp. 415–449). New York: Kluwer.

Saxton, M., Curry, M., Powers, L. E., Maley, S., Eckels, K., & Gross, J. (2001). Bring my scooter so I can leave you: A study of disabled women handling abuse by personal assistance providers. *Violence Against Women, 7,* 393–417.

Scarr, S., & McCartney, K. (1983). How people make their own environments: A theory of genotype-environment effects. *Child Development, 54,* 424–435.

Scerbo, A., & Raine, A. (1992). *Neurotransmitters and antisocial behavior: A meta-analysis.* Unpublished manuscript, University of Southern California.

Schiamberg, L. B., & Gans, D. (2000). Elder abuse by adult children: An applied ecological framework for understanding contextual risk factors and the inter-generational character of quality of life. *International Journal of Aging and Human Development, 50,* 329–335.

Schiff, M., & Cavaiola, A. A. (1993). Child abuse, adolescent substance abuse, and "deadly violence." *Journal of Adolescent Chemical Dependency, 2,* 131–141.

Schilit, R., Lie, G., & Montagne, M. (1990). Substance use as a correlate of violence in intimate lesbian relationships. *Journal of Homosexuality, 19,* 51–65.

Schmidt, M. (1995). Anglo Americans and sexual child abuse. In L. A. Fontes (Ed.), *Sexual abuse in nine North American cultures: Treatment and prevention* (pp. 156–175). Thousand Oaks, CA: Sage.

Schutter, L. S., & Brinker, R. P. (1992). Conjuring a new category of disability from prenatal cocaine exposure: Are the infants unique biological or caretaking casu-alties? *Topics in Early Childhood Special Education, 11,* 84–111.

Scoggin, F., Beall, C., Bynum, J., Stephens, G., Grote, N. P., Baumhover, L. A., & Bolland, J. M. (1989). Training for abusive caregivers: An unconventional approach to an intervention dilemma. *Journal of Elder Abuse and Neglect, 1,* 73–86.

Sedlak, A. J., & Broadhurst, D. D. (1996). *Executive summary of the Third National Incidence Study of child abuse and neglect.* Retrieved January 14, 2003, from http://www.calib.com/nccanch/pubs/statinfo/nis3.cfm#national

Sedney, M., & Brooks, B. (1984). Factors associated with a history of childhood sexual experience in a nonclinical female population. *Journal of the American Academy of Child Psychiatry, 23,* 215–218.

Sen, A. (1998). Universal truths: Human rights and the westernizing illusion. *Harvard International Review, 20,* 40–43.

Shapiro, J. P., Leifer, M., Martone, M. W., & Kassem, L. (1992). Cognitive func-tioning and social competence as predictors of maladjustment in sexually abused girls. *Journal of Interpersonal Violence, 7,* 156–164.

Sheldon, J. P., & Parent, S. L. (2002). Clergy's attitudes and attributions of blame toward female rape victims. *Violence Against Women, 8,* 233–256.

Sherman, L., & Berk, R. (1984). The specific deterrent effects of arrest for domestic assault. *American Sociological Review, 49,* 261–272.

Sherman, L. W., Schmidt, J. D., Rogan, D. P., Smith, D. A., Gartin, P. R., Cohn, E. G., Collins, D., & Bacich, R. (1992). The variable effects of arrest on criminal careers: The Milwaukee domestic violence experiment. *Journal of Criminal Law and Criminology, 83,* 137–169.

Shields, N. M., & Hanneke, C. R. (1983). Battered wives' reactions to marital rape. In D. Finkelhor, R.J. Gelles, G.T. Hotaling, & M.A. Straus (Eds.), *The dark side of families* (pp. 132–148). Beverly Hills, CA: Sage.

Shope, J. H. (2004). When words are not enough: The search for the effect of pornography on abused women. *Violence Against Women, 10,* 56–72.

Silence Whispers. (2003). Retrieved September 11, 2003, from http://siblingabuse.coolfreepage.com/SAlinkgen.htm

Silverman, A. B., Reinherz, H. Z., & Giaconia, R. M. (1996). The long-term sequelae of child and adolescent abuse: A longitudinal community study. *Child Abuse and Neglect, 8,* 709–723.

Silverstein, E. (2001, June 12). Committee approves report on domestic violence: Document makes recommendations for combating abuse. *Presbyterian Church (U.S.A.) News.* Retrieved May 2003, from http://www.pcusa.org/ga213/news/ga01065.htm

Simonelli, C. J., & Ingram, K. M. (1998). Psychological distress among men experiencing physical and emotional abuse in heterosexual dating relationships. *Journal of Interpersonal Violence, 13,* 667–681.

Simonelli, C .J., Mullis, T., Elliott, A. N., & Pierce, T. W. (2002). Abuse by siblings and subsequent experiences of violence within the dating relationship. *Journal of Interpersonal Violence, 17,* 103–121.

Simons, D., Wurtele, S. K., & Heil, P. (2002). Childhood victimization and lack of empathy as predictors of sexual offending against women and children. *Journal of Interpersonal Violence, 17,* 1291–1307.

Simons, R. L., Whitbeck, L. B., Conger, R. D., & Melby, J. N. (1991). The effect of social skills, values, peers, and depression on adolescent substance use. *Journal of Early Adolescence, 11,* 466–481.

Simons, R. L., Whitbeck, L. B., Conger, R. D., & Wu, C. (1991). Intergenerational transmission of harsh parenting. *Developmental Psychology, 27,* 159–171.

Singer, L. T., Arendt, R., Minnes, S., Farkas, K., Salvator, A., Kirchner, H. L., & Kleigman, R. (2002). Cognitive and motor outcomes of cocaine-exposed infants. *Journal of the American Medical Association, 287,* 1952–1960.

Sirles, E., & Franke, P. J. (1989). Factors influencing mothers' reactions to intrafamilial sexual abuse. *Child Abuse and Neglect, 13,* 131–139.

Smeeding, T. A., Rainwater, L., & Burtless, G. (2000). *United States poverty in a cross-national context* (Luxembourg Income Study Working Paper Series

No. 244). Syracuse, NY: Syracuse University, Maxwell School of Citizenship and Public Affairs.

Smetana, J. G., Toth, S. L., Cicchetti, D., Bruce, J., Kane, P., & Daddis, C. (1999). Maltreated and nonmaltreated preschoolers' conceptions of hypothetical and actual moral transgressions. *Developmental Psychology, 35,* 269–281.

Smith, H., & Israel, E. (1987). Sibling incest: A study of the dynamics of 25 cases. *Child Abuse and Neglect, 11,* 101–108.

Smith, M. D., & Bennett, N. (1985). Poverty, inequality, and theories of forcible rape. *Crime and Delinquency, 31,* 295–305.

Smith, P. H., Moracco, K. E., & Butts, H. D. (1998). Partner homicide in context. *Homicide Studies, 2,* 400–421.

Smith, R., & Loring, M. T. (1994). The trauma of emotionally abused men. *Psychology: A Journal of Human Behavior, 31,* 1–4.

Smith, S. K. (1997). Women's experiences of victimizing sexualization, Part II: Community and longer term personal impacts. *Issues in Mental Health Nursing, 18,* 417–432.

Smith, T. W. (1987). *Classifying Protestant denominations.* GSS Methodological Report No. 43. Retrieved November 2003, from http://www.icpsr.umich.edu/GSS99/report/m-report/meth43.htm

Smith, T. W., & Martos, L. (1999, December). *Attitudes towards and experience with guns: A state-level perspective.* National Opinion Research Center at the University of Chicago. Retrieved February 2004, from http://www.vpc.org/graphics/norcpoll.pdf

Sobsey, D., Randall, W., & Parilla, R. K. (1997). Gender differences in abused children with and without disabilities. *Child Abuse and Neglect, 21,* 707–720.

Soler, H., Vinayak, P., & Quadagno, D. (2000). Biosocial aspects of domestic violence. *Psychoneuroendocrinology, 25,* 721–739.

Somer, E., & Szwarcberg, S. (2001). Variables in delayed disclosure of child sexual abuse. *American Journal of Orthopsychiatry, 71,* 332–341.

Sommer, R., Barnes, G. E., & Murray, R. P. (1992). Alcohol consumption, alcohol abuse, personality and female perpetrated spouse abuse. *Personality and Individual Differences, 13,* 1315–1323.

Spaccarelli, S. (1994). Stress, appraisal, and coping in child sexual abuse: A theoretical and empirical review. *Psychological Bulletin, 116,* 1–23.

Stacey, W. A., Hazlewood, L. R., & Shupe, A. (1994). *The violent couple.* Westport, CT: Praeger Publishers.

Stahl, C., & Fritz, N. (2002). Internet safety: Adolescents' self-report. *Journal of Adolescent Health, 31,* 7–10.

Star, B. (1978). Comparing battered and non-battered women. *Victimology, 3,* 32–44.

Stark, E., & Flitcraft, A. (1988). Violence among intimates: An epidemiological review. In V. B. Van Hasselt, R. L. Morrison, A. S. Bellack, & M. Hersen (Eds.), *Handbook of family violence* (pp. 293–317). New York: Plenum.

Stark, E., Flitcraft, A., & Frazier, W. (1981). *Wife abuse in the medical setting: An introduction for health personnel* (Domestic Violence Monograph Series

No. 7). Washington, D.C.: U.S. Department of Health and Human Services, National Clearinghouse on Domestic Abuse and Neglect.

Stark, R., & McEvoy, J. (1970). Middle class violence. *Psychology Today, 4,* 52–65.

Stein, J. A., Golding, J. M., Siegel, J. M., Burnam, M. A., & Sorenson, S. B. (1988). Long-term psychological sequelae of child sexual abuse: The Los Angeles epidemiologic catchment area study. In G. E. Wyatt & G. J. Powell (Eds.), *Lasting effects of child sexual abuse* (pp. 135–154). Newbury Park, CA: Sage.

Steinmetz, S. (1977). *The cycle of violence: Aggressive and abusive family interaction.* New York: Praeger.

Steinmetz, S. K. (1977). Wifebeating, husbandbeating—A comparison of the use of physical violence between spouses to resolve marital fights. In M. Roy (Ed.), *Battered women: A psychosociological study of domestic violence* (pp. 63–72). New York: Van Nostrand Reinhold Co., Inc.

Steinmetz, S. K. (1978). Battered parents. *Society, 15,* 54–55.

Steinmetz, S. K. (1988). *Duty bound: Elder abuse and family care.* Newbury Park, CA: Sage.

Steinmetz, S. K. (1993). The abused elderly are dependent: Abuse is caused by the perception of stress associated with providing care. In R. J. Gelles & D. R. Loseke (Eds.), *Current controversies on family violence* (pp. 222–236). Newbury Park, CA: Sage.

Stets, J. E. (1990). Verbal and physical aggression in marriage. *Journal of Marriage and the Family, 52,* 501–514.

Stets, J. E., & Straus, M. A. (1990a). Gender differences in reporting marital violence and its medical and psychological consequences. In M. A. Straus & R. J. Gelles (Eds.), *Physical violence in American families: Risk factors and adaptations to violence in 8,145 families* (pp. 151–166). New Brunswick, NJ: Transaction.

Stets, J. E., & Straus, M. A. (1990b). The marriage license as a hitting license: A comparison of assaults in dating, cohabiting and married couples. In M. A. Straus & R. J. Gelles (Eds.), *Physical violence in American families: Risk factors and adaptations to violence in 8,145 families* (pp. 227–244). New Brunswick, NJ: Transaction.

Stewart, A. L., & Maddren, K. (1997). Police officers' judgment of blame in family violence: The impact of gender and alcohol. *Sex Roles, 37,* 921–934.

Stith, S. M., Crossman, R. K., & Bischof, G. P. (1991). Alcoholism and marital violence: A comparative study of men in alcohol treatment programs and batterer treatment programs. *Alcoholism Treatment Quarterly, 8,* 3–20.

Stop Abuse For Everyone (2003). Retrieved September 11, 2003, from http://www.safe4all.org

Stotland, N. L. (2000). Tug-of-war: Domestic abuse and the misuse of religion. *American Journal of Psychiatry, 157,* 696–702.

Stout, K. D. (1989). Intimate femicide: Effects of legislation and social services. *Affilia, 4,* 21–27.

Straus, M. A. (1973). A general systems theory approach to a theory of violence between family members. *Social Science Information, 12,* 105–125.

Straus, M. A. (1980). Wife-beating: How common and why? In M. A. Straus & G. T. Hotaling (Eds.), *The social causes of husband-wife violence*. Minneapolis: University of Minnesota Press.

Straus, M. A. (1990a). Injury and frequency of assault and the 'representative sample fallacy' in measuring wife beating and child abuse. In M. A. Straus & R. J. Gelles (Eds.), *Physical violence in American families: Risk factors and adaptations to violence in 8,145 families* (pp. 75–89). New Brunswick, NJ: Transaction.

Straus, M. A. (1990b). Ordinary violence, child abuse, and wife beating: What do they have in common? In M. A. Straus & R. J. Gelles (Eds.), *Physical violence in American families: Risk factors and adaptations to violence in 8,145 families* (pp. 403–424). New Brunswick, NJ: Transaction.

Straus, M. A. (1990c). Social stress and marital violence in a national sample of American families. In M. A. Straus & R. J. Gelles (Eds.), *Physical violence in American families: Risk factors and adaptations to violence in 8,145 families* (pp. 181–202). New Brunswick, NJ: Transaction.

Straus, M. A. (1990d). Measuring intrafamily conflict and violence: The Conflict Tactics (CT) Scales. In M. A. Straus & R. J. Gelles (Eds.), *Physical violence in American families: Risk factors and adaptations to violence in 8,145 families* (pp. 29–47). New Brunswick, NJ: Transaction.

Straus, M. A. (1994). *Beating the devil out of them: Corporal punishment in American families*. New York: Lexington Books.

Straus, M. A. (1996). Spanking and the making of a violent society. *Pediatrics, 98*, 837–842.

Straus, M. A. (2001). *Beating the devil out of them: Corporal punishment in American families and its effects on children*. New Brunswick, NJ: Transaction.

Straus, M. A. (2004). Women's violence toward men is a serious social problem. In D. R. Loseke, R. J. Gelles, & M. M. Cavanaugh (Eds.), *Current controversies on family violence* (2nd ed.) (pp. 55–77). Thousand Oaks, CA: Sage.

Straus, M. A., & Gelles, R. J. (1986). Societal change and change in family violence from 1975 to 1985 as revealed by two national surveys. *Journal of Marriage and the Family, 48,* 465–479.

Straus, M. A., & Gelles, R. J. (1988). How violent are American families? Estimates from the national family violence resurvey and other studies. In G. T. Hotaling, D. Finkelhor, J. T. Kirkpatrick, & M. A. Straus (Eds.), *Family abuse and its consequences: New directions in research* (pp. 14–36). Beverly Hills, CA: Sage.

Straus, M. A., & Gelles, R. J. (1990a). How violent are American families? Estimates from the National Family Violence Resurvey and other studies. In M. A. Straus & R. J. Gelles (Eds.), *Physical violence in American families: Risk factors and adaptations to violence in 8,145 families* (pp. 95–112). New Brunswick, NJ: Transaction.

Straus, M. A., & Gelles, R. J. (1990b). Societal change and change in family violence from 1975 to 1985 as revealed by two national surveys. In M. A. Straus & R. J. Gelles (Eds.), *Physical violence in American families: Risk factors and adaptations to violence in 8,145 families* (pp. 113–132). New Brunswick, NJ: Transaction.

Straus, M. A., Gelles, R. J., & Steinmetz, S. (1980). *Behind closed doors: Violence in the American family.* Garden City, NJ: Anchor.

Straus, M. A., & Hotaling, G. T. (Eds.). (1980). *The social causes of husband-wife violence.* Minneapolis: University of Minnesota Press.

Straus, M. A., Kaufman Kantor, G., & Moore, D. W. (1997). Change in cultural norms approving marital violence from 1968 to 1994. In G. Kaufman Kantor & J. L. Jasinski (Eds). *Out of darkness: Contemporary perspectives on family violence* (pp. 3–16). Thousand Oaks, CA: Sage.

Straus, M. A., & Runyan, D. K. (1997). Physical abuse. In S. B. Friedman, M. M. Fisher, S. K. Schonberg, & E. M. Alderman (Eds), *Comprehensive adolescent health care* (2nd ed.). St. Louis: Mosby-Year Book.

Straus, M. A., & Smith, C. (1990). Family patterns and child abuse. In M. A. Straus & R. J. Gelles (Eds.), *Physical violence in American families: Risk factors and adaptations to violence in 8,145 families* (pp. 245–262). New Brunswick, NJ: Transaction.

Straus, M. A., & Stewart, J. H. (1999). Corporal punishment by American parents: National data on prevalence, chronicity, severity, and duration in relation to child and family characteristics. *Clinical Child and Family Psychology Review, 2,* 55–70.

Straus, M. A., & Sweet, S. (1992). Verbal/symbolic aggression in couples: Incidence rates and relationships to personal characteristics. *Journal of Marriage and the Family, 54,* 346–357.

Straus, M. A., & Yodanis, C. L. (1996). Corporal punishment in adolescence and physical assaults on spouses in later life: What accounts for the link? *Journal of Marriage and the Family, 58,* 825–841.

Strawbridge, W. J., Shema, S. J., Cohen, R. C., & Kaplan, G. A. (2001). Religious attendance increases survival by improving and maintaining good health behaviors, mental health, and social relationships. *Society of Behavioral Medicine, 23,* 68–74.

Streib, V. L. (2003). *The juvenile death penalty today: Death sentences and executions for juvenile crimes, January 1, 1973–June 30, 2003.* Retrieved March 2004, from http://www.law.onu.edu/faculty/streib/documents/JuvDeathJune302004NewTables.pdf

Streissguth, A. P., Barr, H. M., & Olson, H. C. (1994a). Drinking during pregnancy decreases word attack and arithmetic scores on standardized tests: Adolescent data from a population-based prospective study. *Alcoholism: Clinical and Experimental Research, 18,* 248–255.

Streissguth, A. P., Sampson, P. D., & Olson, H. C. (1994b). Maternal drinking during pregnancy: Attention and short term memory in 14 year old offspring: A longitudinal prospective study. *Alcoholism: Clinical and Experimental Research, 18,* 202–218.

Streissguth, A. P., Barr, H. M., & Sampson, P. D. (1990). Moderate prenatal alcohol exposure: Effects on child IQ and learning problems at age 7.5 years. *Alcoholism: Clinical and Experimental Research, 14,* 662–669.

Strube, M. J. (1988). The decision to leave an abusive relationship: Empirical evidence and theoretical issues. *Psychological Bulletin, 104,* 236–250.

Struckman-Johnson, C. (1988). Forced sex on dates: It happens to men, too. *Journal of Sex Research, 24,* 234–241.

Struckman-Johnson, C., & Struckman-Johnson, D. (1998). The dynamics and impact of sexual coercion of men by women. In P. B. Anderson & C. Struckman-Johnson (Eds.), *Sexually aggressive women: Current perspectives and controversies* (pp. 121–143). New York: Guilford Press.

Sugarman, D. B., & Frankel, S. L. (1996). Patriarchal ideology and wife-assault: A meta-analytic review. *Journal of Family Violence, 11,* 13–40.

Suitor, J. J., Pillemer, K., & Straus, M. A. (1990). Marital violence in a life course perspective. In M. A. Straus & R. J. Gelles (Eds.), *Physical violence in American families: Risk factors and adaptations to violence in 8,145 families* (pp. 305–320). New Brunswick, NJ: Transaction.

Sullivan, C. M., Tan, C., Basta, J., Rumptz, M., & Davidson, W. S. (1992). An advocacy intervention program for women with abusive partners: Initial evaluation. *American Journal of Community Psychology, 20,* 309–332.

Sullivan, P. M., & Knutson, J. F. (2000a). Maltreatment and disabilities: A population-based epidemiological study. *Child Abuse and Neglect, 24,* 1257–1273.

Sullivan, P. M., & Knutson, J. F. (2000b). The prevalence of disabilities and maltreatment among runaway children. *Child Abuse and Neglect, 24,* 1275–1288.

Sutherland, K. (1996). Spousal abuse: Does anyone care? *Journal of Christian Nursing, 13,* 17–18.

Swan, S. C., & Snow, D. L. (2002). A typology of women's use of violence in intimate relationships. *Violence Against Women, 8,* 286–319.

Swan, S. C., & Snow, D. L. (2003). Behavioral and psychological differences among abused women who use violence in intimate relationships. *Violence Against Women, 9,* 75–109.

Szymanski, D. M., & Chung, Y. B. (2001). The Lesbian Internalized Homophobia Scale: A rational/theoretical approach. *Journal of Homosexuality, 41,* 37–52.

Szymanski, D. M., & Chung, Y. B. (2003). Internalized homophobia in lesbians. *Journal of Lesbian Studies, 7,* 115–125.

Szymanski, D. M., Chung, Y. B., & Balsam, K. F. (2001). Psychosocial correlates of internalized homophobia in lesbians. *Measurement and Evaluation in Counseling and Development, 34,* 27–38.

Tajima, E. A. (2000). The relative importance of wife abuse as a risk factor for violence against children. *Child Abuse and Neglect, 24,* 1383–1398.

Taylor, C., & Fontes, L. A. (1995). Seventh Day Adventists and sexual child abuse. In L. A. Fontes (Ed.), *Sexual abuse in nine North American cultures: Treatment and prevention* (pp. 176–199). Thousand Oaks, CA: Sage.

Testa, M., & Leonard, K. E. (2001). The impact of marital aggression on women's psychological and marital functioning in a newlywed sample. *Journal of Family Violence, 16,* 115–130.

Thomas, A., & Chess, S. (1977). *Temperament and development.* New York: Bruner-Mazel.

Thompson-Haas, L. (1987). *Marital rape: Methods of helping and healing.* Austin, TX: Austin Rape Crisis Center.

Thurman v. City of Torrington (1984). United States District Court. D. Connecticut, October 23, 1984, 595 F. Supp. 1521. Retrieved January 2004 from http://cyber .law.harvard.edu/vaw00/thurman.html

Tjaden, P., & Thoennes, N. (2000). *Extent, nature, and consequences of intimate partner violence: Findings from the National Violence Against Women Survey.* Retrieved September 9, 2003, from http://www.ncjrs.org/pdffiles1/nij/181867.pdf

Tolman, R. M. (1989). The development of a measure of psychological maltreatment of women by their male partners. *Violence and Victims, 4,* 159–178.

Torres, S. (1991). A comparison of wife abuse between two cultures: Perceptions, attitudes, nature, and extent. *Issues in Mental Health Nursing, 12,* 113–131.

Toth, S. L., Cicchetti, D., Macfie, J., & Emde, R. N. (1997). Representations of self and other in narratives of neglected, physically abused, and sexually abused preschoolers. *Development and Psychopathology, 9,* 781–796.

Tuel, B. D., & Russell, R. K. (1998). Self-esteem and depression in battered women: A comparison of lesbian and heterosexual survivors. *Violence Against Women, 4,* 344–362.

Turell, S. C. (2000). A descriptive analysis of same-sex relationship violence for a diverse sample. *Journal of Family Violence, 15,* 281–293.

Tyiska, C. G. (2001). *Working with victims of crime with disabilities.* Retrieved May 4, 2004, from http://www.vaw.umn.edu/documents/ovcdisable/ovcdisable.html

Ullman, S. E., & Siegel, J. M. (1993). Victim-offender relationship and sexual assault. *Violence and Victims, 8,* 121–134.

Ulman, A., & Straus, M. A. (2003). Violence by children against mothers in relation to violence between parents and corporal punishment by parents. *Journal of Comparative Family Studies, 34,* 41–60.

Underwood, R. C., & Patch, P. C. (1999). Siblicide: A descriptive analysis of sibling homicide. *Homicide Studies, 3,* 333–348.

UNICEF (2003). A league table of child maltreatment deaths in rich nations. *Innocenti Report Cards, 5.*

U.S. Census Bureau (1997). *Americans with disabilities: Household economic studies.* Retrieved August 2003, from http://www.census.gov/prod/2001pubs/p70–73.pdf

U.S. Census Bureau (2000). *Statistical abstract of the United States.* Retrieved October 2003, from http://www.census.gov/prod/2001pubs/statab/sec01.pdf

U.S. Department of Health and Human Services (DHHS) (2001). *12 Years of Reporting Child Maltreatment, 2001.* Administration for Children and Families, Administration on Children Youth and Families, Children's Bureau.

Usborne, D. (2003). *Starvation horror story sparks review of foster and adoption system.* Retrieved November 17, 2003, from http://www.cyc-net.org/features/ ft-horrorstory.html

Van Hook, J. (2003, December 1). *Poverty grows among children of immigrants in U.S.* Migration Information Source. Retrieved March 2004, from http://www .migrationinformation.org/USfocus/display.cfm?ID=188

Van Wyk, J. A., Benson, M. L., Fox, G. L., & DeMaris, A. (2003). Detangling individual-, partner- and community-level correlates of partner violence. *Crime & Delinquency, 49,* 412–438.

Vest, J. R., Catlin, T. K., Chen, J. J., & Brownson, R. C. (2002). Multistate analysis of factors associated with intimate partner violence. *American Journal of Preventive Medicine, 22,* 156–164.

Vinton, L. (1991). Factors associated with refusing services among maltreated elderly. *Journal of Elder Abuse and Neglect, 3,* 89–103.

Vinton, L. (2002). *Questions and answers about older abused women.* Retrieved December 1, 2003, from http://ssw.fsu.edu/qaolderwomen/qaolderwomen.pdf

Violence Policy Center. (2000). *Facts on firearms and domestic violence.* Retrieved January 2004, from http://www.vpc.org/fact_sht/domviofs.htm

Vissing, Y. M., Straus, M. A., Gelles, R. J., & Harrop, J. W. (1991). Verbal aggression by parents and psychosocial problems of children. *Child Abuse and Neglect, 15,* 223–238.

Vivian, D., & Langhinrichsen-Rohling, J. (1994). Are bi-directionally violent couples mutually victimized? A gender sensitive comparison. *Violence and Victims, 9,* 107–124.

Vogeltanz, N. D., Wilsnack, S. C., Harris, T. R., Wilsnack, R. W., Wonderlich, S. A., & Kristjanson, A. F. (1999). Prevalence and risk factors for childhood sexual abuse in women: National survey findings. *Child Abuse and Neglect, 23,* 579–592.

Wald, R. L., & Knutson, J. F. (2000). Childhood disciplinary experiences reported by adults with craniofacial anomalies. *Child Abuse and Neglect, 24,* 1623–1627.

Waldner-Haugrud, L. K., & Gratch, L. V. (1997). Sexual coercion in gay/lesbian relationships: Descriptives and gender differences. *Violence and Victims, 12,* 87–98.

Walker, E., Downey, G., & Bergman, A. (1989). The effects of parental psychopathology and maltreatment on child behavior: A test of the diathesis-stress model. *Child Development, 60,* 15–24.

Walker, L. E. A. (1979). *The battered woman.* New York: Harper & Row.

Walker, L. E. A. (1990). Psychology and domestic violence around the world. *American Psychologist, 54,* 21–29.

Walker, L. E. A. (1993). The battered woman syndrome is a psychological consequence of abuse. In R. J. Gelles & D. R. Loseke (Eds.), *Current controversies on family violence* (pp. 133–153). Newbury Park, CA: Sage.

Walker, L. E. A. (2000). *The battered woman syndrome* (2nd ed.). New York: Springer.

Walker, N. E., Brooks, C. M., & Wrightsman, L. S. (1999). *Children's rights in the United States: In search of a national policy.* Thousand Oaks, CA: Sage.

Wallach, V. A., & Lister, L. (1995). Stages in the delivery of home-based services to parents at risk of child abuse: A Healthy Start experience. *Scholarly Inquiry for Nursing Practice: An International Journal, 9,* 159–173.

Walters, A. S., Barrett, R. P., Knapp, L. G., & Borden, M. C. (1995). Suicidal behavior in children and adolescents with mental retardation. *Research in Developmental Disabilities, 16,* 85–96.

Warner, J. D., Malinosky-Rummell, R., Ellis, J. T., & Hansen, D. J. (1990, November). *An examination of demographic and treatment variables associated with session attendance of maltreating families.* Paper presented at the annual conference of the Association for Advancement of Behavior Therapy, San Francisco.

Waterman, C. K., Dawson, L. J., & Bologna, M. J. (1989). Sexual coercion in gay male and lesbian relationships: Predictors and implications for support services. *Journal of Sex Research, 26,* 118–124.

Wauchope, B. A., & Straus, M. A. (1990). Physical punishment and physical abuse of American children: Incidence rates by age, gender, and occupational class. In M. A. Straus & R. J. Gelles (Eds.), *Physical violence in American families: Risk factors and adaptations to violence in 8,145 families* (pp. 133–150). New Brunswick, NJ: Transaction.

Wealin, J. M., Davies, S., Shaffer, A. E., Jackson, J. L., & Love, L. C. (2002). Family context and childhood adjustment associated with intrafamilial unwanted sexual attention. *Journal of Family Violence, 17,* 151–165.

Wekerle, C., & Wolfe, D. A. (1993). Prevention of child physical abuse and neglect: Promising new directions. *Clinical Psychology Review, 13,* 501–540.

Wellesley Center for Women. (1998). *The wife rape information page.* Retrieved May 20, 2004, from http://www.wcwonline.org/partnerviolence/mrape.html

Whipple, E. E., & Webster-Stratton, C. (1991). The role of parental stress in physically abusive families. *Child Abuse and Neglect, 15,* 279–291.

Whiteford, L. M., & Vitucci, J. (1997). Pregnancy and addiction: Translating research into practice. *Social Science and Medicine, 44,* 1371–1380.

Whittaker, J., Kinney, J., Tracy, E. M., Booth, C. (1990). *Reaching high-risk families: Intensive family preservation in human services.* New York: Aldine.

Widom, C. S. (1989). Does violence beget violence? A critical examination of the literature. *Psychological Bulletin, 106,* 3–28.

Widom, C. S., & Kuhns, J. B. (1996). Childhood victimization and subsequent risk for promiscuity, prostitution, and teenage pregnancy: A prospective study. *American Journal of Public Health, 86,* 1607–1612.

Wiehe, V. R. (1990). Religious influence on parental attitudes toward the use of corporal punishment. *Journal of Family Violence, 5,* 173–187.

Wiehe, V. R. (1997). *Sibling abuse: Hidden physical, emotional, and sexual trauma* (2nd ed.). Thousand Oaks, CA: Sage.

Wilcox, W. B. (1998). Conservative Protestant parenting: Authoritarian or authoritative? *American Sociological Review, 63,* 796–809.

Wille, R., & Beier, K. M. (1989). Castration in Germany. *Annals of Sex Research, 2,* 103–134.

Willett, J. B., Ayoub, C. C., & Robinson, D. (1991). Using growth modeling to examine systematic differences in growth: An example of change in the functioning of families at risk of maladaptive parenting, child abuse, or neglect. *Journal of Consulting and Clinical Psychology, 59,* 38–47.

Williams, L. M. (1994). Recall of childhood trauma: A prospective study of women's memories of child sexual abuse. *Journal of Consulting and Clinical Psychology, 62,* 1167–1176.

Williams, L. M. (1995). Recovered memories of abuse in women with documented child sexual victimization histories. *Journal of Traumatic Stress, 8,* 649–674.

Williams, O. J., & Griffin, L. W. (1996). Elderly maltreatment and cultural diversity: When laws are not enough. *Journal of Multicultural Social Work, 4,* 1–13.

Wolak, J., & Finkelhor, D. (1998). Children exposed to partner violence. In J. L. Jasinski & L. M. Williams (Eds.), *Partner violence: A comprehensive review of 20 years of research* (pp. 73–112). Thousand Oaks, CA: Sage.

Wolak, J., Finkelhor, D., & Mitchell, K. (2003). *Victims of Internet-initiated sex crimes: Dynamics, risk factors, and implications for prevention.* Manuscript submitted for publication.

Wolf, R. S., Strugnell, C. P., & Godkin, M. A. (1982). *Preliminary findings from three model projects on elder abuse.* Worcester, MA: University of Massachusetts Medical Center, University Center on Aging.

Wolfe, D., Fairbank, J. A., Kelly, J. A., & Bradlyn, A. S. (1983). Child abusive parents' physiological responses to stressful and non-stressful behavior in children. *Behavioral Assessment, 5,* 363–371.

Wood, W., Wong, F. Y., & Chachere, J. G. (1991). Effects of media violence on viewers' aggression in unconstrained social interaction. *Psychological Bulletin, 109,* 371–383.

Woody, G. E., Van Etten-Lee, M. L., McKirnan, D., Donnell, D., Metzger, D., Seage, G., & Gross, M. (2001). Substance use among men who have sex with men: Comparison with a national household survey. *Journal of Acquired Immune Deficiency Syndromes, 27,* 86–90.

World Health Organization. (1997). *Intimate partner violence.* Retrieved May 20, 2004, from http://www.who.int/violence_injury_prevention/violence/global_campaign/en/ipvfacts.pdf

Worling, J. (1995). Adolescent sibling-incest offenders: Differences in family and individual functioning when compared to adolescent nonsibling sex offenders. *Child Abuse and Neglect, 19,* 633–643.

Wozencraft, T., Wagner, W., & Pellegrin, A. (1991). Depression and suicidal ideation in sexually abused children. *Child Abuse and Neglect, 15,* 505–510.

Wright, P., Nobrega, J., Langevin, R., & Wortzman, G. (1990). Brain density and symmetry in pedophilic and sexually aggressive offenders. *Annals of Sex Research, 3,* 319–328.

Wright, L. M., Watson, W. L., & Bell, J. M. (1996). *The heart and healing in families and illness.* New York: Basic Books.

Xu, X., Tung, Y. Y., & Dunaway, R. G. (2000). Cultural, human and social capital as determinants of corporal punishment: Toward an integrated theoretical model. *Journal of Interpersonal Violence, 15,* 603–630.

Yick, A., & Agbayani-Siewert, P. (1997). Perceptions of domestic violence in a Chinese-American community. *Journal of Interpersonal Violence, 12,* 832–846.

Yllo, K. (1993). Through a feminist lens: Gender, power, and violence. In R. J. Gelles & D. R. Loseke (Eds.), *Current controversies in family violence,* (pp. 47–62). Newbury Park, CA: Sage.

Yllo, K., & Straus, M. A. (1990). Patriarchy and violence against wives: The impact of structural and normative factors. In M. A. Straus & R. J. Gelles (Eds.), *Physical violence in American families: Risk factors and adaptations to violence in 8,145 families* (pp. 383–402). New Brunswick, NJ: Transaction.

Yoshioka, M. R., DiNoia, J., & Ullah, K. (2001). Attitudes toward marital violence: An examination of four Asian communities. *Violence Against Women, 7,* 900–926.

Young, M. E., Nosek, M. A., Howland, C. A., Chanpong, G., & Rintala, D. H. (1997). Prevalence of abuse of women with physical disabilities. *Archives of Physical Medicine and Rehabilitation, 78,* S34–S38.

Youniss, J., McLellan, J. A., & Yates, M. (1999). Religion, community service, and identity in American youth. *Journal of Adolescence, 22,* 243–253.

Zambrano, M. Z. (1985). *Mejor sola que real acomoanada: For the Latina in an abusive relationship.* Seattle: Seal Press.

Zanarini, M. C., Ruser, T. F., Frankenburg, F. F., Hennen, J., & Gunderson, J. G. (2000). Risk factors associated with the dissociative experiences of borderline patients. *Journal of Nervous and Mental Disease, 188,* 26–30.

Zellman, G. L., & Fair, C. C. (2002). Preventing and reporting abuse. In J. E. B. Myers, L. Berliner, J. Briere, C. T. Hendrix, C. Jenny, & T. A. Reid (Eds.), *The APSAC handbook on child maltreatment* (2nd ed.) (pp. 449–478). Thousand Oaks, CA: Sage.

Zierler, S., Cunningham, W. E., Andersen, R., Shapiro, M. F., Bozzette, S. A., Nakazono, T., Morton, S., Crystal, S., Stein, M., Turner, B., & St. Clair, P. (2000). Violence victimization after HIV infection in a U.S. probability sample of adult patients in primary care. *American Journal Public Health, 90,* 208–215.

Zingraff, M., Leiter, J., Johnson, M. C., & Myers, K. A. (1994). The mediating effect of school performance on the maltreatment-delinquency relationship. *Journal of Research in Crime and Delinquency, 31,* 62–91.

Zolondek, S. C., Abel, G. G., Northey, W. R. Jr., & Jordan, A. D. (2001). The self-reported behaviors of juvenile sexual offenders. *Journal of Interpersonal Violence, 16,* 73–85.

Zuravin, S. J. (1989). The ecology of child abuse and neglect: Review of the literature and presentation of data. *Violence and Victims, 4,* 101–120.

Zuriff, G. (1988). A quick solution to the psychologist's problem of defining "psychological maltreatment." *American Psychologist, 43,* 201.

Zwerling, E. (2003). *Suit presses for "gender symmetry" in shelters.* Retrieved November 13, 2003, from http://www.womensenews.org/article.cfm/dyn/aid/1457/context/archive

Author Index

Subject Index

About the Authors

Denise A. Hines, PhD, completed her doctoral degree in the Human Development Program in the Psychology Department at Boston University. Her dissertation, a behavioral genetic study of aggression in intimate relationships, was supported by an individual National Research Service Award from the National Institute of Mental Health. Her primary research interests include genetic influences on aggressive behaviors in family relationships, female-perpetrated family aggression, and cultural influences on family violence. She has several publications on these topics, including a book published by Sage, *Family Violence in a Cultural Perspective,* and she has made numerous conference presentations relating to issues in family violence. She is currently a postdoctoral research fellow at the Family Research Laboratory and Crimes Against Children Research Center at the University of New Hampshire with Murray Straus and David Finkelhor.

Kathleen Malley-Morrison, EdD, is a Professor of Psychology at Boston University. She has conducted considerable research on family violence since 1980 when she was a postdoctoral fellow on the family violence team at Children's Hospital in Boston. She also regularly teaches undergraduate and graduate courses focusing on family violence. She is the first author of several books, including *Treating Child Abuse and Family Violence in Hospitals* with Eli Newberger, Richard Bourne, and Jane Snyder, and *Family Violence in a Cultural Perspective* with Denise Hines. She is the editor of the book *International Perspectives on Family Violence and Abuse: A Cognitive Ecological Approach,* and she is the second author, with Anne Copeland, of the Sage book *Studying Families.* Her current focus is primarily on cross-cultural and international perspectives on family violence and abuse.